# Adventures in Group Theory

# Adventures in Group Theory

*Rubik's Cube, Merlin's Machine, and Other Mathematical Toys*

**David Joyner**

The Johns Hopkins University Press
Baltimore and London

© 2002 The Johns Hopkins University Press
All rights reserved. Published 2002
Printed in the United States of America on acid-free paper
9 8 7 6 5 4 3 2

The Johns Hopkins University Press
2715 North Charles Street
Baltimore, Maryland 21218-4363
www.press.jhu.edu

**Library of Congress Cataloging-in-Publication Data**
Joyner, David, 1959—
    Adventures in group theory : Rubik's Cube, Merlin's machine, and other mathematical
toys / David Joyner.
        p. cm.
    Includes bibliographical references and index.
    ISBN 0-8018-6945-5 (acid-free paper)—ISBN 0-8018-6947-1 (pbk. : acid-free paper)
        1. Group theory. 2. Mathematical recreations. I. Title.
    QA174.2 .J69 2002
    512'.2—dc21                                                         2001050252

A catalog record for this book is available from the British Library.

The Rubik's Cube, Pyraminx, Megaminx, Masterball, Lights Out, and other
puzzles names mentioned frequently are all trademarked. We shall omit the
symbol ™ after each occurrence for ease of reading.

*In mathematics you don't understand things.*
*You just get used to them.*

—JOHANN VON NEUMANN

# Contents

CONTENTS

# Preface

This book grew out of a combined fascination with games and mathematics, from a desire to marry 'play' with 'work' in a sense. It pursues playing with mathematics and working with games. In particular, abstract algebra is developed and used to study certain toys and games from a mathematical modeling perspective. All the abstract algebra needed to understand the mathematics behind the Rubik's cube, Lights Out, and many other games, is developed here. If you believe in the quote by von Neumann on the previous page, I hope you will enjoy getting used to the mathematics developed here.

Why is it that these games, developed for amusement by non-mathematicians, can be described so well using mathematics? To some extent, I believe it is because many aspects of our experience are universal, crossing cultural boundaries. One can view the games considered in this book, the Rubik's cube, Lights Out, etc., to be universal in this sense. Mathematics provides a collection of universal analytical methods, which is I believe why it works so well to model these games.

This book began as some lecture notes designed to teach discrete mathematics and group theory to students who, though certainly capable of learning the material, had more immediate pressures in their lives than the long-term discipline required to struggle with the abstract concepts involved. My strategy, to tempt them with something irresistible such as the Rubik's cube, worked and students loved it. Since the original class notes were put on the web, they have received between 20 and 30 hits per day. Based on that and my email, my students were not the only ones to enjoy the notes, which have since been expanded considerably. I've tried to write the book to be interesting to people with a wider variety of backgrounds but at the same time I've tried to make it useful as a reference.

This book was truly a labor of love in the sense that I enjoyed every minute of it. I hope the reader derives some fun from it too.

I've tried to ensure no mistakes remain. However, if you find any, I'd appreciate hearing about them so they can be fixed in a possible future edition. My email is wdj@usna.edu.

To continue the universal theme mentioned above, all the royalties from this book will go directly to the Earth Island Institute, a non-profit organization dedicated to environmental projects all over the world, http://www. earthisland.org/. To paraphrase David Brower (1912-2000), the founder of EII, there is but one ocean, one atmosphere, one Earth, and there are no replacements.

# Acknowledgments

This book owes much to the very interesting books of Christoph Bandelow [B], Professor of Mathematics at the Ruhr-University Bochum in Germany, and David Singmaster [Si], Professor of Mathematics in the Department of Computing, Information Systems and Mathematics, South Bank University, London. I thank my fine editor at Johns Hopkins University Press, Trevor Lipscombe, for his encouragement and for his many excellent suggestions. Without them as a source of ideas, this book would not exist. This book has benefited greatly from the stimulating discussions, encouraging correspondence, and collaborations with many people. These people include Christoph Bandelow, Dan Hoey, Ann Luers (now Ann Casey—for help with §§ 13.4, 13.3, and parts of chapter 14), Michael Dunbar (for help with § 13.3), Mark Longridge, Jim McShea (for help with § § 7.4, 13.5). Justin Montague and G. Gomes (for help with § 15.5.2, 13.2), Michael Reid, Spencer Robinson (for help with § 15.3), Andy Southern (for help with § 15.4), Dennis Spellman (for help with § 10.5.1), and many others.

Many of the graphs given below were produced with the help of MAPLE [Maple]. Some of the group-theoretical calculations were determined with the help of GAP [Gap] or MAGMA [Mag]. Some of the biographical information included in this book was borrowed from the award-winning internet site, the MacTutor History of Mathematics archive [MT].

I would also like to thank Hasbro Toys (who market Lights Out) for the opportunity to consult for them.

Last, but certainly not least, I thank my wonderful wife Elva for more things than I can mention here.

# Where to begin . . .

Speaking personally, I am always fascinated to discover that one topic is connected in a surprising way with another completely different topic. The Rubik's Cube, a mechanical toy that can be solved using pure mathematics (and no 'strategy'), is a case in point. It's difficult to know where to begin to explain this connection: as Wodehouse's character Bertie Wooster said, 'I don't know if you have the same experience, but the snag I come up against when I'm telling a story is the dashed difficult problem of where to begin it'.

Erno Rubik was born in the air-raid shelter of a Budapest hospital during World War II. His mother was a published poet, his father an aircraft engineer who started a company to build gliders. Rubik himself, far from being a mathematician, studied architecture and design at the Academy of Applied Arts and Design, remaining there as a professor, teaching interior design. In the mid 1970's, he patented a cube-shaped mechanical puzzle that has since captured the imagination of millions of people worldwide. The Rubik's Cube was born. By 1982, 'Rubik's Cube' was a household term, and became part of the Oxford English Dictionary. More than 100 million cubes have been sold worldwide.

About 150 years earlier, in the late 1820's and early 1830's, a French teenager named Evariste Galois developed a new branch of mathematics: group theory was born. Galois' life story is also interesting (see, for example, Calinger [Ca] and the article by Rothman [Ro]) and we shall read more of it later. His work on group theory solved what was perhaps the main unsolved mathematical problem of the day. In high-school, students learn that

$$x = \frac{-b \pm \sqrt{b^2 - 4ac}}{2a}$$

are the roots of the quadratic equation $ax^2 + bx + c = 0$. There are similar but more complicated formulas for the cubic and quartic polynomials discovered during the middle ages (see for example [JKT] for more details). But does there exist an analogous algebraic formula involving radicals only in the coefficients for an equation of fifth degree or higher? This problem had remained unsolved for centuries despite the efforts of the best and brightest mathematical minds. Galois succeeded where others had failed by inventing new mathematical tools.

Groups measure symmetry. (Galois, for example, studied the 'symmetry group' of the roots of a polynomial, now called Galois groups.) As the mathematician Hermann Weyl said in his wonderful book [We], symmetry is 'one idea by which man has tried throughout the ages to comprehend and create order, beauty, and perfection'. Groups play a key role in the study of roots of polynomials, crystallography, elementary particle physics, campanology (or 'bell-ringing', see chapter 3 below), cryptography, and the Rubik's Cube, among others. Surprisingly enough, it turns out that it is possible to use group theory (and only the 'group-theoretical definition' of the cube—no knowledge of strategy or special moves) to solve the Rubik's Cube (see § 10.2 below).

In this book, we'll develop the basics of group theory and create group-theoretical models of Rubik's Cube-like puzzles. On the practical side, we also discuss the solution strategy for the Rubik's Cube in some detail. (For those wanting to see a solution strategy now, see section 15.1.) Some solution strategies are briefly discussed for similar puzzles (the '15 Puzzle', the 'Rubik tetrahedron' or Pyraminx, the 'Rubik dodecahedron' or Megaminx, the Skewb, the 'Hockeypuck', and the 'Masterball') as well.

The important point to remember, though, is that group theory is a powerful tool with many real-world applications. Solving puzzles happens to be just one of them. Because of our Rubik's Cube focus, the approach in this book is different from some texts:

(a) there are a lot of non-standard, though relatively elementary, group theory topics;

(b) we emphasize permutation groups via examples over general theory (such as Sylow theory);

(c) we present some of the basic notions algorithmically (as in [Bu]); and

(d) we include material which is interesting, from both the mathematical and puzzlists perspective, while keeping the level as low as possible for as long as possible.

Moreover, most group theory texts prove everything. Here a lot of statements are proven, sometimes only a hint or sketch is provided, and the proof is left to the interested reader, other statements are supported only by an example. When a proof is not provided, a reference for a proof in the literature is given. To keep things as clear as possible, the start of a proof is denoted **Proof:** and the end by □. I've used the number system where a result or example in section a.b has a labeling of the form a.b.c.

Chapters 1 and 2 give some basic mathematical background. Chapter 3 introduces some of the puzzles and some notation we use for them. Except for the chapter on solutions, the remaining chapters discuss these puzzles using group theory, graph theory, linear algebra, or automata (finite state machines) theory. It must be confessed that while the earlier chapters can probably be followed by a good high school student, chapters 12 and 14 are relatively advanced. They are, in my opinion, the most interesting, and illustrate some remarkable ways in which the Rubik's Cube is connected with other branches of mathematics, which is a continuing source of my fascination with the subject.

The last chapter, Coda, gives some indications of the present state of our, or at least my own, ignorance in this area. As it only gives some of the problems and questions that I don't have the answer to, it is not intended to be complete!

A note to teachers: Based on my experience teaching at the U.S. Naval Academy, a reasonable semester course based on this book could aim for the First Fundamental Theorem of Cube Theory, in chapter 9, with lots of time for 'side trips' and other material. It is possible to cover the Second Fundamental Theorem of Cube Theory, in chapter 11, in one semester but it is a race against time. For example, chapter 6 on 'Merlin's Machine', or any uncovered sections, might be used for some very interesting term projects.

Adventures in Group Theory

# Chapter 1

# Elementary, my dear Watson

'If logic is the hygiene of the mathematician, it is not his source of food; the great problems furnish the daily bread on which he thrives.'
*Andre Weil, 'The future of mathematics'*, **American Mathematical Monthly,** *May, 1950*

Think of a scrambled Rubik's Cube as a car you want to fix on your own. You not only need some tools but you need to know how to use them. This chapter, among others, provides you with some of the tools needed to get the job done. As one of our goals is to discuss the mathematics of the Rubik's Cube, and other games, we start with some fundamentals. The basic purpose of this chapter is to introduce standard set theory notation and some basic notions of mathematical logic.

Logic and set theory are as basic to mathematics as light is to the 'real world'. The background presented here hopefully will make some of the terminology and notation introduced later a little easier to follow for those who either haven't seen or may have forgotten the mathematical notation. In any case, this is not intended to be a 'serious' introduction to mathematical logic nor to set theory.

## 1.1   You have a logical mind if...

The sentence in the section title, intended to be somewhat whimsical, will be finished later in this section!

A **statement** is a logical assertion, which is either true or false. (Of course we assume that this admittedly circular 'definition' is itself a statement.) Sometimes the truth or falsity of a statement is called its **Boolean value**. One can combine several statements into a single statement using the **connectives** 'and' $\wedge$, 'or' $\vee$, and 'implies' $\Rightarrow$. The Boolean value of a statement is changed using the 'negation' $\sim$. We shall also use 'if and only if' $\iff$ and 'exclusive or' $\underline{\vee}$

1

(this is defined in the table below) but these can be defined in terms of negation $\sim$ and the other three connectives ($\vee$, $\wedge$, and $\Rightarrow$).

**Example 1.1.1.** *Today is Monday if and only if today is the day before Tuesday. An example of 'exclusive or': Either today is Monday or today is not Monday (but not both).*

**Ponderable 1.1.1.** *Express $\_\vee$ and $\Longleftrightarrow$ in terms of $\sim$, $\vee$, $\wedge$, and $\Rightarrow$.*

Solution: $p \Longleftrightarrow q$ is the same as $(p \Rightarrow q) \wedge (q \Rightarrow p)$. $p \_ \vee q$ is the same as $(p \vee q) \wedge \sim (p \Longleftrightarrow q)$. $\square$

**Notation**: Let p and q be statements.

| statement | notation | terminology |
|---|---|---|
| p and q | p $\wedge$ q | 'conjunction' |
| p or q | p $\vee$ q | 'disjunction' |
| p implies q | p $\Rightarrow$ q | 'conditional' |
| ~q implies ~p | ~q $\Rightarrow$ ~p | 'contrapositive' of p $\Rightarrow$ q |
| negate p | ~p | 'negation' |
| p if and only if q | p $\Longleftrightarrow$ q | 'if and only if' |
| either p or q (not both) | p $\_\vee$ q | 'exclusive or' |

The contrapositive is the essence of any 'proof by contradiction', or 'reductio ad absurdum', argument.

**Truth tables**: Given the Boolean values of the statements $p, q$, we can determine the values of the statements $p \wedge q$, $p \vee q$, $p \Rightarrow q$, $p \Longleftrightarrow q$, $p \_ \wedge q$ using the truth tables:

| p | q | p $\wedge$ q |
|---|---|---|
| T | T | T |
| T | F | F |
| F | T | F |
| F | F | F |

| p | q | p $\vee$ q |
|---|---|---|
| T | T | T |
| T | F | T |
| F | T | T |
| F | F | F |

| p | q | p $\Rightarrow$ q |
|---|---|---|
| T | T | T |
| T | F | F |
| F | T | T |
| F | F | T |

| p | q | p $\Longleftrightarrow$ q |
|---|---|---|
| T | T | T |
| T | F | F |
| F | T | F |
| F | F | T |

Note that $\Longleftrightarrow$ is analogous to the $=$ sign.

For example, let p be the statement 'I am a millionaire' and q the statement 'I will buy my wife a brand new car.' I claim that 'if p then q' is true. If I was a millionaire I really would buy my wife a new car. But I am not and cannot afford a car, so p and q are *both* false. This supports the idea that 'F $\Rightarrow$ F' is true. Now let q be the statement 'I will buy my wife some flowers.' I claim that 'if p then q' is true. The trouble is, I plan to buy my wife flowers in any case, i.e., q is true no matter what p is. This supports the idea that 'F $\Rightarrow$ T' is true. In particular, be careful about drawing conclusions from false premises because they imply both true and false statements!

| p | q | p $\vee$ q |
|---|---|---|
| T | T | F |
| T | F | T |
| F | T | T |
| F | F | F |

| p | $\sim$ p |
|---|---|
| T | F |
| F | T |

**Ponderable 1.1.2.** *What can you (logically) conclude from the following Chinese proverb?*

> *If there is light in the soul,*
> *   then there will be beauty in the person.*
> *If there is beauty in the person,*
> *   then there will be harmony in the house.*
> *If there is harmony in the house,*
> *   then there will be order in the nation.*
> *If there is order in the nation,*
> *   then there will be peace in the world.*
> Chinese Proverb

**Definition 1.1.1.** *'For all', written* $\forall$*, is the* **universal quantifier**. *'There exists', written* $\exists$*, is the* **existential quantifier**.

**Ponderable 1.1.3.** *(M. Gardner) Determine which of the following statements is true.*

- *Exactly one of these statements is false.*

- *Exactly two of these statements are false.*

- *Exactly three of these statements are false.*

- *Exactly four of these statements are false.*

- *Exactly five of these statements are false.*

Though a 'set' is defined later (see §1.2), we shall give an example of what can be done with a set containing only two elements.

**Definition 1.1.2.** *Let B be the set containing only the two elements 0 and 1: this is expressed in symbols as* $B = \{0, 1\}$*. Think of 0 as 'off' and 1 as 'on'. (In applications, B models the two possible states of an electrical circuit). Give B the two operations* **addition** *(+) and* **multiplication** *(\*) defined by the following tables*

3

| $x$ | $y$ | $x + y$ |
|---|---|---|
| 1 | 1 | 0 |
| 1 | 0 | 1 |
| 0 | 1 | 1 |
| 0 | 0 | 0 |

| $x$ | $y$ | $x * y$ |
|---|---|---|
| 1 | 1 | 1 |
| 1 | 0 | 0 |
| 0 | 1 | 0 |
| 0 | 0 | 0 |

*(Note how these mimic the truth tables of 'exclusive or' ($\veebar$) and 'and' ($\wedge$).)* *We call B the* **Boolean algebra.** *(In applications, + and * model two commonly occurring 'gates' in an electrical circuit.)*

With this, we can start to do some mathematics. After all, as Bertrand Russell said, 'The rules of logic are to mathematics what those of structure are to architecture.'

We can begin with

**De Morgan's laws:**

(a) $p \wedge (q \vee r) \iff (p \wedge q) \vee (p \wedge r)$ ,

(b) $p \vee (q \wedge r) \iff (p \vee q) \wedge (p \vee r)$,

and the **laws of negation:**

(c) $\sim (p \wedge q) \iff (\sim p) \vee (\sim q)$ ,

(d) $\sim (p \vee q) \iff (\sim p) \wedge (\sim q)$.

(Hint: It is possible, though tedious, to verify this using truth tables. To check this yourself, you may want to do (a), (c), (d) first, then deduce (b) from these. It is easier to prove the set theory analog, given below, using Venn diagrams, then deduce the above version from that.)

De Morgan's laws are analogous to the distributive law $a \cdot (b + c) = a \cdot b + a \cdot c$ for the real numbers. A. De Morgan (1806-1871) started out training to be a mathematician but was refused a fellowship at Cambridge University due to his objections over the mandatory theological test. He then began to study for the bar but also (at the age of 21) applied for the chair of mathematics at the University College London. He was appointed and began a full professor there a year later. He resigned his chair several times on matters of principle.

**Definition 1.1.3.** *Of course, to think logically, we need to produce logical arguments. But what are they?*

*A* **logical argument** *is a sequence of statements (called* **hypotheses***)* $p_1, p_2, ..., p_n$ *which imply a statement q (called the* **conclusion***). Such a logical argument is also called a* **proof** *of q.*

In other words, a logical argument is a true statement of the form

$$(p_1 \wedge p_2 \wedge ... \wedge p_n) \Rightarrow q.$$

**Ponderable 1.1.4.** *Using truth tables, verify the logical argument*

$$((p \Rightarrow q) \wedge (q \Rightarrow r)) \Rightarrow (p \Rightarrow r).$$

4

## 1.1.1 'You talking to me?'

**Definition 1.1.4.** *A* **variable** *is a letter denoting some (possibly unknown) object. A* **constant** *is a letter denoting some specific, well-defined object. A* **term** *is a variable or constant.*
*A* **predicate** *is a rule which associates to each term a logical statement.*

**Example 1.1.2.** *'Daffy Duck' is a constant. 'x is one of Daffy Duck's favorite songs' is a predicate involving a variable and a constant. Here x might be a Miles Davis song, for example.*

To finish the whimsical sentence in the section title, you have a logical mind if when you see the famous saying 'Time waits for no man' and 'No man is an island', you think 'therefore 'time waits for an island'!' If we let $x$ denote the constant 'no man' and $y$ denote the constant 'an island' then these two sayings can be written 'Time waits for $x$' and '$x = y$'. Substituting, we therefore obtain 'Time waits for $y$.' This may seem crazy, but it's logical.

When creating a model of the Rubik's Cube, we shall need to convert some 'everyday statements' into symbolical form in order to reduce them to a form suitable for mathematical analysis. Let's illustrate this with some examples.

**Example 1.1.3.** *(a) Consider the statement 'Each friend of mine can solve the Rubik's Cube'. Let*

$$M(x) = x \text{ can solve the Rubik's Cube}, \qquad S(x) = x \text{ is a friend of mine}.$$

*The symbolic form is*

$$\forall x, \ S(x) \Rightarrow M(x).$$

*(b) The Rubik's Cube is subdivided into 27 smaller cubes, which we call* **subcubes**. *Only 26 of these are visible since the center of the Rubik's Cube is completely surrounded. Consider the statement 'Each of the 26 visible subcubes of the Rubik's Cube can be moved into any other visible subcube using some move of the Rubik's Cube.' This statement is false, since an edge subcube cannot be moved into a corner subcube, but let us convert this statement into more symbolic notation (which we can more easily analyze) anyway. Let $M$ be the set of all possible (sequences of) moves of the Rubik's Cube. Let $S$ be the set of all 26 visible subcubes of the Rubik's Cube. The statement above says that '$\forall x, y \in S$, $\exists m \in M$ such that $m$ sends $x$ to $y$.' Though, as we already noticed, this is false, if we let $S_C$ denote the set of all corner subcubes of the Rubik's Cube then the analogous statement '$\forall x, y \in S_C$, $\exists m \in M$ such that $m$ sends $x$ to $y$' is true, given any two corner subcubes there is a Rubik's Cube move which sends one to the other (and which, of course, may scramble other subcubes as well).*

**Ponderable 1.1.5.** *Convert 'Some friend of mine can solve the Rubik's Cube' to symbolic form.*

**Ponderable 1.1.6.** *(M. Gardner [Gar1]) Professor White, Professor Brown and Professor Black were lunching together. 'Isn't it remarkable', said the lady,* *'that our names are White, Black, and Brown and one of us has black hair, one has brown hair, and one has white hair.'*

*'It is indeed', answered the one with the black hair as Professor Black bit into his sandwich, 'and have you noticed that not one has hair color to match our name?'*

*The lady's hair is not brown. What is the color of Professor Black's hair?*

|        | white | brown | black | lady |
|--------|-------|-------|-------|------|
| White  |       |       |       |      |
| Brown  |       |       |       |      |
| Black  |       |       |       |      |
| lady   |       |       |       | ×    |

**Ponderable 1.1.7.** *Retrograde chess problem by Christoph Bandelow. Let upper case letter denote white pieces (e.g., K is a white king, Q is a white queen, ...) and lower case letters denote black pieces (e.g., n is a black knight, b is a black bishop, ...). Consider the following chess position*

|   |   |   |   |   |   |   |   |
|---|---|---|---|---|---|---|---|
|   | × |   | × |   | × | $K$ | × |
| × |   | × |   | × |   | × |   |
|   | × |   | × | $P$ | $k$ |   | × |
| × |   | × |   | × |   | × |   |
|   | × |   | × |   | × | $B$ | × |
| × |   | × |   | × |   | × |   |
| $b$ | × |   | × | $Q$ | × |   | × |
| $B$ | $n$ | × |   | × |   | × |   |

*What were the last 6 single moves?*

## 1.2  Elements, my dear Watson

A **set** is a 'well-defined' collection of objects. The objects belonging to a set are the **elements** of the set. If $x$ is an element of a set $S$ then we write $x \in S$. If not, we write $x \notin S$. Let $S, T$ be sets. If all the elements of $S$ also belong to $T$ then we write $S \subset T$ and we say that $S$ is a **subset** of $T$. If a set $S$ contains only a finite number of distinct elements then we call this number the **cardinality** or **size** of $S$ and denote it by $|S|$. Two sets are said to have the **same cardinality** if there is a one-to-one correspondence between them. An **infinite set** is a set which is not a subset of a finite set. A set containing exactly one element is called a **singleton**.

There are two common ways to describe a set:

(a) Listing all its elements in between brackets $S = \{a, b, ...\}$ if the set is finite. For example, assume you have a Rubik's Cube lying on a table in front of you. We denote the set of it's **basic moves** by $\{U, D, L, R, F, B\}$, where

- $U$ denotes the move of the Rubik's cube where you turn the upward face clockwise (as you look it from above) one-quarter turn,

- $D$ denotes the move where you turn the downward face clockwise (as you look it from below) one-quarter turn,

- $L$ turns the left face clockwise one-quarter turn,

- $R$ turns the right face clockwise one-quarter turn,

- $F$ turns the front face clockwise one-quarter turn,

- $B$ turns the back face clockwise one-quarter turn.

This shorthand is called the **Singmaster notation**.

(b) Describe the set using properties of its elements using bracket notation: $S = \{x \mid x \text{ has property } P\}$ is pronounced '$S$ is the set consisting of all $x$ such that $x$ satisfies the property $P$'. For example, you know a Rubik's Cube has 6 faces (or sides) and each face is subdivided into 9 **facets** (and each facet has a colored sticker glued on it). We could define $S = \{x \mid x \text{ is a facet of the Rubik's cube}\}$. In this case, $|S| = 54$.

**Remark 1.2.1.** *(Russell's paradox) We must be a little careful when describing sets using properties since some 'self-referential' properties lead to contradictions: let*

$$R = \{x \mid x \notin x\}.$$

*In other words, for all $x$, $x \in R \iff x \notin x$. In particular, if we take $x = R$ then this becomes*

$$R \in R \iff R \notin R,$$

*an obvious contradiction. The problem is that $R$ is not 'well-defined' (in the sense that it does not satisfy the set theory axioms which we will skip here, see [St] for example).*

**Example 1.2.1.** *Here's another way to think of Russell's paradox: A (male) Navy barber must only shave those (male) Naval personnel who do not shave themselves. Does he shave himself? If yes, then he shaves himself, which is impossible. If no, then he must shave himself, another impossibility. How is this poor barber to avoid being court-martialed for not following orders?!*

The **empty set** is the set containing no elements, denoted $\emptyset$.

**Notation**: Let $S$ and $T$ be sets. Assume that $S \subset X$. The table below defines the terms **intersection**, **union**, **symmetric difference**, and **complement**.

| statement | notation | terminology |
|---|---|---|
| set of elements in $S$ and $T$ | $S \cap T$ | intersection |
| set of elements in $S$ or $T$ | $S \cup T$ | union |
| set of elements in $S$ or in $T$ (not both) | $S \triangle T$ | symmetric difference |
| set of elements not in $S$ | $S^c$ | complement |
| $S$ is a subset of $T$ | $S \subset T$ | subset |

We say that two sets $S, T$ are **equal** if $S \subset T$ and $T \subset S$. For example, we can visualize the intersection with the help of a Venn diagram, as in the Figure below.

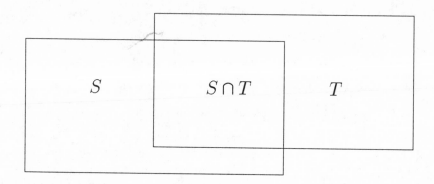

Venn diagrams are named after John Venn (1834-1923), an ordained priest. He lectured at Cambridge University as a lecturer in logic, probability, and set theory.

**Ponderable 1.2.1.** *Using Venn diagrams, verify the* **De Morgan laws***:*
*(a)* $S \cap (T \cup U) = (S \cap T) \cup (S \cap U)$ ,
*(b)* $S \cup (T \cap U) = (S \cup T) \cap (S \cup U)$ ,
*and the* **laws of negation***:*
*(c)* $(S \cap T)^c = S^c \cup T^c$,
*(d)* $(S \cup T)^c = S^c \cap T^c$.

De Morgan's laws for sets are analogous to De Morgan's laws for logic.

**Definition 1.2.1.** *We call two sets $S, T$* **disjoint** *if they have no elements in common, i.e., if*

$$S \cap T = \emptyset.$$

*If*

$$S = \cup_{i=1}^{n} S_i,$$

*where the $S_1, S_2, ..., S_n$ are pairwise disjoint sets, then we call this union a* **partition** *of $S$.*

**Example 1.2.2.** *Let $S$ denote the set of all 54 facets of the Rubik's Cube. Let $S_C$ denote all the facets belonging to a corner subcube, $S_E$ denote all the facets belonging to an edge subcube, and $S_Z$ denote all the facets belonging to a center subcube (we use the subscript $Z$ since the German word for center, zentrum, begins with a z). Then $S = S_C \cup S_E \cup S_Z$ is a partition.*

The set of all integers will be denoted by $\mathbb{Z}$. By the way, the letter $Z$ (for the German word Zahlen, meaning numbers) for the integers was apparently first introduced by N. Bourbaki in the 1930's, a pseudonym for a group of French mathematicians aiming to write a thorough unified account of all mathematics. Earlier, the notation $\overline{3}$ was used by the German number theorist E. Landau (1877-1938) for the integers.

**Example 1.2.3.** *If*
$$S = \mathbb{Z},$$
$$S_1 = \{..., -2, 0, 2, ...\} = \text{even integers},$$
$$S_2 = \{..., -3, -1, 1, 3, ...\} = \text{odd integers},$$
*then $S = S_1 \cup S_2$ is a partition of the integers into the set of even and odd ones.*

**Logic/set theory analogs:** Just as one can use connectives to form new statements from old statements, there are analogous ways to form new sets from old ones using 'intersection' (the analog of 'and'), 'union' (the analog of 'or'), and 'complement' (the analog of 'negation'). The analog of 'implies' is 'subset'. The analog of 'if and only if' is 'equals'.

| set theory | logic |
|---|---|
| sets | statements |
| union | or |
| intersection | and |
| subset | implies |
| symmetric difference | exclusive or |
| equal | if and only if |
| Venn diagrams | truth tables |
| complement | negation |

We shall often make use of the following facts: if $S = \{x \mid x \text{ has property } P\}$ and if $T = \{x \mid x \text{ has property } Q\}$ then

- $S \cap T = \{x \mid x \text{ has property P } and \text{ property Q}\}$,

- $S \cup T = \{x \mid x \text{ has property P } or \text{ property Q (or both)}\}$,

- if $P \Rightarrow Q$ then $S \subset T$,

- if $P \wedge Q$ is false then $S \cap T = \emptyset$,

- if $P \Longleftrightarrow Q$ then $S = T$.

We shall prove the first property and leave the other two to the interested reader.

**Proof:** To show $S \cap T = \{x \mid x \text{ has property P and property Q}\}$, we must show $S \cap T \subset \{x \mid x \text{ has property P and property Q}\}$, and

$$\{x \mid x \text{ has property P and property Q}\} \subset S \cap T.$$

9

Here are the gory details: First pick an arbitrary element $x \in S \cap T$. $x$ satisfies property $P$ since $x \in S$ and $x$ satisfies property $Q$ since $x \in T$. Thus $x \in \{y \mid y$ has property P and property Q$\}$. This shows that each element of $S \cap T$ belongs to

$$\{x \mid x \text{ has property P and property Q}\},$$

which we can write symbolically (by definition of $\subset$) as,

$$S \cap T \subset \{x \mid x \text{ has property P and property Q}\}.$$

Next, pick an arbitrary element

$$x \in \{y \mid y \text{ has property P and property Q}\}.$$

$x$ has property $P$, so $x \in S$, and $x$ has property $Q$, so $x \in T$. Therefore (by definition of $\cap$), $x \in S \cap T$. This shows that each element of

$$\{y \mid y \text{ has property P and property Q}\}$$

belongs to $S \cap T$, so

$$\{y \mid y \text{ has property P and property Q}\} \subset S \cap T.$$

Putting these together, we have shown that

$$S \cap T \subset \{y \mid y \text{ has property P and property Q}\}.$$

This forces the two sets to be equal, and the proof of the first property is complete. $\square$

**Ponderable 1.2.2.** *(Joiner's problem, E. I. Ignjat'ev [Pe]) Cut a $6 \times 6$ chessboard and an $8 \times 8$ chessboard each into two pieces and join the four pieces into a $10 \times 10$ chessboard.*

For more on logic or set theory, see for example [C] or [St].

# Chapter 2

# 'And you do addition?'

> 'And you do Addition?' the White Queen asked. 'What's one and one and one and one and one and one and one and one and one and one?'
> **Through the looking glass**, *Lewis Carroll*

Just as the White Queen asked Alice about addition, we need to ask ourselves how well we know how to count. Just remember, there are three kinds of people in the world, those who can count and those who can't!

The mathematics of the Rubik's Cube involves understanding combinatorial ideas, such as permutations and counting arguments. This chapter will build up some background to introduce some of these notions, and we will end with the binomial theorem.

## 2.1  Functions

The type of function we will run across here most frequently is a 'permutation', defined precisely later, that is roughly speaking a rule which mixes up and swaps around the elements of a finite set.

Let $S$ and $T$ be finite sets.

**Definition 2.1.1.** : *A* **function** *(sometimes also called a* **map** *or* **transformation***) $f$ from $S$ to $T$ is a rule that associates to each element $s \in S$ exactly one element $t \in T$. We will use the following notation and terminology for this:*

$$f : S \to T \quad (f \text{ is a function from } S \text{ to } T),$$
$$f : s \longmapsto t \quad (f \text{ sends } s \text{ in } S \text{ to } t \text{ in } T),$$
$$t = f(s) \quad (t \text{ is the image of } s \text{ under } f).$$

We call $S$ the **domain** of $f$, $T$ the **range** of $f$, and the set

$$f(S) = \{f(s) \in T \mid s \in S\}$$

the **image** of $f$. The Venn diagram depicting this setup is:

11

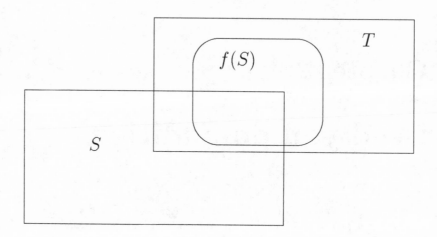

**Example 2.1.1.**

- *One typical example for us will be when $S$ is the set of all 54 facets of the Rubik's Cube (which we introduced in the last chapter) and $f$ is a rule for associating to each facet some other facet determined by a Rubik's Cube move. This example will be given more precisely later.*

- *Let $\mathbb{N} = \{1, 2, 3, ...\}$ denote the set of all* **natural numbers.** *If $n$ is any natural number, let $\phi(n)$ denote the number of natural numbers less than or equal to $n$ which have no prime factor in common with $n$. (A* **prime number** *is a natural number greater than 1 which has no factors less than itself except 1. For example, 2, 3, 5, are primes but $10 = 2 \cdot 5$ is not. By the way, a natural number greater than 1 which is not a prime is called a* **composite.**) *The function $\phi : \mathbb{N} \to \mathbb{N}$ is sometimes called* **Euler's phi function.** *We have $\phi(1) = 1$, $\phi(2) = 1$, $\phi(3) = 2$.*

- *If $n$ is any natural number, let $\pi(n)$ denote the number of prime numbers less than or equal to $n$. The function $\pi : \mathbb{N} \to \mathbb{N}$ is simply called the $\pi$* **function.** *We have $\pi(1) = 0$, $\pi(2) = 1$, $\pi(3) = 2$, $\pi(4) = 2$, and so on. Though the rough asymptotic behavior of $\pi(n)$ is known, there are still many unsolved problems regarding $\pi(n)$. For example, it is not yet known if there is an $n > 1$ such that $\pi(n^2 + 2n + 1) = \pi(n^2)$ (i.e., if there is always a prime between any two consecutive squares).*

**Ponderable 2.1.1.** *What is $\phi(16)$? What is $\pi(16)$?*

If $f : S \to T$ is any function and $t \in f(S)$ then we write the **preimage of $t$** as

$$f^{-1}(t) = \{s \in S \mid f(s) = t\}.$$

This is simply the set of all $s$'s that get sent to $t$. For example, if $f : S = \mathbb{R} \to T = \mathbb{R}$ is the function $f(x) = x^2$, and if $t = 4$ then $f^{-1}(4) = \{2, -2\}$. If $T \subset \mathbb{R}$ then the preimage of 0 is called the **zeros** of $f$.

**Ponderable 2.1.2.** : *If $f : S = \mathbb{R} \to T = \mathbb{R}$ is the map $f(x) = x^4 - 16$ and if $t = 0$ then find $f^{-1}(0)$.*

The **Cartesian product** of two sets $S$, $T$ is the set of pairs of elements taken from these sets:
$$S \times T = \{(s, t) \mid s \in S, t \in T\}.$$
An element of $S \times T$ is simply a list of two things, the first one from $S$ and the second one from $T$. This construction of a new set from two given sets is named for the French philosopher René Descartes (1596-1650) whose work **La géométrie** includes the application of algebra to geometry.

More generally, given any collection of $m$ sets, $S_1$, $S_2$, ..., $S_m$, we can define the $m$-**fold Cartesian product**, to be the set
$$S_1 \times ... \times S_m = \{(s_1, ..., s_m) \mid s_i \in S_i, \ 1 \leq i \leq m\}.$$
Elements of the Cartesian product $S_1 \times ... \times S_m$ are called $m$-**tuples**.

**Example 2.1.2.** *If $\mathbb{R}$ denotes the set of all real numbers then the Cartesian product $\mathbb{R} \times \mathbb{R}$ is simply the set of all pairs or real numbers. In other words, this is the Cartesian plane we are all familiar with from high school.*

*More generally, we may repeat this process and take the Cartesian product of the set of real numbers with itself $n$ times (where $n > 1$ is any integer) to get $\mathbb{R} \times ... \times \mathbb{R}$ (n times). This n-fold Cartesian product is denoted more conveniently by $\mathbb{R}^n$.*

The set $\mathbb{R}^n$ has some extra structure, a 'vector space structure', which we shall make use of frequently in this book. The general definition given next.

**Definition 2.1.2.** *An element of the set $\mathbb{R}^n$ is simply an ordered list of n real numbers. Such a list will be called a* **vector** *or an n-**vector** to be specific.*

*More generally, a* **real vector space** *is a set $V$ (whose elements are called vectors) having two operations,* **vector addition** $+$ *and* **scalar multiplication** $\cdot$, *satisfying the following conditions: (a) if $v, w \in V$ are any two vectors then $v + w = w + v$ is also a vector in $V$, (b) there is a zero vector $\vec{0}$ such that for any vector $v \in V$, $v + \vec{0} = v$, (c) for any real number $c \in \mathbb{R}$ (called a* **scalar**) *and any $v \in V$, the product $c \cdot v$ is a vector in $V$, (d) the distributive laws hold: $(a + b)v = av + bv$ and $c(v + w) = cv + cw$, for all $a, b, c \in \mathbb{R}$ and $v, w \in V$.*

Sometimes it is convenient to 'picture' a function as the set of its values inside the Cartesian product $S \times T$. The **graph** of a function $f : S \to T$ is the subset
$$\{(s, f(s)) \mid s \in S\} \subset S \times T.$$
In high-school, if $y = x^2$, one plotted $x^2$ as a function of $x$. Here $f(x) = x^2$ and you plotted $\{(x, f(x)) \mid x \in \mathbb{R}\}$. It is not possible that every subset of $S \times T$ is the graph of some function from $S$ to $T$. The following fact classifies exactly which subsets of $S \times T$ can arise as the graph of a function from $S$ to $T$.

13

**Ponderable 2.1.3.** : *Let $X \subset S \times T$. Show that $X$ is the graph of a function from $S$ to $T$ if and only if, for all $(s_1, t_1) \in X$ and $(s_2, t_2) \in X$, whenever $t_1 \neq t_2$ we also have $s_1 \neq s_2$.*

(Hint: Let

$$pr_1 : S \times T \to S$$
$$(s, t) \longmapsto s$$

be projection onto the 1st component. Recall that the graph of a function $f$ has the property that $f^{-1}(s)$ is always a singleton, i.e., one gets messed up if one $s$ gives rise to two different $f(s)$ values.)

**Definition 2.1.3.** *Let $f : S \to T$ and $g : T \to U$ be two functions. We can compose them to get another function, the* **composition**, *denoted $fg : S \to U$:*

$$
\begin{array}{ccccc}
x & \longmapsto & f(x) & \longmapsto & g(f(x)) \\
S & \to & T & \to & U
\end{array}
$$

**Definition 2.1.4.** *If the image of the function $f : S \to T$ is all of $T$, i.e., if $f(S) = T$, then we call $f$* **surjective** *(or 'onto', or 'is a surjection'). Equivalently, a function $f$ from $S$ to $T$ is surjective if every $t \in T$ is the image of some $s \in S$ under $f$.*

For example, the map $f : \mathbb{R} \to \mathbb{R}$ defined by $f(x) = 2x$, for any real number $x$, is surjective. Another example, let $S$ be the set of all 54 facets of the Rubik's Cube. Let $f : S \to S$ be the map which sends a facet to the facet which is diametrically opposite (for instance, the upward center facet would be mapped to the downward center facet). This function is also surjective.

**Ponderable 2.1.4.** *Suppose that $|S| < |T|$. Is there a surjective function $f : S \to T$? Explain.*

**Definition 2.1.5.** : *A function $f : S \to T$ is called* **injective** *(or 'one-to-one' or 'an injection') if each element $t$ belonging to the image $f(S)$ is the image of exactly one $s$ in $S$.*

In other words, $f$ is an injection if the condition $f(s_1) = f(s_2)$ (for some $s_1, s_2 \in S$) always forces $s_1 = s_2$.

**Ponderable 2.1.5.** *Suppose that $|S| > |T|$. Is there an injective function $f : S \to T$? Explain.*

**Ponderable 2.1.6.** *Suppose that $|S| = |T|$. Show that a function $f : S \to T$ is surjective if and only if it is injective.*

**Definition 2.1.6.** *A function $f : S \to T$ is called a* **bijection** *if it is both injective and surjective.*

Equivalently, a bijection from $S$ to $T$ is a function $f : S \to T$ for which each $t \in T$ is the image of exactly one $s \in S$.

**Definition 2.1.7.** *A set $S$ is called* **countable** *if there exists a bijection $f : S \to \mathbb{Z}$ to the set of integers $\mathbb{Z}$.*

**Example 2.1.3.**

- *The set of all odd integers $S = \{..., -3, -1, 1, 3, ...\}$ is countable since the map $f : S \to \mathbb{Z}$ defined by $f(x) = (x+1)/2$ is a bijection.*

- *The set of all rational numbers will be denoted by $\mathbb{Q}$. The set $S$ of all rational numbers within the unit interval $0 < r < 1$ is countable since you can define $f : S \to \mathbb{Z}$ as follows: $f(1) = 1/2$, $f(2) = 1/3$, $f(3) = 2/3$, $f(4) = 1/4$, $f(5) = 3/4$, $f(6) = 1/5$, $f(7) = 2/5$, $f(8) = 3/5$, and so on. (There are $\phi(n)$ terms of the form $m/n$, where $m$ is relatively prime to $n$ and $\phi(n)$ denotes the number of positive integers less than or equal to $n$ which are relatively prime to $n$, i.e., have no common prime divisors.).*

**Example 2.1.4.** *Let $C$ be the cube in 3-space having vertices at the points $O = (0,0,0), P_1 = (1,0,0), P_2 = (0,1,0), P_3 = (0,0,1), P_4 = (1,1,0), P_5 = (1,0,1), P_6 = (0,1,1), P_7 = (1,1,1)$. We shall also (to use a notation which will be used more later) denote these by dbl, dfl, dbr, ubl, dfr, ufl, ubr, ufr, respectively. Let $C_0 = \{O, P_1, ..., P_7\}$ be the set of the eight vertices of $C$, let $C_1 = \{uf, ur, ub, ul, fr, br, bl, fl, df, dr, db, dl\}$ be the set of the 12 edges of $C$, and let $C_2 = \{F, B, U, D, L, R\}$ be the set of the 6 faces of $C$. Let $r$ be the rotation of a point $(x, y, z)$ by 180 degrees about the line passing through the points $(1/2, 1/2, 0)$, $(1/2, 1/2, 1)$. Note that $r : \mathbb{R}^3 \to \mathbb{R}^3$ is a function which sends the cube $C$ onto itself.*
*The cube $C$ is pictured as follows:*

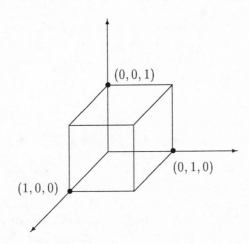

*This function $r$ induces three functions*

$$f_0 : C_0 \to C_0, \qquad f_1 : C_1 \to C_1, \qquad f_2 : C_2 \to C_2.$$

15

where $f_i$ is the function which sends $x \in C_i$ to its image $r(x)$ under $r$ (which is again in $C_i$), for $i = 0, 1, 2$. Each $f_i$ is a bijection.

**Ponderable 2.1.7.** *Finish the labeling of the vertices and the faces in the above picture and describe $f_1$ and $f_2$ explicitly by filling out the tables*

| $v$ | $f_0(v)$ |
|-----|----------|
| $P_1 = dfl$ | $P_2 = dbr$ |
|  |  |
|  |  |

| $e$ | $f_1(e)$ |
|-----|----------|
| $fl$ | $br$ |

| $f$ | $f_2(f)$ |
|-----|----------|
| $F$ | $B$ |

**Definition 2.1.8.** *Suppose $f : S \to T$ is a bijection. Define $f^{-1}$ to be the rule which associates to $t \in T$ the element $s \in S$,*

$$f^{-1}(t) = s \qquad \text{if and only if } f(s) = t.$$

*This rule defines a function $f^{-1} : T \to S$. This function satisfies $f(f^{-1}(t)) = t$, for all $t \in T$, and $f^{-1}(f(s)) = s$, for all $s \in S$. In other words, the composition $f \circ f^{-1} : T \to T$ sends each element of $T$ to itself, and $f^{-1} \circ f : S \to S$ sends each element of $S$ to itself.*

*The function $f^{-1}$ constructed above is called the **inverse function** of $f$.*

## 2.2 Functions on vectors

Later on, we shall need some particular types of functions. This section presents, very briefly, a few basic facts about matrices, which we regard as a certain type of function sending vectors to vectors. For further details, see, for example, [JN].

### 2.2.1 History

The mathematician who first published a major work that seriously studied matrices and matrix algebra in the western world was Arthur Cayley (1821-1895) of Cambridge, England. He wrote a memoir on the theory of linear transformations which was published in 1858 and is often thought of as one of the 'fathers' of matrix theory. In fact, it was his friend and colleague James Sylvester (1814-1897) who first coined the term 'matrix'. (Incidentally, Sylvester taught at Johns Hopkins University from 1877 to 1883 and founded the first mathematics journal in the United States, the American Journal of Mathematics, put out by the same press that published the book you are reading now.) Though most people learn of matrices then determinants, historically determinants were studied long before matrices arose.

One of the earliest examples that motivated Cayley's study of matrices was the following. If we have three coordinate systems $(x, y), (x', y')$, and $(x'', y'')$, connected by

$$\begin{cases} x' = x + y \\ y' = x - y \end{cases} \text{ and } \begin{cases} x'' = -x' - y' \\ y'' = -x' + y' \end{cases}$$

then the relationship between $(x, y)$ and $(x'', y'')$ is given

$$\begin{cases} x'' = -x' - y' = -(x + y) - (x - y) = -2x \\ y'' = -x' + y' = -(x + y) + (x - y) = -2y \end{cases}$$

If we do as Cayley did and abbreviate the three change of coordinates by the square array of the coefficients then we obtain

$$\begin{bmatrix} 1 & 1 \\ 1 & -1 \end{bmatrix}, \begin{bmatrix} -1 & -1 \\ -1 & 1 \end{bmatrix}, \begin{bmatrix} -2 & 0 \\ 0 & -2 \end{bmatrix}.$$

Cayley noticed that there was an algebraic rule that allows us to determine the third array from the first two, without doing the substitution we did above to relate $(x'', y'')$ to $(x, y)$. In doing so, Cayley invented 'matrix multiplication'. It corresponds to composing the function sending $(x, y) \longmapsto (x', y')$ with the function sending $(x', y') \longmapsto (x'', y'')$.

## 2.2.2  $3 \times 3$ matrices

A $3 \times 3$ matrix is a $3 \times 3$ table of real numbers

$$A = \begin{pmatrix} a_{11} & a_{12} & a_{13} \\ a_{21} & a_{22} & a_{23} \\ a_{31} & a_{32} & a_{33} \end{pmatrix}.$$

This may be thought of as a function (which we shall still denote by $A$) which sends 3-vectors to 3-vectors. This function is defined by the **matrix-vector multiplication formula**,

$$A\vec{v} = \begin{pmatrix} a_{11} & a_{12} & a_{13} \\ a_{21} & a_{22} & a_{23} \\ a_{31} & a_{32} & a_{33} \end{pmatrix} \begin{pmatrix} x \\ y \\ z \end{pmatrix} = \begin{pmatrix} a_{11}x + a_{12}y + a_{13}z \\ a_{21}x + a_{22}y + a_{23}z \\ a_{31}x + a_{32}y + a_{33}z \end{pmatrix}$$

In other words, $A$ sends the 3-vector $(x, y, z)$ (which we have written above as a column, as is the convention) to the 3-vector $(a_{11}x + a_{12}y + a_{13}z, a_{21}x + a_{22}y + a_{23}z, a_{31}x + a_{32}y + a_{33}z)$. In general, any such $3 \times 3$ matrix gives rise to a function $A : \mathbb{R}^3 \to \mathbb{R}^3$.

**Example 2.2.1.** *If $A = I_3$, the $3 \times 3$ identity matrix,*

$$\begin{pmatrix} 1 & 0 & 0 \\ 0 & 1 & 0 \\ 0 & 0 & 1 \end{pmatrix}$$

*then $I_3\vec{v} = \vec{v}$ for all $\vec{v} \in \mathbb{R}^3$.*

17

**Ponderable 2.2.1.** *Show that*

$$\begin{pmatrix} 0 & 1 & 0 \\ 1 & 0 & 0 \\ 0 & 0 & 1 \end{pmatrix}$$

*has the effect of swapping the x- and y-axes.*

### 2.2.3  $m \times n$ matrices

An $m \times n$ **matrix** (of real numbers) is a rectangular array or table of numbers arranged with $m$ rows and $n$ columns. It is usually written:

$$A = \begin{pmatrix} a_{11} & a_{12} & \cdots & a_{1n} \\ a_{21} & a_{22} & \cdots & a_{2n} \\ \vdots & & & \vdots \\ a_{m1} & a_{m2} & \cdots & a_{mn} \end{pmatrix}.$$

The $(i,j)^{th}$ **entry** of $A$ is $a_{ij}$. The $i$th **row of** $A$ is

$$\begin{bmatrix} a_{i1} & a_{i2} & \cdots & a_{in} \end{bmatrix} \qquad (1 \le i \le m)$$

The $j$th **column of** $A$ is

$$\begin{bmatrix} a_{1j} \\ a_{2j} \\ \vdots \\ a_{mj} \end{bmatrix} \qquad (1 \le j \le n)$$

A matrix having as many rows as it has columns $(m = n)$ is called a **square matrix**. The entries $a_{ii}$ of an $m \times n$ matrix $A = (a_{ij})$ are called the **diagonal entries**, the entries $a_{ij}$ with $i > j$ are called the **lower diagonal entries**, and the entries $a_{ij}$ with $i < j$ are called the **upper diagonal entries**. An $m \times n$ matrix $A = (a_{ij})$ all of whose lower diagonal entries are zero is called an **upper triangular matrix**. This terminology is logical if the matrix is a square matrix but both the matrices below are called upper triangular

$$\begin{pmatrix} 1 & 2 & 3 & 4 \\ 0 & 5 & 6 & 7 \\ 0 & 0 & 8 & 9 \\ 0 & 0 & 0 & 10 \end{pmatrix}, \quad \begin{pmatrix} 1 & 2 & 3 & 4 \\ 0 & 5 & 6 & 7 \\ 0 & 0 & 8 & 9 \\ 0 & 0 & 0 & 10 \\ 0 & 0 & 0 & 0 \end{pmatrix}$$

whether they look triangular of not! A similar definition holds for lower triangular matrices. The square $n \times n$ matrix with 1's on the diagonal and 0's elsewhere,

$$\begin{pmatrix} 1 & & \cdots & 0 \\ 0 & \ddots & & \vdots \\ \vdots & & & 0 \\ 0 & \vdots & 0 & 1 \end{pmatrix},$$

is called the $n \times n$ **identity matrix** and denoted $I$ or $I_n$. This is both upper triangular and lower triangular. (In general, any square matrix which is both upper triangular and lower triangular is called a **diagonal matrix**.)

A square $n \times n$ matrix with exactly one 1 in each row and each column, and 0's elsewhere, is called an $n \times n$ **permutation matrix**. The identity $I_n$ is a permutation matrix. We shall discuss these types of matrices in detail in the next chapter.

A square $n \times n$ matrix with exactly one *non-zero entry* in each row and each column, and 0's elsewhere, is called an $n \times n$ **monomial matrix**. We shall discuss these types of matrices later in the book. They have many properties similar to permutation matrices. Monomial matrices occur in the explicit description of the Rubik's Cube group, which we shall give later.

### 2.2.4 Multiplication and inverses

Fortunately, we shall not be forced to deal in this book too much with computations of matrix multiplications of large matrices. *Roughly speaking*, we shall eventually show how each move of the Rubik's Cube can be expressed in terms of matrices (more precisely, as a pair of matrices - an $8 \times 8$ matrix corresponding to the movement of the 8 corners and a $12 \times 12$ matrix corresponding to the movement of the 12 edges). Therefore, a little bit of brief background on matrix multiplication is appropriate.

When you multiply an $m \times n$ matrix $A$ by a $n \times p$ matrix $B$, you get an $m \times p$ matrix $AB$. The $(i,j)^{th}$ entry of $AB$ is computed as follows:

1. Let $k = 1$ and $c_0 = 0$.

2. If $k = m$, you're done and $a_{ij} = c_m$. Otherwise proceed to the next step.

3. Take the $k^{th}$ entry of the $i^{th}$ row of $A$ and multiply it by the $k^{th}$ entry of the $j^{th}$ row of $B$. Let $c_k = c_{k-1} + a_{ik}b_{kj}$.

4. Increment $k$ by 1 and go to step 2.

In other words, multiply each element of row $i$ in $A$ with the corresponding entry of column $j$ in $B$, add them up, and put the result in the $(i,j)$ position of the array for $AB$.

If $A$ is a square $n \times n$ matrix and if there is a matrix $B$ such that $AB = I_n$ then we call $B$ the **inverse** matrix of $A$, denoted $A^{-1}$. If you think of $A$ as a function $A : \mathbb{R}^n \to \mathbb{R}^n$ then $A^{-1}$ is the inverse function. As a practical matter, if $n$ is 'small' (say, $n \le 3$) then matrix inverses can be computed by pencil and paper using known techniques (see for example [JN]). For most larger matrices, computers are needed. We shall discuss easier techniques for inverting permutation matrices in the next chapter and this is the type of matrix we will have to deal with most frequently anyway.

**Ponderable 2.2.2.** *Imagine a chessboard in front of you. You can place at most 8 non-attacking rooks on the chessboard. (Rooks move only horizontally and vertically.) Now imagine you have done this and let $A = (a_{ij})$ be the $8 \times 8$ matrix of 0's and 1's (called a $(0, 1)$-matrix) where $a_{ij} = 1$ if there is a rook on the square belonging to the $i^{th}$ horizontal down and the $j^{th}$ vertical from the left. Call such a matrix a* **rook matrix***. If there are exactly 8 1's in A then we shall call A a* **full rook matrix***. For example,*

$$\begin{pmatrix} 0 & 0 & 0 & 0 & 0 & 0 & 1 & 0 \\ 0 & 0 & 0 & 0 & 0 & 0 & 0 & 1 \\ 0 & 0 & 0 & 0 & 0 & 1 & 0 & 0 \\ 1 & 0 & 0 & 0 & 0 & 0 & 0 & 0 \\ 0 & 0 & 0 & 0 & 1 & 0 & 0 & 0 \\ 0 & 0 & 1 & 0 & 0 & 0 & 0 & 0 \\ 0 & 0 & 0 & 1 & 0 & 0 & 0 & 0 \\ 0 & 1 & 0 & 0 & 0 & 0 & 0 & 0 \end{pmatrix}$$

*is a full rook matrix. Show that*
*(a) any full rook matrix is invertible,*
*(b) the product of any two rook matrices is a rook matrix.*

## 2.2.5  Determinants

The determinant of a matrix $A$, denoted $\det(A)$, is only defined when $A$ is a square matrix, i.e., the number of rows is equal to the number of columns. If $A$ is an $n \times n$ real matrix then $A$ may be regarded as a function of $\mathbb{R}^n$, sending points to points. It turns out it will send the unit hypercube in $\mathbb{R}^n$ to a parallelepiped (the $n$-dimensional analog of a parallelogram). It also turns out that the *absolute value* of the determinant of $A$ measures the volume of that parallelpiped.

If $m = n = 2$ the determinant is easy to define:

$$\det \begin{pmatrix} a & b \\ c & d \end{pmatrix} = ad - bc.$$

The proof that $|ad-bc|$ is the area of the parallelogram with vertices $(0, 0)$, $(a, c)$, $(b, d)$, $(a + b, c + d)$ may be found in [JN], for example. The easiest *constructive* way to define the determinant of an arbitrary $n \times n$ matrix $A = (a_{ij})$ is to use the **Laplace cofactor expansion**: for any $1 \leq i \leq n$, we have

$$\det(A) = \sum_{j=1}^{n} (-1)^{i+j} \det(A_{ij}), \qquad (2.1)$$

20

where $A_{ij}$ is the $(n-1) \times (n-1)$ 'submatrix of $A$' obtained by omitting all the entries in the $i^{th}$ row or the $j^{th}$ column. The matrix $A_{ij}$ is called the $(i,j)$-**minor** of $A$ and $(-1)^{i+j} \det(A_{ij})$ is called the $(i,j)$-**cofactor**.

Pierre-Simon Laplace (1749-1827) was one of the most influential scientists in French history. However, modesty was not his strong point and he announced in 1780 that he was the greatest mathematician in France. He was right of course. In fact, Laplace went on to make major contributions to probability and celestial mechancs (the mathematics of the motion of planets is mostly due to him).

A square matrix $A$ is **singular** if $\det(A) = 0$. Otherwise, $A$ is called **non-singular** or **invertible**.

**Example 2.2.2.** *Taking $i = 2$,*

$$\det \begin{pmatrix} 1 & 2 & 3 \\ 4 & 5 & 6 \\ 7 & 8 & 9 \end{pmatrix}$$

$$= (-1) \cdot 4 \cdot \det \begin{pmatrix} 2 & 3 \\ 8 & 9 \end{pmatrix} + (+1) \cdot 5 \cdot \det \begin{pmatrix} 1 & 3 \\ 7 & 9 \end{pmatrix} + (-1) \cdot 6 \cdot \det \begin{pmatrix} 1 & 2 \\ 7 & 8 \end{pmatrix}$$

$$= (-4)(18 - 24) + (5)(9 - 21) + (-6)(8 - 14) = 0.$$

*This implies $A$ is singular. Indeed, the parallelepiped generated by $(1, 2, 3)$, $(4, 5, 6)$, $(7, 8, 9)$, must be flat (2-dimensional, hence have 0 volume) since $(4, 5, 6) = (1, 2, 3) + (1, 1, 1)$ and $(7, 8, 9) = (1, 2, 3) + 2(1, 1, 1)$.*

There is an analogous Lagrange cofactor expansion for columns as well. See [JN] (or any other book on linear algebra), for example.

An important fact about singular matrices, and one that we will use later, is the following.

**Lemma 2.2.1.** *Suppose $A$ is an $n \times n$ matrix with real entries. $\det(A) = 0$ if and only if there is a non-zero vector $v$ such that $Av = 0$, where $0$ is the zero vector in $\mathbb{R}^n$.*

Here's a sketch: $\det(A) = 0$ if and only if $A$ is singular if and only if $A$ is not 1-1. $A$ is not 1-1 if and only if there are vectors $v_1 \neq v_2$ for which $Av_1 = Av_2$, i.e., $Av = 0$, where $v = v_1 - v_2 \neq 0$.

We end this section with one last key property of determinants.

**Lemma 2.2.2.** *If $A, B$ are any two $n \times n$ matrices having real entries then $\det(AB) = \det(A) \det(B)$.*

More details on all this material can be found in any text book on linear algebra, for example [JN].

## 2.3   Relations

A relation on a set is a generalization of the concept of a function from S to itself. Relations include $=$, $<$, and many other types of 'comparisons'.

**Definition 2.3.1.** : *Let $S$ be a set. If $R$ is a subset of $S \times S$ then we call $R$ a* **relation on** *$S$. If $(x, y) \in R$ then we say that $x$ is* **related** *to $y$.*

We may also regard a relation $R$ on $S$ as a function

$$R : S \times S \to \{0, 1\}.$$

In this form, we say $x$ is related to $y$ (for $x, y \in S$) if $f(x, y) = 1$.

This is a very general notion. There are lots and lots of relations in mathematics - inequality symbols, functions, subset symbols are all common examples of relations.

**Example 2.3.1.** *Let $S$ be any set and let $f$ be a function from $S$ to itself. This function gives rise to the relation $R$ on $S$ defined by the graph of $f$:*

$$R = \{(x, y) \in S \times S \mid y = f(x), \text{ for } x \in S\}.$$

*(It is through this correspondence that we may regard a function as a relation.)*

**Example 2.3.2.** *Let $S$ be the set of all subsets of $\{1, 2, ..., n\}$. Let $R$ be defined by*

$$R = \{(S_1, S_2) \mid S_1 \subset S_2, \ S_1 \in S, \ S_2 \in S\}.$$

*Note that $R$ is a relation.*

**Definition 2.3.2.** *Let $R$ be a relation on a set $S$. We call $R$ an* **equivalence relation** *if*

*(a) any element $s \in S$ is related to itself ('reflexive'),*

*(b) if $s$ is related to $t$ (i.e., $(s, t)$ belongs to $S$) then $t$ is related to $s$ ('symmetry'),*

*(c) if $s_1$ is related to $s_2$ and $s_2$ is related to $s_3$ then $s_1$ is related to $s_3$ ('transitivity').*

**Example 2.3.3.** *The equality symbol $=$ provides an equivalence relation on the real line: let $D = \{(x, x) \mid x \text{ real}\}$. This is an equivalence relation on the real line: note $x = y$ if and only if $(x, y)$ belongs to $D$.*

**Notation**: If $R$ is an equivalence relation on $S$ then we often write $x \sim y$ or $x \equiv y$ in place of $(x, y) \in R$, for simplicity.

**Example 2.3.4.** *Fix an integer $n > 1$. For integers $x, y$, define $x \equiv y$ if and only if $n$ divides $x - y$. In this case, we say that $x$ is* **congruent** *to $y$ mod $n$. The equivalence class of $x$ is sometimes (for historical reasons) called the* **residue class** *(or* **congruence class***) of $x$ mod $n$. We shall sometimes abuse notation and denote residue class of $x$ mod $n$ simply by $x$.*

*This notation was first introduced by Carl F. Gauss (1777-1855), who is regarded by many as one of the top mathematicians of all time. At the age of 21 he wrote* **Disquisitiones Arithmeticae***, which started a new era of number theory and introduced this notation.*

**Ponderable 2.3.1.** *(a) Let* $f(x) = x^2$, *let* $S$ *be the real line, and let* $R$ *be the corresponding relation as in the first example. Is* $R$ *an equivalence relation?*

*(b) Let* $f(x) = 2x$, *let* $S$ *be the real line, and let* $R$ *be the corresponding relation as in the first example. Is* $R$ *an equivalence relation?*

**Ponderable 2.3.2.** *Let* $R$ *be the corresponding relation as in the second example. Is* $R$ *an equivalence relation?*

**Definition 2.3.3.** *Let* $R$ *be an equivalence relation on a set* $S$. *For* $s \in S$, *we call the subset*

$$[s] = \{t \in S \mid s \sim t\}$$

*the* **equivalence class** *of* $s$ *in* $S$.

**Example 2.3.5.** *For integers* $x, y$, *define* $x \equiv y$ *if and only if* $3$ *divides* $x - y$. *the equivalence classes are*

$$[0] = \{..., -6, -3, 0, 3, 6, ...\},$$
$$[1] = \{..., -5, -2, 1, 4, 7, ...\},$$
$$[2] = \{..., -4, -1, 2, 5, 8, ...\}.$$

**Ponderable 2.3.3.** *Show that for any* $s_1$ *and* $s_2$ *in* $S$, *we have either*
*(a)* $[s_1] = [s_2]$, *or*
*(b)* $[s_1]$ *is disjoint from* $[s_2]$.

As a consequence of this problem, we see that if $R$ is an equivalence relation on a set S then the equivalence classes of $R$ partition $S$ into disjoint subsets. We state this separately for future reference (we also assume $S$ is finite for simplicity):

**Theorem 2.3.1.** *If* $S$ *is a finite set and* $R$ *is an equivalence relation on* $S$ *then there are subsets*

$$S_1 \subset S, \ S_2 \subset S, ..., S_k \subset S,$$

*satisfying the following properties:*
*(1)* $S$ *is the union of all the* $S_i$ *'s:*

$$S = S_1 \cup S_2 \cup ... \cup S_k = \cup_i S_i$$

*(2) the* $S_i$ *'s are disjoint: for* $1 \leq i \leq k$, $1 \leq j \leq k$, *if* $i \neq j$ *then* $S_i \cap S_j = \emptyset$.
*(These last two properties say that the* $S_i$ *'s partition* $S$ *in the sense of the previous chapter.)*
*(3) the* $S_i$ *'s exhaust the collection of equivalence classes of* $R$: *for each* $1 \leq i \leq k$, *there is an* $s_i \in S$ *such that*

$$S_i = [s_i].$$

*(This element* $s_i$ *is called a* **representative** *of the equivalence class* $S_i$.*)*

**Example 2.3.6.** *For real numbers $x, y$, define $x \equiv y$ if and only if $x - y$ is an integer. The equivalence classes are of the form*

$$[x] = \{..., x - 2, x - 1, x, x + 1, x + 2, ...\},$$

*for $x$ real. Each equivalence class has exactly one representative in the half open interval $[0, 1)$.*

**Remark 2.3.1.** *Conversely, given a partition as in (1), there is an equivalence relation $R$ such that $S_i = [s_i]$, for some some $s_i \in S$, where*

$$[s] = \{x \in S \mid s \sim x\}$$

*is the equivalence class of $s$ with respect to $R$. Indeed, we define*

$$R = \cup_{i=1}^{k} S_i \times S_i.$$

*This is an equivalence relation and $s \sim t$ if and only if $s, t \in S_i$, for some $i = 1, 2, ..., k$.*

**Ponderable 2.3.4.** *Let $S$ be the set $\mathbb{Z}$ of all integers. Let $R$ be the relation defined by $(x, y) \in R$ if and only if $x - y$ is an even number (i.e., an integer multiple of 2).*
*(1) Show that this is an equivalence relation,*
*(2) Find the sets $S_i$ in the above lemma which partition $S$,*
*(3) Find a representative of each equivalence class $S_i$.*

# 2.4 Counting and mathematical induction

'About binomial theorem I'm teeming with a lot o' news...'
*Major General Stanley,* **Pirates of Penzance,** *W. S. Gilbert and A. Sullivan*

This section quickly surveys the few basic counting principles we shall use later.

Suppose you have 15 monkeys in one cage and 20 orangutans in another cage. Now put them all in one big cage. After the commotion dies down, count up the total. What do you get? 35 of course. This is the basic idea behind the next result.

**Theorem 2.4.1.** *(Addition Principle) Let $S_1, ..., S_n$ denote disjoint finite sets. Then*

$$|S_1 \cup ... \cup S_n| = |S_1| + ... + |S_n|.$$

The proof proceeds by the method of **mathematical induction** (or simply **induction**, not to be confused with the logical reasoning of induction). This shall be presented from a general viewpoint and then applied to proving the Addition Principle.

**Definition 2.4.1.** *Let $P(1), P(2), ...$ be a sequence of logical statements which you want to prove to be true (i.e., $P(k)$ is a 'predicate' with 'term' $k = 1, 2, ...$).*

*(1) Show that the case $k = 1$ is true, i.e., prove $P(1)$.*

*(2) Assuming the truth of the case $k = n - 1$, i.e., assuming $P(n-1)$ is true (the* **induction hypothesis***), prove $P(n)$.*

*Once we have these two facts, we can conclude $P(1)$ is true (case (1)), $P(1) \Rightarrow P(2)$ is true (case (2)), so $P(2)$ is true (by the truth tables in the previous chapter), $P(2) \Rightarrow P(3)$ is true (case (2)), so $P(3)$ is true (by the truth tables in the previous chapter), and so on. We deduce that the logical statements $P(k)$ are all true. This is the method of* **mathematical induction***.*

De Morgan, who we have met already, was the first to define the term mathematical induction (though the basic idea was known before him).

As a humorous example, we 'prove', using mathematical induction, that horses have an odd number of feet. Let $P(k)$ be the statement that $k$ horses have an odd number of feet. Case $k = 1$ is the statement that one horse has an odd number of feet. Well, any horse with a broken leg has an odd number of feet, so $P(1)$ is true. Now assume $P(k)$ is true and we try to prove $P(k+1)$. Any $k$ of the $k + 1$ horses have an odd number of feet, so all $k + 1$ do as well, hence $P(k + 1)$ holds as well.

As a more serious example, let $P(k)$ be the logical statement $|S_1 \cup ... \cup S_k| = |S_1| + ... + |S_k|$, $1 \le k \le n$. We need to:

(1) Prove the claim for $k = 1$, i.e., prove $P(1)$.

(2) Assuming the truth of the case $k = n - 1$, prove $P(n)$.

**Proof:** Let $P(k)$ be the logical statement $|S_1 \cup ... \cup S_k| = |S_1| + ... + |S_k|$, $1 \le k \le n$.

Case $k = 1$. $P(1)$ is the statement $|S_1| = |S_1|$, which is of course true.

Case $k = n - 1$. Assume $|S_1 \cup ... \cup S_{n-1}| = |S_1| + ... + |S_{n-1}|$ is true.

Let $S = S_1 \cup ... \cup S_{n-1}$ and let $T = S_n$. Each element of $S \cup T$ is either an element of $S$ or an element of $T$ but not both since they are disjoint. How many ways can we pick an element from $S \cup T$? $|S| + |T|$ ways, since may pick one from either $S$ or from $T$.

The induction hypothesis implies $|S| = |S_1 \cup ... \cup S_{n-1}| = |S_1| + ... + |S_{n-1}|$. Since $S_1 \cup ... \cup S_n = S \cup T$, this and the previous paragraph all together implies $|S_1 \cup ... \cup S_{n-1} \cup S_n| = |S \cup T| = |S| + |T| = |S_1| + ... + |S_{n-1}| + |S_n|$. This proves the case $k = n$.

By mathematical induction, the proof of the addition principle is complete. $\square$

**Example 2.4.1.** *If there are $n$ bowls, each containing some distinguishable marbles and if $S_i$ is the set of marbles in the $i^{th}$ bowl then the number of ways to pick a marble from exactly one of the bowls is $|S_1| + ... + |S_n|$, by the addition principle.*

As a corollary of the above addition principle, we have the following result.

**Proposition 2.4.1.** *(***Dirichlet Pigeonhole Principle***) Let $d, m, n$ be positive integers. If there are $n$ objects (pigeons) which must be placed in $m$ boxes (these* *'pigeonholes' as assumed to be disjoint), where $md < n$, then there is at least one box with at least $d + 1$ objects.*

**Proof:** We prove this using the **reductio ad absurdum** argument: assume the statement claimed to be true is actually false and derive a contradiction.

Suppose that the $i^{th}$ box $S_i$ has less than or equal to $d$ objects in it, for every $1 \le i \le m$. Since the boxes are assumed to be disjoint, the hypothesis of the addition principle holds. Therefore, the total number of objects in all the boxes is $|S_1| + ... + |S_m| \le d + ... + d = md$. But there are $n$ such objects, so we must therefore have $n \le md$. This contradicts the assumption that $md < n$. □

**Example 2.4.2.** *If you are in a room with 9 other people then there must be either at least 5 people you know or 5 people you don't know (not counting yourself). This is obvious enough. It is also what the Dirichlet Pigeonhole Principle tells you: In this case, there are $n = 9$ objects (the people in the room) and $m = 2$ boxes (the friend box and the stranger box) so we may take $d = 4$ in the pigeonhole principle.*

The Three Stooges, Curly, Moe, and Larry, each know 10 different insults (probably more, but let's keep things simple!). For example, Moe knows 'You knucklehead!', and 9 others. Each one of them insults you exactly once. How many insults are possible? $10^3$, because each one of the 3 can insult you each one of 10 ways. The following principle generalizes this idea.

**Theorem 2.4.2. (Multiplication Principle)** *Let $S_1, ..., S_n$ denote finite sets. Then*

$$|S_1 \times ... \times S_n| = |S_1| \cdot ... \cdot |S_n|.$$

The proof proceeds by the method of induction. Let $P(k)$ be the logical statement $|S_1 \times ... \times S_k| = |S_1| \cdot ... \cdot |S_k|$, $1 \le k \le n$. Singing our chorus once again, we must:

(1) Prove the claim for $k = 1$, i.e., prove $P(1)$.

(2) Assuming the truth of the case $k = n - 1$, prove $P(n)$.

**Proof:** Let $P(k)$ be the logical statement $|S_1 \times ... \times S_k| = |S_1| \cdot ... \cdot |S_k|$, $1 \le k \le n$.

Case $k = 1$. $P(1)$ is the statement $|S_1| = |S_1|$, which is of course true.

Case $k = n - 1$. Assume $|S_1 \times ... \times S_{n-1}| = |S_1| \cdot ... \cdot |S_{n-1}|$ is true.

Let $S = S_1 \times ... \times S_{n-1}$ and let $T = S_n$.

Each element of $S \times T$ is a list of 2 elements, the first from $S$, the second from $T$. How many ways can we pick an element from $S$ to put in the first coordinate? $|S|$ ways. Suppose that the first coordinate has been choosen. How many ways can we pick an element from $T$ to put in the second coordinate? $|T|$ ways. Therefore, the number of ways to pick elements for the first coordinate and the second coordinate is $|S| \cdot |T|$.

The induction hypothesis implies $|S| = |S_1 \times ... \times S_{n-1}| = |S_1| \cdot ... \cdot |S_{n-1}|$. Since $S_1 \times S_n = S \times T$, this and the previous paragraph all together implies $|S_1 \times ... \times S_{n-1} \times S_n| = |S \times T| = |S| \cdot |T| = |S_1| \cdot ... \cdot |S_{n-1}| \cdot |S_n|$. This proves the case $k = n$.

By mathematical induction, the proof of the multiplication principle is complete. □

**Example 2.4.3.** *If there are n bowls, each containing some distinguishable marbles and if $S_i$ is the set of marbles in the $i^{th}$ bowl then the number of ways to pick exactly one marble from each of the bowls is $|S_1| \cdot ... \cdot |S_n|$, by the multiplication principle.*

**Corollary 2.4.1.** *The number of ordered selections, taken with repetition, of m objects from a set of n objects (m < n) is $n^m$.*

An ordered selection is sometimes called an **arrangement**.

**Example 2.4.4.** *From the set of majors,*

*mathematics, computer science, chemistry, engineering, physics,*

*select 3 (with repetition allowed). This is the same as selecting 3 object, with replacement, from 5 jars each containing the 5 words above. There are $5^3$ possible choices.*

**Corollary 2.4.2.** *The number of ordered selections, taken without repetition, of m objects from a set of n objects (m < n) is*

$$\frac{n!}{(n-m)!} = n \cdot (n-1) \cdot ... \cdot (n-m+1).$$

**Proof:** Strictly speaking this is not a corollary of the multiplication principle, so a proof is given.

Since the objects selected are ordered, by hypothesis, we may speak of a first selection, second selection, and so on. There are $n$ ways to make the first selection. Since there is no repetition allowed, this object is removed from the set used to make the second selection. There are $n-1$ ways to make the second selection, leaving $n-2$ objects remaining. This process can be continued $m$ times since $m < n$, by hypothesis. Multiplying the integers $n$, $n-1$, ..., $n-m$ gives the result. □

**Example 2.4.5.** *The number of poker hands, 5-tuples, without repetition, of objects from the set $\{1, 2, ..., 52\}$ is*

$$\frac{52!}{(47)!} = 52 \cdot 51 \cdot 50 \cdot 49 \cdot 48 \cdot 47 = 14658134400 \cong 1.4 \times 10^{10}.$$

**Ponderable 2.4.1.** *Let C be a set of 6 distinct colors. Fix a cube in space (imagine it sitting in front of you on a table). We call a **coloring** of the cube a choice of exactly one color per side. Let S be the set of all colorings of the cube. We say $x, y \in S$ are **equivalent** if x and y agree after a suitable rotation of the cube.*
*(a) Show that this is an equivalence relation.*
*(b) Count the number of equivalence classes in S.*

27

The next two results are corollaries of the multiplication principle as well.

**Corollary 2.4.3.** *The number of unordered selections, taken without repetition, of m objects from a set of n objects (m < n) is*

$$C(n, m) = \frac{n!}{(n-m)!m!} = \frac{n \cdot (n-1) \cdot \ldots \cdot (n-m+1)}{m \cdot (m-1) \cdot \ldots \cdot 1}.$$

An unordered selection is sometimes called an **combination**.

**Corollary 2.4.4.** *The number of unordered selections, taken with repetition allowed, of m objects from a set of n objects (m < n) is $C(n+m-1, m)$*

**Proof:** The proof is rather clever, so we include a sketch.

Imagine $m$ 1's in a row. Order the $n$ objects you will be selecting from as object 1, object 2, ... . Starting from the leftmost 1, count the number of object 1's you will select, then put a | mark to the right of the last 1. Put the | to the left of all the 1's if you don't select any from object 1. If you selected $m_1$ elements from object 1 then you have $m_1$ 1's to the right of the | and $m - m_1$ 1's to the left. Do the same for object 2, then object 3, ... . You must have inserted $n$ |'s. There are $C(n+m-1, m)$ ways to do this. □

This symbol $C(n, m)$, also commonly written $\binom{n}{m}$, is called the **combinatorial** (or **binomial**) **symbol** and pronounced '$n$ choose $m$'. It occurs in the binomial theorem and in Pascal's triangle,

$$
\begin{array}{ccccccc}
 & & & 1 & & & \\
 & & 1 & & 1 & & \\
 & 1 & & 2 & & 1 & \\
1 & & 3 & & 3 & & 1 \\
\end{array}
$$

$$\cdots\cdots\cdots\cdots\cdots,$$

where each entry is obtained by adding the two immediately above it. In other words, the defining property of Pascal's triangle is the following general fact, whose proof is left as an exercise for the interested reader.

**Lemma 2.4.1.** *If $0 < m < n$ then*

$$C(n, m) + C(n, m+1) = C(n+1, m+1).$$

**Theorem 2.4.3.** *(Binomial Theorem)*

$$(x + y)^n = \sum_{m=0}^{n} C(n, m) x^m y^{n-m}$$
$$= x^n + nx^{n-1}y + \ldots + C(n, m)x^m y^{n-m} + \ldots + nxy^{n-1} + y^n.$$

We shall not prove this theorem here. For the interested reader: the key ideas in the proof involve using mathematical induction and the defining property of Pascal's triangle - try to find it! By the way, Pascal's triangle was known long before Blaise Pascal (1623-1662) wrote his book **Treatise on the Arithmetical Triangle** on it in the 1600's.

# Chapter 3

# Bell ringing and other permutations

'Mathematics, springing up from the soil of basic human experience with numbers and data and space and motion, builds up a far-flung architectural structure composed of theorems which reveal insights into the reasons behind appearances and of concepts which relate totally disparate concrete ideas.'

*Sanders MacLane*, **American Mathematical Monthly**, *1954*

In the last chapter we discussed counting arguments and some combinatorial notions. This chapter introduces more combinatorial ideas and techniques - more tools for your toolbox, so to speak. The objective is to keep track of rearrangements of objects, such as the facets of a Rubik's Cube which get rearranged each time you make a move. Some new notation, symbols, and terms must be introduced.

Talk about relating totally disparate ideas - at the end of the chapter, we will see how combinatorial ideas already occurred in (church) bell ringing over 300 years ago!

## 3.1 Definitions

Suppose you shuffle a pack of cards. You are simply rearranging all 52 of them. The mathematical equivalence of shuffling is a permutation. A more precise general definition is as follows. Let $\mathbb{Z}_n = \{1, 2, ..., n\}$ be the set of integers from 1 to a fixed positive integer $n$. A **permutation** of $\mathbb{Z}_n$ is a bijection from $\mathbb{Z}_n$ to itself. (A bijection was defined in Definition 2.1.6.) More generally, if $T$ is any finite set then a permutation of $T$ is a bijection from $T$ to itself. In the case when $T$ has $n$ elements, we shall often label the elements of $T$ by $T = \{t_1, ..., t_n\}$ and regard a permutation $f : T \to T$ as a permutation $\phi : \mathbb{Z}_n \to \mathbb{Z}_n$, where $f(t_i) = t_j$ if and only if $\phi(i) = j$. In other words, once we label the elements of $T$, there is a 1-1 correspondence between the permutations of $T$ and those of

$\mathbb{Z}_n$. Since the elements of $\mathbb{Z}_n$ are easier to write (fewer subscripts!), we often just work with $\mathbb{Z}_n$.

As an example, on the $3 \times 3$ Rubik's Cube there are $9 \cdot 6 = 54$ facets. If you label them $1, 2, ..., 54$ (in any way you like), then any move of the Rubik's Cube corresponds to a permutation of $\mathbb{Z}_{54}$. In this chapter we present some basic notation and properties of permutations.

**Notation**: We may denote a permutation $f : \mathbb{Z}_n \to \mathbb{Z}_n$ by a $2 \times n$ array:

$$f \leftrightarrow \begin{pmatrix} 1 & 2 & ... & n \\ f_1 & f_2 & ... & f_n \end{pmatrix},$$

where $f_1, f_2, ..., f_n$ is simply a rearrangement (in some order depending on $f$) of the integers $1, 2, ..., n$. This notation means that $f$ sends 1 to $f_1$, $f$ sends 2 to $f_2$, ..., $f$ sends $n$ to $f_n$. In other words, $f_1 = f(1)$, $f_2 = f(2)$, ..., $f_n = f(n)$, and $f_i \neq f_j$ unless $i = j$.

**Example 3.1.1.** *(a) The* **identity** *permutation, denoted by I, is the permutation which doesn't do anything:*

$$I = \begin{pmatrix} 1 & 2 & ... & n \\ 1 & 2 & ... & n \end{pmatrix}$$

*(b) The permutation $f : \mathbb{Z}_3 \to \mathbb{Z}_3$ defined by*

$$f = \begin{pmatrix} 1 & 2 & 3 \\ 3 & 2 & 1 \end{pmatrix}$$

*simply swaps 1 and 3 and leaves 2 alone.*

*(a) The* **n-cycle** *is a permutation which cyclically permutes the values:*

$$\begin{pmatrix} 1 & 2 & ... & n \\ 2 & 3 & ... & 1 \end{pmatrix}$$

*Imagine an analog clock with the numbers 1, 2, ..., n arranged around the dial. An n-cycle simply moves each number forward (clockwise) by 1 unit. The permutation*

$$\begin{pmatrix} 1 & 2 & ... & n \\ n & 1 & ... & n-1 \end{pmatrix}$$

*is also called an n-cycle.*

**Definition 3.1.1.** *Let $f : \mathbb{Z}_n \to \mathbb{Z}_n$ be a permutation and let*

$$e_f(i) = \#\{j > i \mid f(i) > f(j)\}, \qquad 1 \le i \le n - 1.$$

*Let*

$$swap(f) = e_f(1) + ... + e_f(n - 1).$$

*We call this the* **swapping number** *(or* **length***) of the permutation $f$ since it counts the number of times $f$ swaps the inequality in $i < j$ to $f(i) > f(j)$. If*

30

*we plot a bar-graph of the function $f$ then $swap(f)$ counts the number of times the bar at $i$ is higher than the bar at $j$. We call $f$ **even** if $swap(f)$ is even and we call $f$ **odd** otherwise.*

*The number*

$$sign(f) = (-1)^{swap(f)}$$

*is called the **sign** (or **signum** function) of the permutation $f$.*

The reader may verify that the sign function satisfies the following property.

**Lemma 3.1.1.** *Let $f : \mathbb{Z}_n \to \mathbb{Z}_n$ be a permutation. Then*

$$sign(f) = \prod_{i < j \le n} \frac{f(i) - f(j)}{i - j}.$$

**Remark 3.1.1.** *Let $S \subset \mathbb{Z}_n$ be a subset. Let $f : \mathbb{Z}_n \to \mathbb{Z}_n$ be a permutation for which $f(i) = i$, for all $i \in S$. In other words, $f$ permutes the elements of $S^c = \mathbb{Z}_n - S$ but not the elements of $S$. Define a new permutation $g$ of $S^c$ (not of $\mathbb{Z}_n$) by $g(i) = f(i)$ for all $i \in S^c$. The above Lemma implies that the sign of the permutation $f : \mathbb{Z}_n \to \mathbb{Z}_n$ is the same as the sign of the permutation $g : S^s \to S^c$. This will turn out to be a useful fact to keep in mind later.*

**Example 3.1.2.** *We define the permutation $f : \mathbb{Z}_3 \to \mathbb{Z}_3$ for which $f(1) = 2, f(2) = 1, f(3) = 3$ by a $2 \times 3$ array*

$$\begin{pmatrix} 1 & 2 & 3 \\ 2 & 1 & 3 \end{pmatrix}$$

*This swaps 1 and 2 and leaves 3 alone. There are 3 inequalities for $\mathbb{Z}_3$: $1 < 2$, $2 < 3$, and $1 < 3$. Applying $f$ to these, we get: $f(1) = 2 > f(2) = 1$, $f(2) = 1 < f(3) = 3$, and $f(1) = 2 < f(3) = 3$. Only the first inequality is changed, so the swapping number is $swap(f) = 1$. Another way to picture $f$ is by a 'crossing diagram', where $f$ sends the left-hand column to the right-hand column:*

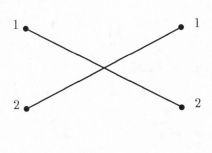

*The number of crosses in this diagram is the swapping number of $f$, from which we can see that $f$ is an odd permutation.*

31

**Ponderable 3.1.1.** *Express $f : \mathbb{Z}_3 \to \mathbb{Z}_3$ given by $f(1) = 3, f(2) = 1, f(3) = 2$, as (a) a $2 \times 3$ array, (b) a crossing diagram. Find its swapping number and sign.*

**Definition 3.1.2.** *Let $f : \mathbb{Z}_n \to \mathbb{Z}_n$ and $g : \mathbb{Z}_n \to \mathbb{Z}_n$ be two permutations. We can compose them to get another permutation, the* **composition**, *denoted $fg : \mathbb{Z}_n \to \mathbb{Z}_n$:*

$$
\begin{array}{ccccc}
k & \longmapsto & f(k) & \longmapsto & g(f(k)) \\
\mathbb{Z}_n & \to & \mathbb{Z}_n & \to & \mathbb{Z}_n
\end{array}
$$

**Notation/Warning!** We shall follow standard convention and write our compositions of permutations **left-to-right**. (This contrasts to the *right-to-left* composition of functions you may have seen in calculus.) When a possible ambiguity may arise, we call this type of composition 'composition as permutations' and call 'right-to-left composition' the 'composition as functions'.

When $f = g$ then we write $ff$ as $f^2$. In general, we write the n-fold composition $f...f$ ($n$ times) as $f^n$. Every permutation $f$ has the property that there is some integer $N > 0$, which depends on $f$, such that $f^N = 1$. That is to say, if you repeatedly apply a permutation enough times you will eventually obtain the identity permutation. In particular, if you can perform the same exact shuffle on a deck of cards repeatedly, you will eventually get back to the original ordering of the card deck.

This is vaguely reminiscent of a theorem due to P. Diaconis which roughly speaking says that you don't need more than 7 shuffles to scramble a card deck. The two situations are almost completely different though, because in one case you are trying to scramble the ordering up and in the other case you are trying to get back to the original ordering.

**Lemma 3.1.2.** *Let $f$ and $g$ be permutations $\mathbb{Z}_n \to \mathbb{Z}_n$. $sign(fg) = sign(f)sign(g)$.*

**Ponderable 3.1.2.** *For $f, g$ as in Lemma 3.1.1, we have*

$$
\prod_{i<j\leq n} \frac{fg(i) - fg(j)}{i - j} = \prod_{i<j\leq n} \frac{g(i) - g(j)}{i - j} \prod_{i<j\leq n} \frac{fg(i) - fg(j)}{g(i) - g(j)}.
$$

*Use this to prove Lemma 3.1.2?*

**Definition 3.1.3.** *The smallest integer $N > 0$ such that $f^N = 1$ is called the* **order** *of $f$.*

For an example from science fiction, suppose you take off in a space ship which goes faster than the speed of light. Suppose it is so fast that in one trip you come back exactly one hour earlier, according to your wrist watch, than everyone else! In other words, your hourly times are the same as everyone elses, except that they have been shifted by 1. (Incidentally, this type of permutation is called a 'cyclic permutation'.) How many trips do you need to take before your watch reads the same as everyone elses again? 24 trips total (assuming everyone is using 'military time').

**Example 3.1.3.** *Let*

$$f = \begin{pmatrix} 1 & 2 & 3 \\ 2 & 1 & 3 \end{pmatrix} \qquad g = \begin{pmatrix} 1 & 2 & 3 \\ 3 & 1 & 2 \end{pmatrix}$$

*be permutations of* $\mathbb{Z}_n$. *We have*

$$fg = \begin{pmatrix} 1 & 2 & 3 \\ 1 & 3 & 2 \end{pmatrix}, \qquad f^2 = 1, \qquad g^3 = 1.$$

**Ponderable 3.1.3.** *Compute (a)* $fg$ *and (b) the order of* $f$ *and the order of* $g$, *where*

$$(a) \qquad f = \begin{pmatrix} 1 & 2 & 3 \\ 3 & 2 & 1 \end{pmatrix} \qquad g = \begin{pmatrix} 1 & 2 & 3 \\ 3 & 1 & 2 \end{pmatrix}$$

$$(b) \qquad f = \begin{pmatrix} 1 & 2 & 3 \\ 3 & 1 & 2 \end{pmatrix} \qquad g = \begin{pmatrix} 1 & 2 & 3 \\ 2 & 1 & 3 \end{pmatrix}$$

**Ponderable 3.1.4.** *If* $f, g, h$ *are permutations of* $\mathbb{Z}_n$, *is* $(fg)h = f(gh)$? *In other words, we have three permutations, I have* $f, g$ *and you have* $h$. *First, you perform* $h$ *and then I perform* $g$ *then* $f$. *Second, I give you* $g$ *and you perform* $h$ *then* $g$ *and then I perform* $f$. *Are these the same? Explain.*

## 3.2 Inverses

> 'I don't understand you,' said Alice. 'It's dreadfully confusing!'
> 'That's the effect of living backwards,' the Queen said kindly:
> 'it always makes one a little giddy at first –'
> 'Living backwards!' Alice repeated in great astonishment. 'I never heard of such a thing!'
> ' – but there's one great advantage in it, that one's memory works both ways.'
> **Through the looking glass,** *Lewis Carroll*

Try to visualize in your mind the graph of a function $f : \mathbb{Z}_n \to \mathbb{Z}_n$. We can think of this as either a point of points $(i, f(i))$, $i = 1, 2, ..., n$, or as a bar graph. In any case, we can look at this graph of $f$ and determine

(a) if $f$ is injective,

(b) if $f$ is surjective,

(c) the inverse $f^{-1}$, if it exists.

How? Well, from the graph of $f$ we can determine the image $f(\mathbb{Z}_n)$ and this determines if $f$ is surjective or not. The inverse exists only if f is injective (hence, in our case, surjective by Ponderable 2.1.6). It's graph is determined by reflecting the graph of f about the diagonal, $x = y$.

**Lemma 3.2.1.** *The following statements are equivalent:*

*(1)* $f : \mathbb{Z}_n \to \mathbb{Z}_n$ *is injective,*

*(2)* $f : \mathbb{Z}_n \to \mathbb{Z}_n$ *is surjective,*

*(3)* $|f(\mathbb{Z}_n)| = |\mathbb{Z}_n|$.

33

**Proof:** The equivalence of the first two statements is left to the interested reader as an exercise (hints: first show (1) implies (2) using reductio ad absurdum, then show (2) implies (1), again using reductio ad absurdum). (2) is equivalent to (3), by definition of surjectivity. □

**Example 3.2.1.** *The inverse of*

$$\begin{pmatrix} 1 & 2 & 3 \\ 3 & 1 & 2 \end{pmatrix}$$

*is obtained by reflecting its graph about the $x = y$ line.*

**Ponderable 3.2.1.** *Graph and determine the inverses of the following permutations:*

$$\begin{pmatrix} 1 & 2 & 3 \\ 2 & 1 & 3 \end{pmatrix}, \begin{pmatrix} 1 & 2 & 3 & 4 \\ 2 & 3 & 4 & 1 \end{pmatrix}, \begin{pmatrix} 1 & 2 & 3 & 4 & 5 \\ 2 & 1 & 5 & 3 & 4 \end{pmatrix}$$

There are two more commonly used ways of expressing a permutation. The first is the 'matrix notation':

**Definition 3.2.1.** *To a permutation $f : \mathbb{Z}_n \to \mathbb{Z}_n$, given by*

$$f = \begin{pmatrix} 1 & 2 & \dots & n \\ f(1) & f(2) & \dots & f(n) \end{pmatrix}$$

*we associate to it the matrix $P(f)$ of 0's and 1's defined as follows: the $ij^{th}$ entry of $P(f)$ is 1 if $j = f(i)$ and is 0 otherwise.*

**Example 3.2.2.** *The matrix of the permutation $f$ given by*

$$f = \begin{pmatrix} 1 & 2 & 3 \\ 1 & 2 & 3 \end{pmatrix}$$

*is*

$$P(f) = \begin{pmatrix} 1 & 0 & 0 \\ 0 & 1 & 0 \\ 0 & 0 & 1 \end{pmatrix}.$$

**Definition 3.2.2.** *A square matrix which has exactly one 1 per row and per column (as $P(f)$ does) is called a* **permutation matrix** *.*

**Lemma 3.2.2.** *There are $n!$ distinct $n \times n$ permutation matrices and there are $n!$ distinct permutations of the set $\{1, 2, ..., n\}$.*

**Ponderable 3.2.2.** *Prove this using the multiplication principle from Theorem 2.4.2 in the previous chapter.*

**Example 3.2.3.** *The matrix of the permutation $f$ given by*

$$f = \begin{pmatrix} 1 & 2 & 3 \\ 2 & 1 & 3 \end{pmatrix}$$

34

*is*

$$P(f) = \begin{pmatrix} 0 & 1 & 0 \\ 1 & 0 & 0 \\ 0 & 0 & 1 \end{pmatrix}$$

*Note that matrix multiplication gives*

$$\begin{pmatrix} 0 & 1 & 0 \\ 1 & 0 & 0 \\ 0 & 0 & 1 \end{pmatrix} \begin{pmatrix} 1 \\ 2 \\ 3 \end{pmatrix} = \begin{pmatrix} 2 \\ 1 \\ 3 \end{pmatrix}$$

*from which we can recover the 2 × 3 array.*

How does the permutation function relate to the permutation matrix? The theorem below explains parts of this relationship.

**Theorem 3.2.1.** *If $f : \mathbb{Z}_n \to \mathbb{Z}_n$ is a permutation then*

$$(a) \qquad P(f) \begin{pmatrix} 1 \\ 2 \\ \vdots \\ n \end{pmatrix} = \begin{pmatrix} f(1) \\ f(2) \\ \vdots \\ f(n) \end{pmatrix}$$

*(b) $P(f)^{-1} = P(f^{-1})$, (the inverse of the permutation matrix is the matrix of the inverse of the permutation),*

*(c) $P(fg) = P(f)P(g)$ (the permutation matrix of the product is the product of the permutation matrices).*

**Proof:** If $\vec{v}$ is the column vector with entries $v_1, v_2, ..., v_n$ (the $v_i$ are arbitrary real numbers) then $P(f)\vec{v}$ is the column vector whose $i^{th}$ coordinate is equal to $v_j$ if $f$ sends $i$ to $j$. Since, in this case, $j = f(i)$ (here we write $f(i)$ to denote the image of $i$ under the permutation $f$, even though $i$ really gets plugged into $f$ on the left), this implies that $P(f)\vec{v}$ is the column vector with entries $v_{f(1)}, v_{f(2)}, ..., v_{f(n)}$. This proves (a).

Note (b) is a consequence of (c) so we need only prove (c). We compute $P(fg)\vec{v}$ and $P(f)P(g)\vec{v}$. By the same reasoning as in (a), we find that the $i^{th}$ coordinate of $P(fg)\vec{v}$ is $v_{(fg)(i)}$. Similarly, the $i^{th}$ coordinate of $P(g)\vec{v}$ is $v_i' = v_{g(i)}$. Therefore, the $i^{th}$ coordinate of $P(f)(P(g)\vec{v})$ is $v_{f(i)}' = v_{g(f(i))} = v_{(fg)(i)}$. This implies $P(fg)\vec{v} = P(f)P(g)\vec{v}$. Since the $v_i$ were arbitrary real numbers, this implies the theorem. $\square$

**Example 3.2.4.** *Let*

$$f = \begin{pmatrix} 1 & 2 & 3 \\ 2 & 1 & 3 \end{pmatrix}, \qquad g = \begin{pmatrix} 1 & 2 & 3 \\ 3 & 2 & 1 \end{pmatrix}, \qquad h = \begin{pmatrix} 1 & 2 & 3 \\ 2 & 3 & 1 \end{pmatrix},$$

*so $f = f^{-1}, g = g^{-1}, h = fg$. Moreover,*

$$P(g) = \begin{pmatrix} 0 & 0 & 1 \\ 0 & 1 & 0 \\ 1 & 0 & 0 \end{pmatrix}, \qquad P(h) = \begin{pmatrix} 0 & 1 & 0 \\ 0 & 0 & 1 \\ 1 & 0 & 0 \end{pmatrix}.$$

*A direct matrix calculation verifies that* $P(f)P(g) = P(fg) = P(h)$ *and* $P(h^{-1}) = P(g^{-1}f^{-1}) = P(g^{-1})P(f^{-1}) = P(g)^{-1}P(f)^{-1}$, *as predicted by the above theorem.*

The matrix can be determined from the graph of the function $f : \mathbb{Z}_n \to \mathbb{Z}_n$ as follows: in the $n \times n$ grid of integral points $(x, y)$, with $x$ and $y$ integers between 1 and $n$ inclusive, fill in all the plotted points with 1's and all the unplotted points with 0's The resulting $n \times n$ array is the matrix $P(f)$.

Rubik cubers will often, without knowing it perhaps, use the following lemma to solve their cube.

**Lemma 3.2.3.** *Let* $r \in S_n$ *denote any permutation and let* $i, j$ *denote distinct integers belonging to* $\{1, 2, ..., n\}$. *Let* $s$ *denote a permutation sending* $i$ *to* $j$:

$$s(i) = j.$$

*Then* $s^r = r^{-1}sr$ *is a permutation sending* $r(i)$ *to* $r(j)$:

$$s^r(r(i)) = r(j).$$

*More specifically (and this is the case which this lemma is most often applied in this book): let* $i_1, i_2, ..., i_k$ *denote distinct integers belonging to* $\{1, 2, ..., n\}$. *Let* $s$ *denote the permutation sending* $i_j$ *to* $i_{j+1}$:

$$s(i_j) = i_{j+1}, \qquad 1 \le j < k, \qquad s(i_k) = i_1, \qquad s(m) = m, \ \forall m \notin \{i_1, ..., i_k\}.$$

*Then* $s^r = r^{-1}sr$ *is the permutation sending* $r(i_j)$ *to* $r(i_{j+1})$:

$$s^r(r(i_j)) = r(i_{j+1}), \qquad 1 \le j < k, \qquad s^r(r(i_k)) = r(i_1),$$

$$s^r(m) = m, \ \forall m \notin \{r(i_1), ..., r(i_k)\}.$$

**Example 3.2.5.** *Let us label the 12 edges of the Rubik's Cube using the* **Singmaster notation***:*

- *uf denotes the 'up, front edge',*

- *ul denotes the 'up, left edge',*

- *ur denotes the 'up, right edge',*

- *ub denotes the 'up, back edge',*

- *df denotes the 'down, front edge',*

- *dl denotes the 'down, left edge',*

- *dr denotes the 'down, right edge',*

- *db denotes the 'down, back edge',*

- *fl denotes the 'front, left edge',*

- *fr denotes the 'front, right edge',*
- *bl denotes the 'back, left edge',*
- *br denotes the 'back, right edge'.*

*If you have a Rubik's Cube move s which is a 3-cycle on 3 particular edges, say*

$$uf \longmapsto ul \longmapsto ur \longmapsto uf,$$

*and another move r which sends these edges somewhere else, say $r = F^2$ so that $r : uf \longmapsto df$ but leaves the other edges alone, then $r^{-1}sr$ is the permutation*

$$df \longmapsto ul \longmapsto ur \longmapsto df.$$

*Try it!*

A proof of this lemma will be given in chapter 8.

## 3.3 Cycle notation

The most common notation for a permutation is the 'cycle notation', due to Cauchy in the mid 1800's. This notation is more compact that the array notation we've been using and from this point on we will switch over to the cycle notation. If $a_1, a_2, ..., a_r \in \mathbb{Z}_n$ then the symbol

$$(a_1, a_2, ..., a_r) \qquad \text{(some r less than or equal to n)}$$

denotes the permutation $f$ of $\mathbb{Z}_n$ which sends $a_1$ to $a_2$, sends $a_2$ to $a_3$, ..., sends $a_{r-1}$ to $a_r$, sends $a_r$ to $a_1$, and leaves all the other numbers in $\mathbb{Z}_n$ alone. (By the way, when there is no possible confusion with the notation for r-tuples, we sometimes use the notation $(a_1, a_2, ..., a_r)$ in place of $(a_1 \ a_2 \ ... \ a_r)$.) In other words,

$$f(a_1) = a_2, \ f(a_2) = a_3, \ ..., f(a_r) = a_1,$$

and $f(i) = i$, if $i$ is not equal to one of the $a_1, ..., a_r$. Such a permutation is called **cyclic**. The number $r$ is called the **length** of the cycle.

We call two such cycles $(a_1, a_2, ..., a_r)$ and $(b_1, b_2, ..., b_t)$ **disjoint** if the sets $\{a_1, a_2, ..., a_r\}$ and $\{b_1, b_2, ..., b_t\}$ are disjoint.

**Lemma 3.3.1.** *If $f$ and $g$ are disjoint cyclic permutations of $\mathbb{Z}_n$ then $fg = gf$.*

**Proof:** This is true because the permutations $f$ and $g$ of $\mathbb{Z}_n$ affect disjoint collections of integers, so the permutations may be performed in either order. □

**Lemma 3.3.2.** *The cyclic permutation $(a_1, a_2, ..., a_r)$ has order $r$.*

**Proof:** Note $f(a_1) = a_2$, $f^2(a_1) = a_3$, ..., $f^{r-1}(a_1) = a_r$, $f^r(a_1) = a_1$, by definition of $f$. Likewise, for any $i = 1, ..., r$, we have $f^r(a_i) = a_i$. □

**Definition 3.3.1.**

*A* **transposition** *is a cycle* $(i, j)$ *of length* 2 *which interchanges* $i$ *and* $j$.

**Theorem 3.3.1.** *Every permutation* $f : \{1, 2, ..., n\} \to \{1, 2, ..., n\}$ *is the product of disjoint cyclic permutations. More precisely, if* $f$ *is a permutation of* $\{1, 2, ..., n\}$ *(with* $n > 1$*) then there are non-empty disjoint subsets of distinct integers*

$$A_1 = \{a_{11}, ..., a_{1,r_1}\} \subset \{1, 2, ..., n\},$$
$$A_2 = \{a_{21}, ..., a_{2,r_2}\} \subset \{1, 2, ..., n\},$$
$$\vdots$$
$$A_k = \{a_{k1}, ..., a_{k,r_k}\},$$

*such that*

$$\{1, 2, ..., n\} = A_1 \cup ... \cup A_k, \qquad n = r_1 + r_2 + ... + r_k,$$

*and*

$$f = (a_{11}, ..., a_{1,r_1})...(a_{k1}, ..., a_{k,r_k}).$$

This product is called a **cycle decomposition** of $f$ . If we rearrange the cardinalities $r_i$ of these sets $S_i$ in decreasing order, say we write this as

$$r'_1 \geq r'_2 \geq ... \geq r'_k,$$

then the $k$-tuple $(r'_1, ..., r'_k)$ is called the **cycle structure** of $f$ and $f$ is called a $(r'_1, ..., r'_k)$-**cycle**. For example, $(1, 2)(3, 4, 5)$ is a $(3, 2)$-cycle in $S_5$.
   **Proof:** The proof is constructive.
   Let $f : \mathbb{Z}_n \to \mathbb{Z}_n$ be a permutation. List all the *distinct* elements in $\mathbb{Z}_n$ you get by repeatedly applying $f$ to 1. Call them (for lack of a better notation!), $a_{10}, a_{11}, a_{12}, ..., a_{1,r_1}$, where,

$$O_1 = \{a_{10} = 1, a_{11} = f(1), a_{12} = f^2(1), ..., a_{1,r_1} = f^{r_1}(1)\}.$$

(Incidentally, this is called the **orbit of 1 under** $f$. We shall discuss orbits later.) Now list the elements in the orbit of 2:

$$O_2 = \{a_{20} = 2, a_{21} = f(2), a_{22} = f^2(2), ..., a_{2,r_2} = f^{r_2}(2)\},$$

and so on until we get to the 'orbit of $n$':

$$O_n = \{a_{n0} = n, a_{n1} = f(n), a_{n2} = f^2(n), ..., a_{n,r_n} = f^{r_n}(n)\}.$$

   An example: let $f : \mathbb{Z}_3 \to \mathbb{Z}_3$ be the 3-cycle permutation, $f(1) = 2, f(2) = 3, f(3) = 1$. In this case, $O_1 = \{1, 2, 3\}, O_2 = \{2, f(2) = 3, f(f(2)) = f(3) = 1\}$, and $O_3 = \{3, f(3) = 1, f(f((3)) = f(1) = 2\}$, so all three orbits are equal to each other in this example.

In general, if you pick any two of these $n$ orbits $O_1, ..., O_n$, they will either be the same or disjoint. Denote all the *distinct* orbits by $O'_1, ..., O'_k$. (The $O'_1, ..., O'_k$ are a subsequence of the $O_1, ..., O_n$. It doesn't matter what order you write the $O'_i$'s in or in what order you write the elements in each individual orbit.) Suppose that

$$O'_1 = \{b_{11}, ..., b_{1,s_1}\} \quad \text{so } |O'_1| = s_1,$$
$$O'_2 = \{b_{21}, ..., b_{2,s_2}\} \quad \text{so } |O'_2| = s_2,$$
$$.$$
$$.$$
$$.$$
$$O'_k = \{b_{k1}, ..., b_{k,s_k}\} \quad \text{so } |O'_k| = s_k.$$

(The $a_{ij}$'s have been relabeled as $b_{ij}$'s to try to simplify the notation.) In this case,

$$\mathbb{Z}_n = \cup_{i=1}^k O'_i = O'_1 \cup ... \cup O'_k,$$

and $s_1 + s_2 + ... + s_k = n$. The restriction of $f$ to $O'_1$, denoted $f_{O'_1} : O'_1 \to O'_1$, is equal to the $s_1$-cycle $(b_{11}, b_{12}, ..., b_{1,s_1})$. In general, the restriction of $f$ to $O'_j$, denoted $f_{O'_j} : O'_1 \to O'_j$, is equal to the $s_j$-cycle $(b_{j1}, b_{j2}, ..., b_{j,s_j})$. Since the $O'_j$'s partition $\mathbb{Z}_n$, the definition of $f$ and our construction implies that

$$f = (b_{11}, b_{12}, ..., b_{1,s_1})...(b_{k1}, b_{k2}, ..., b_{k,s_k}).$$

□

**Example 3.3.1.** • *The cycle notation for*

$$\begin{pmatrix} 1 & 2 & 3 \\ 2 & 1 & 3 \end{pmatrix}$$

*is $(1,2)(3)$ or simply $(1,2)$. In general, if any of the orbits $O_j$ in the above construction is a singleton, it is often omitted from the notation, with the implicit understanding that $f$ doesn't permute the omitted numbers.*

• *The cycle notation for*

$$\begin{pmatrix} 1 & 2 & 3 \\ 2 & 3 & 1 \end{pmatrix}$$

*is $(1, 2, 3)$.*

• *The cycle notation for*

$$\begin{pmatrix} 1 & 2 & 3 & 4 \\ 3 & 4 & 1 & 2 \end{pmatrix}$$

*is $(1, 3)(2, 4) = (2, 4)(1, 3)$.*

• *The cycle notation for*

$$\begin{pmatrix} 1 & 2 & 3 & 4 & 5 \\ 3 & 4 & 1 & 5 & 2 \end{pmatrix}$$

*is $(1, 3)(2, 4, 5) = (4, 5, 2)(1, 3)$.*

**Ponderable 3.3.1.** *The disjoint cycle decomposition of* $(2,3,7)(3,7,10)$ *is either* $(2,3)(7,10)$ *or* $(2,7)(3,10)$. *Which one is it?*

**Ponderable 3.3.2.** *Divide a square into 4 subsquares ('facets') and label them* $1, 2, 3, 4$. *For example,*

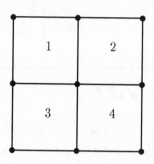

    *Let* $r$ *denote counterclockwise rotation by 90 degrees. Then, as a permutation on the facets,* $r = (1,\ 3,\ 4,\ 2)$. *Let* $f_x$ *denote reflection about the horizonal line dividing the square in two, let* $f_y$ *denote reflection about the vertical line dividing the square in two. Use the cycle notation to determine the permutations of the facets (a)* $r^2$, *(b)* $r^3$, *(c)* $f_x$, *(d)* $f_y$, *(e)* $f_x r f_x$, *(f)* $f_x f_y$.

**Ponderable 3.3.3.** *Label the 24 facets of the* $2 \times 2$ *Rubik's Cube as follows:*

| 1 | 2 |
|---|---|
| $U$ | |
| 3 | 4 |

| 5 | 6 | 9 | 10 | 13 | 14 | 17 | 18 |
|---|---|---|----|----|----|----|----|
| $L$ | | $F$ | | $R$ | | $B$ | |
| 7 | 8 | 11 | 12 | 15 | 16 | 19 | 20 |

| 21 | 22 |
|----|----|
| $D$ | |
| 23 | 24 |

*(You may want to xerox this page then cut this cube out and tape it together for this.) Let $X$ denote rotation clockwise by 90 degrees of the face labeled $x$, where $x \in \{r, l, f, b, u, d\}$ (so, for example, if $x = f$ then $X = F$). Use the cycle notation to determine the permutations of the facets given by (a) R, (b) L, (c) F, (d) B, (e) U, (f) D.*

**Lemma 3.3.3.** *A cyclic permutation is even if and only if the length of its cycle is odd. A general permutation $f : \mathbb{Z}_n \to \mathbb{Z}_n$ is odd if and only if the number of cycles of even length in its disjoint cycle decomposition is odd.*

This follows from the definition of an even/odd permutation (see Definition 3.1.1), the fact that $sign(p_1 p_2 ... p_k) = sign(p_1)sign(p_2)...sign(p_k)$, for permutations $p_i$ (because of Lemma 3.1.2), and and the fact that any $k$-cycle can be written as a product of $k - 1$ transpositions,

$$(a_1, a_2, ..., a_k) = (a_1, a_k)(a_1, a_{k-1})...(a_1, a_2). \qquad (3.1)$$

(Equation (3.1) can be proven by mathematical induction. The argument is left to the reader.) For details, we refer to related results in chapter 1 of Rotman [R] or Theorem 3.3 in Gaglione [G], §3.2, for example.)

**Lemma 3.3.4.** *The order of a permutation is the least common multiple (lcm) of the lengths $r_1, r_2, ..., r_k$ of the disjoint cycles in its cycle decomposition.*

**Example 3.3.2.** *The order of $(1, 3)(2, 4)$ is 2. It is even. The order of $(1, 3)(2, 4, 5)$ is 6. It is odd.*

## 3.4   An algorithm to list all the permutations

In Martin Gardner [Gar1], an algorithm is mentioned that lists all the permutations of $\{1, 2, ..., n\}$. This algorithm gives the fastest known method of listing all permutations of $\{1, 2, ..., n\}$. Rediscovered many times since, the procedure is due originally to the Polish mathematician Hugo Steinhaus (1887-1972). A student of David Hilbert at Göttingen, he did important work on orthogonal series, probability theory, real functions and their applications.

We shall denote each permutation by the second row in its $2 \times n$ array notation. For example, in the case $n = 2$:

$$\begin{matrix} 1 & 2 \\ 2 & 1 \end{matrix}$$

are the permutations.

To see the case $n = 3$, the idea is to

(a) write down each row $n = 3$ times each as follows:

$$\begin{matrix} 1 & 2 \\ 1 & 2 \\ 1 & 2 \\ 2 & 1 \\ 2 & 1 \\ 2 & 1 \end{matrix}$$

(b) 'weave' in a 3 as follows

$$
\begin{array}{ccc}
1 & 2 & 3 \\
1 & 3 & 2 \\
3 & 1 & 2 \\
3 & 2 & 1 \\
2 & 3 & 1 \\
2 & 1 & 3
\end{array}
$$

In case $n = 4$, the idea is to

(a) write down each row $n = 4$ times each as follows (go down the first row then down the second row):

$$
\begin{array}{ccc}
1 & 2 & 3 \\
1 & 2 & 3 \\
1 & 2 & 3 \\
1 & 2 & 3 \\
1 & 3 & 2 \\
1 & 3 & 2 \\
1 & 3 & 2 \\
1 & 3 & 2 \\
3 & 1 & 2 \\
3 & 1 & 2 \\
3 & 1 & 2 \\
3 & 1 & 2
\end{array}
\qquad
\begin{array}{ccc}
3 & 2 & 1 \\
3 & 2 & 1 \\
3 & 2 & 1 \\
3 & 2 & 1 \\
2 & 3 & 1 \\
2 & 3 & 1 \\
2 & 3 & 1 \\
2 & 3 & 1 \\
2 & 1 & 3 \\
2 & 1 & 3 \\
2 & 1 & 3 \\
2 & 1 & 3
\end{array}
$$

(b) now 'weave' a 4 in:

$$
\begin{array}{cccc}
1 & 2 & 3 & 4 \\
1 & 2 & 4 & 3 \\
1 & 4 & 2 & 3 \\
4 & 1 & 2 & 3 \\
4 & 1 & 3 & 2 \\
1 & 4 & 3 & 2 \\
1 & 3 & 4 & 2 \\
1 & 3 & 2 & 4 \\
3 & 1 & 2 & 4 \\
3 & 1 & 4 & 2 \\
3 & 4 & 1 & 2 \\
4 & 3 & 1 & 2
\end{array}
\qquad
\begin{array}{cccc}
4 & 3 & 2 & 1 \\
3 & 4 & 2 & 1 \\
3 & 2 & 4 & 1 \\
3 & 2 & 1 & 4 \\
2 & 3 & 1 & 4 \\
2 & 3 & 4 & 1 \\
2 & 4 & 3 & 1 \\
4 & 2 & 3 & 1 \\
4 & 2 & 1 & 3 \\
2 & 4 & 1 & 3 \\
2 & 1 & 4 & 3 \\
2 & 1 & 3 & 4
\end{array}
$$

In general, we have the following

**Theorem 3.4.1.** *(Steinhaus) There is a sequence of (not necessarily distinct) 2-cycles, $(a_1, b_1),...,(a_N, b_N)$, where $N = n! - 1$, such that each non-trivial permutation $f$ of $\{1, 2, ..., n\}$ may be expressed in the form*

$$
f = \prod_{i=1}^{k} (a_i, b_i),
$$

*for some $k$, $1 \le k \le N$. Furthermore, these transpositions $(a_i, b_i)$, for $k = 1, 2, ..., N$, are all distinct (but not necessarily disjoint).*

42

In other words, each permutation may be written as a product of (not necessarily disjoint) 2-cycles. This will be proven in the next section.

There is an analogous result valid only for *even* permutations: each even permutation may be written as a product of (not necessarily disjoint) 3-cycles. This will be stated more precisely (and proved) in the next section - see Proposition 9.4.1.

### 3.4.1 Why Steinhaus' algorithm works

The argument is by mathematical induction.

1. First, note that Steinhaus' algorithm works for $n = 2, n = 3$.

2. Write down all the elements of $S_{n-1}$ as a list using Steinhaus' algorithm (by the induction hypothesis, each element differs from the previous one by a suitable transposition in this list). We represent each element in the $2 \times (n-1)$ array notation. In our list, we simply record the $2^{nd}$ row in the list. ie, as an $(n-1)$-tuple. There are $(n-1)!$ elements in this list.

3. For the $1^{st}$ $(n-1)$-tuple, write an $n$ at the end, for the $2^{nd}$ $(n-1)$-tuple, write an $n$ at the beginning, for the $3^{rd}$ $(n-1)$-tuple, write an $n$ at the end, for the $4^{th}$ $(n-1)$-tuple, write an n at the beginning, and so on. Note we have a sequence of $(n-1)!$ $n$-tuples.

4. Suppose the $i^{th}$ $n$-tuple is $(n, a_1, a_2, ..., a_{n-1})$. First, act on this by the transposition $(n, a_1)$, then by the transposition $(n, a_2)$, then by $(n, a_3)$, ..., by $(n, a_n)$. The result of the last one is $(a_1, a_2, ..., a_{n-1}, n)$. Suppose the $j^{th}$ $n$-tuple is $(b_1, b_2, ..., b_{n-1}, n)$. (Note that by the induction hypothesis, $(a_1, a_2, ..., a_{n-1}, n)$ and $(b_1, b_2, ..., b_{n-1}, n)$ 'differ' only by a suitable transposition, in $S_{n-1}$.) First, act on this by the transposition $(b_1, n)$, then by the transposition $(b_2, n)$, then by $(b_3, n)$, ..., by $(b_n, n)$. The result of the last one is $(n, b_1, b_2, ..., b_{n-1})$.

5. Note that all these (for $i, j$ ranging over even or odd integers from 1 to $n-1$), the resulting $n!$ $n$-tuples are distinct. This yields a listing of $S_n$ as desired.

### 3.4.2 A side order of dessert: cake cutting

Though somewhat off the topic, another problem connected with Steinhaus is worth mentioning. In 1944 Steinhaus proposed the problem of dividing a cake into $n$ pieces so that it is 'proportional' (i.e., each person is satisfied with their share) and 'envy free' (i.e., each person is satisfied nobody is receiving more than a fair share). For $n = 2$, one person cuts the cake, the other chooses their piece. Steinhaus found a proportional but *not* envy free solution for $n = 3$. An envy free solution to Steinhaus' problem for $n = 3$ was found in 1962 by John H. Conway (who we will run into again later in this book) and, independently, by John Selfridge. For general $n$ the problem was solved by Steven Brams and Alan Taylor in 1995.

## 3.5 Permutations and bell ringing

Since the seventeenth century, and possibly before, cathedral bells in England have been rung by permuting the order of a 'round' of bells. The art and study of such bell ringing is referred to as campanology.

Fabian Stedman provided significant contributions to bell ringing. Born in 1640 to Reverend Francis Stedman, Fabian Stedman's connections with campanology took root at the early age of 15 when he moved to London to work as an apprentice to a Master Printer. While in London Stedman joined a bell-ringing society known as the 'Society of Colledg(sic) Youths,' which has since been renamed the Ancient Society of College Youths and is still in existence to-day. Stedman's major contributions to campanology are reflected in his efforts on **Tintinnalogia** and **Campanologia**, the first two books published on the subject, in 1668 and 1677, respectively.

Below is a glossary of a few essential terms:

- **Change**: the swapping of one or more disjoint pairs of adjacent bells

- **Plain change**: involves swapping one pair of adjacent bells only

- **Cross change**: involves more than one swapping pair of bells

- **Round**: A ordering of the bells (i.e., a permutation of $(1, 2, 3, ..., n)$)

In the beginning, change ringing concerned itself with a single row of bells whose order could be denoted by $(1, 2, 3, ..., n)$. Considering the case where $n = 6$ the concepts of plain and cross changes can be understood more clearly. If we use only plain changes we can generate permutations of the bells as follows:

$$
\begin{array}{cccccc}
1 & 2 & 3 & 4 & 5 & 6 \\
2 & 1 & 3 & 4 & 5 & 6 \\
2 & 1 & 4 & 3 & 5 & 6 \\
2 & 1 & 4 & 3 & 6 & 5,
\end{array}
$$

It should be fairly obvious on inspection that the first plain change swaps 1 and 2, the second swaps 3 and 4, and the third swaps 5 and 6. Considering the same set of six bells acted upon by a cross change, the same result is achieved in one change, as seen below:

$$
\begin{array}{cccccc}
1 & 2 & 3 & 4 & 5 & 6 \\
2 & 1 & 4 & 3 & 6 & 5.
\end{array}
$$

More useful and interesting patterns can be generated by combining plain and cross changes. The **plain lead on four bells** is one of the most simplistic patterns and was devised sometime around 1621 by alternating consecutive cross and plain changes as seen below:

```
1 2 3 4
2 1 4 3
2 4 1 3
4 2 3 1
4 3 2 1
3 4 1 2
3 1 4 2
1 3 2 4
1 2 3 4
```

The pattern that defines the plain lead on four bells is nothing more than a cross change followed by a plain change on the middle two bells until we reach the round, which is where we started.

Generating the permutations contained in the plain lead on four bells can be easily described using the notation for permutations we have developed. We begin by representing the cross change as $a = (1,2)(3,4)$, which swaps the first two and last two bells, and representing the plain change as $b = (2,3)$, which swaps the middle pair. We begin with the first element, $a$. To generate the next permutation we multiply this first element by $b$. To generate the third element we simply multiply this second term, $ab$, by $a$ to get $aba$. Continuing on in this manner we multiply alternately by $a$ then $b$ to generate the set $D_4 = \{a, ab, aba, (ab)^2, a(ab)^2, (ab)^3, (ab)^3a, (ab)^4\}$. This is the set of permutations in the plain lead on four bells. (It is denoted $D_4$ here for reasons which will be explained later.) Since $(ab)^4$ yields the original round, we say $(ab)^4 = 1$ and $D_4 = \{1, a, ab, aba, (ab)^2, a(ab)^2, (ab)^3, (ab)^3a\}$.

Now for a more complex example. We turn our attention now to the composition which is commonly referred to as **Plain Bob Minimus**. Plain Bob Minimus begins at the round and ends at the round $(1,2,3,4)$ and contains all possible permutations of these four bells (in a particular order, described below):

```
1 2 3 4     1 3 4 2     1 4 2 3
2 1 4 3     3 1 2 4     4 1 3 2
2 4 1 3     3 2 1 4     4 3 1 2
4 2 3 1     2 3 4 1     3 4 2 1
4 3 2 1     2 4 3 1     3 2 4 1
3 4 1 2     4 2 1 3     2 3 1 4
3 1 4 2     4 1 2 3     2 1 3 4
1 3 2 4     1 4 3 2     1 2 4 3
                        1 2 3 4
```

We can now describe this composition using the permutation notation as we did for the plain lead on four bells. Let $a = (1,2)(3,4)$ and $b = (2,3)$ represent possible changes between rows. If we look at the first column of the Plain Bob

Minimus composition, we see that it is nothing more than the plain lead on four bells. To generate the second column, we introduce a new permutation $c = (3,4)$ and we simplify our notation by letting $k = (ab)^3 ac$. Multiplying through we generate the second column,

$$\{k, ka, kab, kaba, k(ab)^2, ka(ab)^2, k(ab)^3, k(ab)^3 a\}.$$

Multiplying through by $k$ again, we generate the third column,

$$\{k^2, k^2 a, k^2 ab, k^2 aba, k^2(ab)^2, k^2 a(ab)^2, k^2(ab)^3, k^2(ab)^3 a\}.$$

This generation of Plain Bob Minimus shows that it can be expressed as the disjoint union of 'translations' of the plain lead on four bells! We shall explain this fact using group theory in the next chapter.

Now, since we chose $a = (1,2)(3,4)$, $b = (2,3)$, and $c = (3,4)$, where $b$ and $c$ are obviously by definition 2-cycle or transpositions and $a$ is the product of two such 2-cycles or transpositions, we have shown a further result, that each permutation of $\mathbb{Z}_4$ can be written as a product of 2-cycles. More generally, we can state the following theorem originally due to H. Steinhaus.

**Theorem 3.5.1.** *Let $f$ be a member of $S_n$, i.e., let $f$ be any permutation of degree $n$. Then $f$ can be written as a product of transpositions.*

To sketch a proof of this theorem (following [G]) and hence prove Theorem 3.4.1 as promised, we need only to recall that: Every permutation of $S_n$ can be written uniquely (up to order) as a product of disjoint cycles (Theorem 3.3.1 above).

**Proof:** It is a fact that any $k$-cycle can be written as a product of $k-1$ transpositions, as in equation (3.1). We see that since any permutation can be written in terms of cycles, and any cycle can be written as product of transposition, it follows that every permutation of $\mathbb{Z}_n$ can be written as a product of transpositions. $\square$

It was a knowledge of the permutations in Kent Treble Bob Major, a ringing of 8 bells, which helped Lord Peter Wimsey solve a murder mystery in Dorothy Sayer's **The Nine Tailors** [Sa].

A good reference for this section is A. White [Wh].

# Chapter 4

# A procession of permutation puzzles

'How can it be that mathematics, being after all a product of human thought independent of experience, is so admirably adapted to the objects of reality?'
*Albert Einstein*

We shall describe several permutation puzzles in this chapter. After presenting a lot of mathematics, we shall finally have a chance to have fun attacking some real puzzles! Though Einstein was talking about relativity in his quote, you'll see for yourself how admirably adapted combinatorics (and later group theory) is to describing these puzzles.

First, we nail down for the sake of discussion a definition of what a permutation puzzle is.

A **one person game** is a sequence of moves that follow certain rules:

- there are finitely many moves at each stage,

- there is a finite sequence of moves that yields a solution,

- there are no chance or random moves,

- there is complete information about each move,

- each move depends only on the present position, not on the existence or non-existence of a certain previous move (such as chess, where castling is made illegal if the king has been moved previously).

(The reader will find a fascinating discussion of 'two person games' in the book by Berlekamp, Conway, and Guy [BCG].)

A **permutation puzzle** is a one person game with the following five properties listed below. The five properties of a permutation puzzle are:

1. for some $n > 1$ depending only on the puzzle's construction, each move of the puzzle corresponds to a unique permutation of the numbers in $\mathbb{Z}_n$,

2. if the permutation of $\mathbb{Z}_n$ in (1) corresponds to more than one puzzle move then the two positions reached by those two respective moves must be indistinguishable,

3. each move, say $M$, must be 'invertible' in the sense that there must exist another move, say $M^{-1}$, which restores the puzzle to the position it was at before $M$ was performed,

4. if $M_1$ is a move corresponding to a permutation $f_1$ of $T$ and if $M_2$ is a move corresponding to a permutation $f_2$ of $T$ then $M_1 * M_2$ (the move $M_1$ followed by the move $M_2$) is either

   - not a legal move, or

   - corresponds to the permutation $f_1 * f_2$.

**Notation:** As in step 4 above, we shall always write successive puzzle moves *left-to-right*.

## 4.1  15 Puzzle

This puzzle is often associated with the puzzle inventor and problem composer Sam Loyd (1841-1911). He began publishing chess puzzles at the age of 14, had his own regular magazine column on chess problems by the age of 16, and by the time he died he had composed over 10,000 original problems in chess, mathematics, and logic.

One of the earliest and most popular permutation puzzles is the '15 Puzzle'. The 'solved position' looks like

| 1 | 2 | 3 | 4 |
|----|----|----|----|
| 5 | 6 | 7 | 8 |
| 9 | 10 | 11 | 12 |
| 13 | 14 | 15 | • |

These numbered squares represent sliding blocks which can only move into the blank square. We shall sometimes label the blank square as '16' for convenience. The moves of the puzzle consist of sliding numbered squares (such as 12, for example) into the blank square (e.g., swapping 12 with 16). In this way, each move of this puzzle may be regarded as a permutation of the integers in $\{1, 2, ..., 16\}$.

Though some accounts state that the 15 Puzzle was discovered by Sam Loyd, most authorities dispute this.

**Ponderable 4.1.1.** *Check that the five conditions of a permutation puzzle are satisfied by the 15 Puzzle.*

Not every permutation of the $\{1, 2, ..., 16\}$ corresponds to a possible position of the puzzle. For example, the position

| 1 | 2 | 3 | 4 |
|---|---|---|---|
| 5 | 6 | 7 | 8 |
| 9 | 10 | 11 | 12 |
| 13 | 15 | 14 | • |

cannot be attained from the previous position. (The mathematical reason for this is explained in 7.4 below, for example.)

Apparently, Sam Loyd applied for a U.S. patent for the above 'impossible' puzzle (the one with the 14, 15 swapped, called the '14-15 Puzzle') but since it could not be 'solved' - i.e., put in the correct order $1, 2, ..., 15$ - no working model could be supplied, so his patent was denied. (Patent or no, there were apparently thousands of them on the market already.) Before the Rubik's Cube was invented, the 15 Puzzle was probably the most popular puzzle of all time.

The moves of the 15 Puzzle may be denoted as follows: suppose we are in a position such as

| * | U | * | * |
|---|---|---|---|
| L |   | R | * |
| * | D | * | * |
| * | * | * | * |

where the squares labeled with an asterisk can be anything. The possible moves are

$$R = R_{u,r,d,l} = (r\ 16) = \text{swap r and 16},$$
$$L = L_{u,r,d,l} = (l\ 16) = \text{swap l and 16},$$
$$U = U_{u,r,d,l} = (u\ 16) = \text{swap u and 16},$$
$$D = D_{u,r,d,l} = (d\ 16) = \text{swap d and 16}.$$

**Ponderable 4.1.2.** *Verify that the five defining properties of a permutation puzzle are satisfied by the 15 Puzzle.*

We shall call the 15 Puzzle a **planar puzzle** since all its pieces lie on a flat board.

## 4.2 The Hockeypuck puzzle

The Hockeypuck puzzle is a mechanical puzzle invented by Andràs Vègh. Slice a disk with 6 diametrical cuts into 12 congruent 'pie pieces'. In the solved position, the pieces are denoted $1, 2, ..., 12$ clockwise. You may make the following moves with this puzzle:

- you may flip over any half of the disk bounded by a cut, leaving the other half unmoved (of course this will change the colors of those pie slices which were flipped);

- you may rotate the disk by any multiple of $30°$.

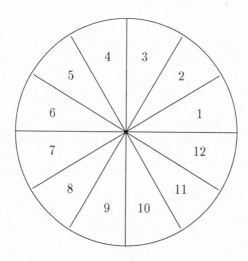

Since the pieces can be flipped over, the top halves of the pieces will be denoted $(1,0), (2,0), ..., (12,0)$ and the corresponding bottom halves by $(1,1), (2,1), ..., (12,1)$.

Basic moves: For the $i-th$ radial segment, let $f_i$ be the transformation which flips over that half of the circle which contains the $i^{th}$ pie piece on its edge. Let r denote the rotation of this circle by $30°$ radians counterclockwise.

**Ponderable 4.2.1.** *Verify that the five defining properties of a permutation puzzle are satisfied by this example.*

## 4.3 Rainbow Masterball

The rainbow Masterball puzzle will simply be referred to as a 'Masterball' in the following. It is a sphere that has been sliced like an apple along its core into 8 congruent wedges, each having a different color. It has also been sliced three times in the orthogonal direction : viewed as a globe - once along the equator, once along the Tropic of Cancer and once along the Tropic of Capricorn, as follows:

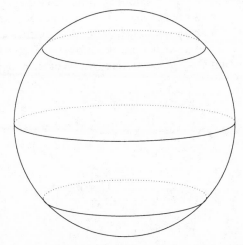

A geodesic path from the north pole to the south pole is called a **longitudinal line** and a closed geodesic path parallel to the equator is called a **latitudinal line**. There are 8 longitudinal lines and 3 latitudinal lines. In spherical coordinates, the longitudinal lines are at the angles which are multiples of $\pi/4$ (i.e., at $\theta = n\pi/4$, $n = 1, .., 8$) and the latitudinal lines are at $\phi = \pi/4, \pi/2, 3\pi/4$. (Here $\pi = 3.141592...$ as usual.)

Without these latitudinal slices, it looks similar to a beachball. Therefore, a Masterball sphere has 32 tiles of 8 distinct colors. We shall assume that the Masterball is in a fixed position in space, centered at the origin.

The sphere shall be oriented by the right-hand rule - the thumb of the right hand wrapping along the polar axis points towards the north pole. We assume that one of the longitudinal lines has been fixed once and for all. This fixed line shall be labeled '1', the next line (with respect to the orientation above) as '2', and so on.

Allowed moves: One may rotate the Masterball east-to-west by multiples of $\pi/4$ along each of the 4 latitudinal bands or by multiples of $\pi$ along each of the 8 longitudinal lines.

A **facet** will be one of the 32 subdivisions of the Masterball created by these geodesics. A facet shall be regarded as immobile positions on the sphere and labeled either by an integer $i \in \{1, ..., 32\}$ or by a pair $(i, j) \in [1, 4] \times [1, 8]$, whichever is more convenient at the time. If a facet has either the north pole or the south pole as a vertex then we call it a **small** (or **polar**) facet. Otherwise, we call a facet **large** (or **middle** or **equatorial**). A **coloring** of the Masterball will be a labeling of each facet by one of the 8 colors in such a way that

(a) each of the 8 colors occurs exactly twice in the set of the 16 small facets,

(b) each of the 8 colors occurs exactly twice in the set of the 16 large facets.

A **move** of the Masterball will be a change in the coloring of the Masterball associated to a sequence of maneuvers as described above.

If we now identify each of the 8 colors with an integer in $\{1,...,8\}$ and identify the collection of facets of the Masterball with a $4 \times 8$ array of integers in this range. To **solve** an array one must, by an appropriate sequence of moves corresponding to the above described rotations of the Masterball, put this array into a 'rainbow' position so that the matrix entries of each column has the same number. Thus the array

$$
\begin{array}{cccccccc}
1 & 2 & 3 & 4 & 5 & 6 & 7 & 8 \\
1 & 2 & 3 & 4 & 5 & 6 & 7 & 8 \\
1 & 2 & 3 & 4 & 5 & 6 & 7 & 8 \\
1 & 2 & 3 & 4 & 5 & 6 & 7 & 8
\end{array}
$$

is 'solved'. The array

$$
\begin{array}{cccccccc}
6 & 7 & 8 & 1 & 2 & 3 & 4 & 5 \\
1 & 2 & 3 & 4 & 5 & 6 & 7 & 8 \\
1 & 2 & 3 & 4 & 5 & 6 & 7 & 8 \\
1 & 2 & 3 & 4 & 5 & 6 & 7 & 8
\end{array}
$$

corresponds to a rotation of the north pole facets by $3\pi/4$.

**Notation.** We use matrix notation to denote the 32 facets of the Masterball. The generators for the latitudinal rotations are denoted $r_1, r_2, r_3, r_4$. For example, $r_1$ sends

$$
\begin{array}{cccccccc}
11 & 12 & 13 & 14 & 15 & 16 & 17 & 18 \\
21 & 22 & 23 & 24 & 25 & 26 & 27 & 28 \\
31 & 32 & 33 & 34 & 35 & 36 & 37 & 38 \\
41 & 42 & 43 & 44 & 45 & 46 & 47 & 48
\end{array}
$$

to

$$
\begin{array}{cccccccc}
12 & 13 & 14 & 15 & 16 & 17 & 18 & 11 \\
21 & 22 & 23 & 24 & 25 & 26 & 27 & 28 \\
31 & 32 & 33 & 34 & 35 & 36 & 37 & 38 \\
41 & 42 & 43 & 44 & 45 & 46 & 47 & 48
\end{array}
$$

As you look down at the ball from the north pole, this move rotates the top of the ball clockwise. The other moves $r_2, r_3, r_4$ rotate the associated band of the ball in the same direction - clockwise as viewed from the north pole.

The generators for the longitudinal rotations are denoted $f_1, f_2, ..., f_8$. For example, $f_1$ sends

$$
\begin{array}{cccccccc}
11 & 12 & 13 & 14 & 15 & 16 & 17 & 18 \\
21 & 22 & 23 & 24 & 25 & 26 & 27 & 28 \\
31 & 32 & 33 & 34 & 35 & 36 & 37 & 38 \\
41 & 42 & 43 & 44 & 45 & 46 & 47 & 48
\end{array}
$$

to

$$
\begin{array}{cccccccc}
44 & 43 & 42 & 41 & 15 & 16 & 17 & 18 \\
34 & 33 & 32 & 31 & 25 & 26 & 27 & 28 \\
24 & 23 & 22 & 21 & 35 & 36 & 37 & 38 \\
14 & 13 & 12 & 11 & 45 & 46 & 47 & 48
\end{array}
$$

With these rules, one can check the relation

$$f_5 = r_1^4 * r_2^4 * r_3^4 * r_4^4 * f_1 * r_1^4 * r_2^4 * r_3^4 * r_4^4.$$

**Ponderable 4.3.1.** *Find similar identities for* $f_6, f_7, f_8$.

Also, one can check that

$$r_1 = (f_3 * f_7)^{-1} * r_4^{-1} * f_3 * f_7.$$

**Ponderable 4.3.2.** *There are similar identities for* $r_2, r_3, r_4$. *Find them.*

Identify the facets of the Masterball with the entries of the array

$$\begin{array}{cccccccc}
8 & 7 & 6 & 5 & 4 & 3 & 2 & 1 \\
16 & 15 & 14 & 13 & 12 & 11 & 10 & 9 \\
24 & 23 & 22 & 21 & 20 & 19 & 18 & 17 \\
32 & 31 & 30 & 29 & 28 & 27 & 26 & 25
\end{array}$$

We may express the generators of the Masterball group in disjoint cycle notation as a subgroup of $S_{32}$ (the symmetric group on 32 letters):

$$r_1^{-1} = (1,2,3,4,5,6,7,8),$$
$$r_2^{-1} = (9,10,11,12,13,14,15,16),$$
$$r_3^{-1} = (17,18,19,20,21,22,23,24),$$
$$r_4^{-1} = (25,26,27,28,29,30,31,32),$$
$$f_1 = (5,32)(6,31)(7,30)(8,29)(13,24)(14,23)(15,22)(16,21),$$
$$f_2 = (4,31)(5,30)(6,29)(7,28)(12,23)(13,22)(14,21)(15,20),$$
$$f_3 = (3,30)(4,29)(5,28)(6,27)(11,22)(12,21)(13,20)(14,19),$$
$$f_4 = (2,29)(3,28)(4,27)(5,26)(10,21)(11,22)(12,23)(13,24),$$
$$f_5 = (1,28)(2,27)(3,26)(4,25)(9,20)(10,19)(11,18)(12,17),$$
$$f_6 = (8,27)(1,26)(2,25)(3,32)(16,19)(9,18)(10,17)(11,24),$$
$$f_7 = (7,26)(8,25)(1,32)(2,31)(15,18)(16,17)(9,24)(10,23),$$
$$f_8 = (6,25)(7,32)(8,31)(1,30)(14,17)(15,24)(16,23)(9,22),$$

**Ponderable 4.3.3.** *Verify that the properties of a permutation puzzle are satisfied for the Masterball.*

More information on this puzzle will be given in §§15.4, 13.7.

# 4.4 Pyraminx

The Pyraminx is one of the puzzles invented by Uwe Méffert (1939-), who also markets several of his inventions (see www.mefferts.com for more information).

A **tetrahedron** is a 4-sided regular platonic solid, all of whose faces are equilateral triangles. In the Pyraminx, each of the 4 faces of the puzzle is divided into 9 triangular facets.

There are a total of $4 \cdot 9 = 36$ facets on the Pyraminx. They will be labeled as in the figure below (the reader may want to xerox this figure (enlarging it), cut it out, then fold the corners and tape it into a tetrahedron).

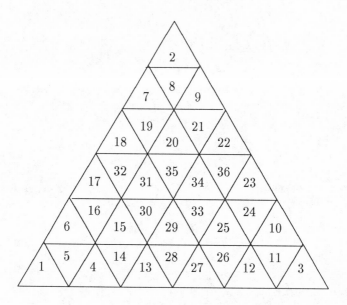

We fix an orientation of the tetrahedron in space so that you are looking at a face which we call the 'front'. We may also speak of a 'right', 'left', and 'down' face. We label the 4 faces as $f$(ront), $r$(ight), $l$(eft), $d$(own). We label the vertices $U$(p), $R$(ight), $L$(eft), and $B$(ack).

The tetrahedron itself has been subdivided into sub-tetrahedrons as follows: to each vertex $X$ (so $X \in \{U, R, L, B\}$) there is an opposing face $F$ of the solid. For each such face, we slice the solid along two planes parallel to the vertex $X$ and lying in between the face and the vertex. We want these planes, along with the face and the vertex to be spaced apart equally. The sub-tetrahedrons in the slice of the face itself will be called the **face slice** associated to the face $F$, denoted $F_1$, the sub-tetrahedrons in the middle slice parallel to the face $F$ will be called the **middle slice** associated to that face, denoted $F_2$, and the sub-tetrahedron containing the vertex $X$ to the face **tip** associated to that vertex, denoted $F_3$.

To each face labeled $F$, we have a clockwise rotation by 120 degrees of the first slice $F_1$ of the face. We shall denote this rotation also by $F_1$. This rotation only moves the facets living on the slice $F_1$. Similarly, we have a clockwise rotation by 120 degrees of the second slice $F_2$ of the face. We shall denote this rotation also by $F_2$. $F_3$ denotes the clockwise rotation by 120 degrees of the opposing sub-tetrahedron containing the vertex $X$. These moves permute the labels for the 36 facets, hence may be regarded as a permutation of the numbers $1, 2, ..., 36$.

For example, the clockwise rotation by 120 degrees (looking at the front face) of the sub-tetrahedron opposite to the front face will be denoted $F_3$. The disjoint cycle notation for this move, regarded as a permutation, is

$$F_3 = (23,\ 22,\ 36).$$

The **basic moves** are given as follows:

$$
\begin{aligned}
F_1 &= (2,32,27)(8,31,26)(7,30,12)(19,29,11)\times \\
&\quad \times(18,28,3)(1,17,13)(6,15,4)(5,16,14) \\
F_2 &= (9,35,25)(21,34,24)(20,33,10) \\
F_3 &= (23,22,36) \\
R_1 &= (3,36,17)(11,34,16)(10,35,6)\times \\
&\quad \times(24,31,5)(23,32,1)(2,22,18)(9,20,7)(8,21,19) \\
R_2 &= (12,33,15)(26,29,14)(25,30,4) \\
R_3 &= (27,28,13) \\
L_1 &= (1,28,22)(5,29,21)(4,33,9)\times \\
&\quad \times(14,34,8)(13,36,2)(3,27,23)(11,26,24)(12,25,10) \\
L_2 &= (6,30,20)(16,31,19)(15,35,7) \\
L_3 &= (17,32,18) \\
D_1 &= (13,18,23)(14,19,24)(15,20,25)\times \\
&\quad \times(16,21,26)(17,22,27)(28,32,36)(29,31,34)(30,35,33) \\
D_2 &= (4,7,10)(5,8,11)(6,9,12) \\
D_3 &= (1,2,3)
\end{aligned}
$$

All other moves are obtained by combining these moves sequentially. Indeed, later, we shall want to use moves of the form $F_2 * F_3$, for each face $F$, but the disjoint cycle notation for these permutations are a little more cumbersome to write down.

**Ponderable 4.4.1.** *Verify that the properties of a permutation puzzle are satisfied for the Pyraminx, and while doing so, remember Shakespeare's phrase, 'Though this be madness, yet there is method in't.'*

# 4.5 Rubik's Cubes

We shall consider briefly the $2 \times 2$ and $3 \times 3$ Rubik's Cubes.

## 4.5.1 $2 \times 2$ Rubik's Cube

The 'pocket' Rubik's Cube has 6 sides, or 'faces', each of which has $2 \cdot 2 = 4$ 'facets', for a total of 24 facets:

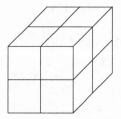

Fix an orientation of the Rubik's Cube in space. Therefore, we may label the 6 sides as $f, b, l, r, u, d$ ('front, back, left, right, up, down'). It has 8 subcubes. Each face of the cube is associated to a 'slice' of 4 subcubes which share a facet with the face. The face, along with all of the 4 cubes in the 'slice', can be rotated by 90 degrees clockwise. We denote this move by the upper case letter associated to the lower case letter denoting the face. For example, F denotes the move which rotates the front face by 90 degrees to clockwise.

We label the 24 facets of the 2 × 2 Rubik's Cube as in Ponderable 3.3.3. The 24 facets will be denoted by xyz where x is the face on which the facet lives and y, z (or z, y - it doesn't matter) indicate the 2 edges of the facet. Written in clockwise order:

| front face | fru | frd | fld | flu |
|---:|---|---|---|---|
| back face | blu | bld | brd | bru |
| right face | rbu | rbd | rfd | rfu |
| left face | lfu | lfd | lbd | lbu |
| up face | urb | urf | ulf | ulb |
| down face | drf | drb | dlb | dlf |

**Ponderable 4.5.1.** *Verify that the properties of a permutation puzzle are satisfied for this puzzle.*

For future reference, we call this system of notation (which we will also use for the 3 × 3 and 4 × 4 Rubik's Cube) the **Singmaster notation**, named after the British mathematician and puzzling enthusiast David Singmaster.

### 4.5.2 $3 \times 3$ Rubik's Cube

In this section we shall, for the most part, simply introduce enough notation (due to Singmaster [Si]) to allow us to check that the puzzle is in fact a permutation puzzle. We shall also introduce a two-person game that is easier to play and learn than solving the cube.

The Rubik's Cube has 6 sides, or 'faces', each of which has $3 \cdot 3 = 9$ 'facets', for a total of 54 facets. Since the center facets are fixed by the basic moves, there are only $54 - 6 = 48$ facets which need labeling. We label these facets $1, 2, ..., 48$ as follows:

|   |   |   | 1 | 2 | 3 |    |    |    |    |    |    |
|---|---|---|---|---|---|----|----|----|----|----|----|
|   |   |   | 4 | $U$ | 5 |    |    |    |    |    |    |
|   |   |   | 6 | 7 | 8 |    |    |    |    |    |    |
| 9 | 10 | 11 | 17 | 18 | 19 | 25 | 26 | 27 | 33 | 34 | 35 |
| 12 | $L$ | 13 | 20 | $F$ | 21 | 28 | $R$ | 29 | 36 | $B$ | 37 |
| 14 | 15 | 16 | 22 | 23 | 24 | 30 | 31 | 32 | 38 | 39 | 40 |
|   |   |   | 41 | 42 | 43 |    |    |    |    |    |    |
|   |   |   | 44 | $D$ | 45 |    |    |    |    |    |    |
|   |   |   | 46 | 47 | 48 |    |    |    |    |    |    |

then the generators, corresponding to the six faces of the cube, may be written in disjoint cycle notation as:

$F = (17, 19, 24, 22)(18, 21, 23, 20)(6, 25, 43, 16)(7, 28, 42, 13)(8, 30, 41, 11),$
$B = (33, 35, 40, 38)(34, 37, 39, 36)(3, 9, 46, 32)(2, 12, 47, 29)(1, 14, 48, 27),$
$L = (9, 11, 16, 14)(10, 13, 15, 12)(1, 17, 41, 40)(4, 20, 44, 37)(6, 22, 46, 35),$
$R = (25, 27, 32, 30)(26, 29, 31, 28)(3, 38, 43, 19)(5, 36, 45, 21)(8, 33, 48, 24),$
$U = (1, 3, 8, 6)(2, 5, 7, 4)(9, 33, 25, 17)(10, 34, 26, 18)(11, 35, 27, 19),$
$D = (41, 43, 48, 46)(42, 45, 47, 44)(14, 22, 30, 38)(15, 23, 31, 39)(16, 24, 32, 40).$

**Ponderable 4.5.2.** *Check this. (It is helpful to xerox the above diagram, cut it out and tape together a paper cube for this exercise.)*

The notation for the facets will be similar to the notation used for the 2 × 2 Rubik's Cube. The corner facets will have the same notation and the edge facets will be denoted by $xy$, where $x$ is the face the facet lives on and $y$ is the face the facet borders to. In clockwise order, starting with the upper right-hand corner of each face:

| front face | fru | fr | frd | fd | fld | fl | flu | fu |
|---|---|---|---|---|---|---|---|---|
| back face | blu | bl | bld | bd | brd | br | bru | bu |
| right face | rbu | rb | rbd | rd | rfd | rf | rfu | ru |
| left face | lfu | lf | lfd | ld | lbd | lb | lbu | lu |
| up face | urb | ur | urf | uf | ulf | ul | ulb | ub |
| down face | drf | dr | drb | db | dlb | dl | dlf | df |

**Ponderable 4.5.3.** *Verify that the properties of a permutation puzzle are satisfied for the Rubik's Cube.*

## 4.5.3  Some two-player Rubik's Cube games

We collect some two-player games associated with the Rubik's Cube, motivated by [BCG], vol II.

**The superflip game**

The position of the Rubik's Cube where every edge is flipped, but all the others subcubes are unaffected, is called the **superflip position**.

To play the game, first choose two particular faces as your up (U) and front (F) face - say white is up and red is front (assuming you have a cube with adjacent white and red faces). Imagine the cube being placed in space with rectangular coordinate axes in such a way that the *bdl* corner is at the origin $(0, 0, 0)$, the *dl* edge is along the $x$-axis and the *bl* edge is along the $z$-axis.

The rules ('slice-superflip game'):

1. Players alternate making moves starting with the cube in the solved position. The first player is determined by (say) a coin toss.

2. A move consists of flipping over exactly two edges. Both edges must lie in a slice. The edge closest to the origin (or, if this is a tie, closest to the $x$-axis) must be flipped from 'solved' to 'wrong'.

3. The first player to reach the superflip position wins.

Of course, to play this game you must know several edge-flipping moves, such as those in §15.1 below.

This game is related to a game that might be called a 'three dimensional acrostic twins game'. (See [BCG], vol II, page 441, for a two dimensional acrostic twins game.)

Several alternate versions of this game may also be played.

'Nonslice-superflip game': The rules are the same except the condition that the two edges belong to the same slice is either dropped altogether or replaced by the condition that the two edges do not belong to the same slice.

'Möbius-superflip game': The rules for this version are the same except the condition that two edges are flipped is to be replaced by any number of edges less than 6 (i.e., exactly 2, 4, or 6) are to be flipped. This game is related to a game that might be called a 'three dimensional Möbius'. (See [BCG], vol II, page 434, for a Möbius game.)

**Ponderable 4.5.4.** *Play a game!*

## 4.6  Skewb

The Skewb is a cube-shaped puzzle invented by a London journalist Tony Durham. The Skewb has been subdivided into regions differently than the Rubik's Cube: it's 8 corners have been sliced in such a way that each edge of the cube is bisected by the slice. The 8 corner pieces are each in the shape of a tetrahedron. First, fix an orientation of the cube in space, so we may talk about a front face, a back face, up, down, left, and right. Each of these 6 square faces are subdivided into 5 facets as follows:

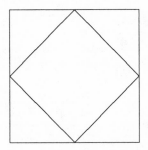

The 4 corner facets are labeled exactly as in the case of the Rubik's cube (as the lower case $xyz$, where $x$ is the label of the face the facet lives on, $y$ and $z$ the two neighboring faces).

The moves of the Skewb are different from the Rubik's Cube as well: Label the corners as $XYZ$, where $xyz$ is the notation for any of the facets belonging to that corner piece. Pick a corner $XYZ$ of the cube and draw a line $L$ passing through that corner vertex and the opposite corner vertex ('skewering the cube'). That line defines a 120 degree rotation in the clockwise direction (viewed from the line looking down onto the corner you picked). One move of the Skewb is defined in terms of this rotation as follows: Of course a 120 degree rotation of the entire cube about the line $L$ will preserve the cube but swap some faces and some vertices. The Skewb has a mechanism so that you can actually rotate half (a 'skewed' half) the Skewb by 120 degrees about $L$ and leave the other half completely fixed. This rotation of half the Skewb about $L$ will also be denoted $XYZ$.

Our labeling of the $5 \cdot 6 = 30$ facets is as follows.

|       |       |   |       |       |       |       |       |       |
|-------|-------|---|-------|-------|-------|-------|-------|-------|
|       |       | 20 |      | 17    |       |       |       |       |
|       |       |    | 16   | U     |       |       |       |       |
|       |       | 19 |      | 18    |       |       |       |       |

| 5  |    | 2 | 10 |   | 7  | 25 |    | 22 | 30 |    | 27 |
|----|----|---|----|---|----|----|----|----|----|----|----|
|    | 1  | L |    | 6 | F  | 21 | R  |    |    | 26 | B  |
| 4  |    | 3 | 9  |   | 8  | 24 |    | 23 | 29 |    | 28 |

|   |   | 15 |    | 12 |
|---|---|----|----|----|
|   |   |    | 11 | D  |
|   |   | 14 |    | 13 |

**Example 4.6.1.** *Consider the rotation $UFR$ associated to the corner $ufr$. This move permutes the facets of the Skewb. As a permutation, the disjoint cycle notation for this move is*

$$UFR = (6,16,21)(7,18,25)(10,17,24)(8,19,22).$$

*Note, in particular $UFR$ does not move the 9-facet.*

The eight **basic moves** are given by

$$FUR = (6,16,21)(7,18,25)(10,17,24)(8,19,22)$$
$$RUB = (21,16,26)(22,17,30)(25,20,29)(23,18,27)$$
$$BUL = (26,16,1)(27,20,5)(28,17,2)(30,19,4)$$
$$LUF = (1,16,6)(2,19,10)(5,18,9)(3,20,7)$$
$$FDR = (11,6,21)(25,13,9)(23,15,7)(24,12,8)$$
$$BDR = (26,11,21)(29,13,23)(27,12,22)(30,14,24)$$
$$FDL = (6,11,1)(9,15,3)(10,12,4)(8,14,2)$$
$$LDB = (1,11,26)(3,13,27)(4,14,28)(5,15,29).$$

All other moves are obtained by combining these moves sequentially.

**Ponderable 4.6.1.** *Verify that the properties of a permutation puzzle are satisfied for the Skewb.*

# 4.7 Megaminx

This puzzle is in the shape of a **dodecahedron**, a 12-sided regular platonic solid for which each of the 12 faces is a pentagon. We call two faces **neighboring** if they share an edge. There are 20 vertices and 30 edges on a dodecahedron. Each vertex meets exactly three edges and exactly three faces.

Each of the puzzle faces has been subdivided into 11 facets by slicing it with a cut that is parallel to an edge and about one-fifth the way to the opposite vertex. Each face has a total of 5 such cuts, creating 11 facets. There are a total of $11 \cdot 12 = 132$ facets on the puzzle. Each face of the solid is parallel to a face on the opposite side. Fix a face of the dodecahedron and consider a plane parallel to that face slicing through the solid and about one-fifth the way to the opposite face. Now imagine this plane as subdividing, or slicing, the dodecahedron in two. Do this for each of the 12 faces, yielding 12 such slices of the dodecahedron in total. These slices subdivide each of the 12 faces into the same 11 facets as mentioned above.

For each such slice associated to a given face $f_i$ there is a **basic move**, still denoted $f_i$, of the Megaminx given by rotation of the slice clockwise of the Megaminx by 72 degrees, leaving the rest of the dodecahedron untouched. Such a move effects 26 facets of the Megaminx and leaves the remaining 106 facets completely fixed.

Label the 12 faces of the solid as $f_1, f_2, ..., f_{12}$ in some fixed way.

Imagine that the dodecahedron is placed in 3-space in such a way that one side lies flat on the $xy$-plane and is centered along the positive $z$-axis. Let $s$ denote its height and $r$ the radius of the inscribed circle for any pentagonal face.

**Ponderable 4.7.1.** *Suppose $r = 1$. Find $s$. (This is fairly hard - see the chapter below on Platonic solids for some ideas.)*

The up face we label as $f_1$. The neighboring faces on the upper half of the puzzle may be labeled according to the picture in Figure 4.1, which represents the dodecahedron as viewed from above. The faces on the bottom are labeled similarly: the bottom face is labeled $f_{12}$ and, moving counterclockwise, label the remaining bottom faces $f_7, ..., f_{11}$, where $f_7$ neighbors both $f_2$ and $f_6$.

A more symmetric way to order the faces of the dodecahedron is as follows (see [B], exercise 18.35):

| $f_1$ | $u$ | | $f_5$ | $u_3$ | | $f_9$ | $d_4$ |
|---|---|---|---|---|---|---|---|
| $f_2$ | $u_0$ | | $f_6$ | $u_4$ | | $f_{12}$ | $d$ |
| $f_3$ | $u_1$ | | $f_7$ | $d_2$ | | $f_{11}$ | $d_1$ |
| $f_4$ | $u_2$ | | $f_8$ | $d_3$ | | $f_{10}$ | $d_0$ |

One property of this labeling is explained in the following:

**Ponderable 4.7.2.** *Suppose that the permutation $(0,1,2,3,4)$ of the numbers $\{0,1,2,3,4\}$ acts on the labels $u_0, ..., u_4$ and $d_0, ..., d_4$ in the obvious way. Show that this permutation of the faces corresponds to a rotation of the dodecahedron.*

61

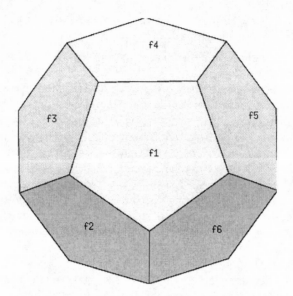

Figure 4.1: The Megaminx labeling

Notice that, like the cube, each vertex is uniquely determined by specifying the three faces it has in common. We use the notation $x.y.z$ for the vertex of the dodecahedron that lies on the three faces $x, y, z$. Note that the order is irrelevent: $x.y.z$ denotes the same vertex as $y.x.z$ or $z.y.x$.

The facets of the Megaminx may be specified as with the Rubik's Cube: a corner facet may be specified as $[x.y.z]$, where $x$ is the face the facet lives on and $y, z$ are the two neighboring faces of the facet. An edge facet may be specified by $[x.y]$, where $x$ is the face the facet lives on and $y$ is the other neighboring face of the facet. The center facet of $f_1$ will simply be denoted by $[f_1]$. We will call this label the **intrinsic label**.

We may label the facets of the up face $f_1$ as follows:

| $f_1$ facet symbol | numerical label | intrinsic label |
|:---:|:---:|:---:|
| a | 1 | $[f_1.f_6.f_2]$ |
| b | 2 | $[f_1.f_2]$ |
| c | 3 | $[f_1.f_2.f_3]$ |
| d | 4 | $[f_1.f_3]$ |
| e | 5 | $[f_1.f_3.f_4]$ |
| f | 6 | $[f_1.f_4]$ |
| g | 7 | $[f_1.f_4.f_5]$ |
| h | 8 | $[f_1.f_5]$ |
| i | 9 | $[f_1.f_5.f_6]$ |
| j | 10 | $[f_1.f_6]$ |
| k | 11 | $[f_1]$ |

For the next face (the $f_2$ face), we label the facets in such a way that the $abc$ edge of $f_1$ joins the $ghi$ edge of $f_2$:

62

| $f_2$ facet symbol | numerical label | intrinsic label |
|---|---|---|
| a | 12 | $[f_2.f_6.f_7]$ |
| b | 13 | $[f_2.f_7]$ |
| c | 14 | $[f_2.f_7.f_8]$ |
| d | 15 | $[f_2.f_8]$ |
| e | 16 | $[f_2.f_8.f_3]$ |
| f | 17 | $[f_2.f_3]$ |
| g | 18 | $[f_2.f_3.f_1]$ |
| h | 19 | $[f_2.f_1]$ |
| i | 20 | $[f_1.f_5.f_6]$ |
| j | 21 | $[f_2.f_6]$ |
| k | 22 | $[f_2]$ |

In general, we can label the remaining facets in such a way that the **basic moves** are, as permutations, given by:

$$f_1 = (1,3,5,7,9)(2,4,6,8,10)(20,31,42,53,64)\times$$
$$\times(19,30,41,52,63)(18,29,40,51,62)$$
$$f_2 = (12,14,16,18,20)(13,15,17,19,21)(1,60,73,84,31)\times$$
$$\times(3,62,75,86,23)(2,61,74,85,32)$$
$$f_3 = (23,25,27,29,31)(24,26,28,30,32)(82,95,42,3,16)\times$$
$$\times(83,96,43,4,17)(84,97,34,5,18)$$
$$f_4 = (34,36,38,40,42)(35,37,39,41,43)(27,93,106,53,5)\times$$
$$\times(28,94,107,54,6)(29,95,108,45,7)$$
$$f_5 = (45,47,49,51,53)(46,48,50,52,54)(38,104,117,64,7)\times$$
$$\times(39,105,118,65,8)(40,106,119,56,9)$$
$$f_6 = (56,58,60,62,64)(57,59,61,63,65)(49,115,75,20,9)\times$$
$$\times(50,116,76,21,10)(51,117,67,12,1)$$
$$f_7 = (67,69,71,73,75)(68,70,72,74,76)(58,113,126,86,12)\times$$
$$\times(59,114,127,7,13)(60,115,128,78,14)$$
$$f_8 = (78,80,82,84,86)(79,81,83,85,87)(71,124,97,23,14)\times$$
$$\times(72,125,98,24,15)(73,126,89,25,16)$$
$$f_9 = (89,91,93,95,97)(90,92,94,96,98)(80,122,108,34,25)\times$$
$$\times(81,123,109,35,26)(82,124,100,36,27)$$
$$f_{10} = (100,102,104,106,108)(101,103,105,107,109)\times$$
$$\times(91,130,119,45,36)(92,131,120,46,37)(93,122,111,47,38)$$
$$f_{11} = (111,113,115,117,119)(112,114,116,118,120)\times$$
$$\times(102,128,67,56,47)(103,129,68,57,48)(104,130,69,58,49)$$
$$f_{12} = (122,124,126,128,130)(123,125,127,129,131)\times$$
$$\times(100,89,78,69,111)(101,90,79,70,112)(102,91,80,71,113)$$

## 4.8 Other permutation puzzles

With regret, I have left out several puzzles. These include such marvelous items as the $4\times4$ Rubik's Cube, the $5\times5$ Rubik's Cube; 'Topspin' and 'Turnstile' (planar puzzles); 'Mozaika' and 'Equator'; '(Adam) Alexander's Star' (a stellated icosahedron), the 'Impossiball' (a spherically shaped icosahedron - see [H]), and 'Mickey's Challenge' (a spherically shaped irregular polyhedron with Mickey Mouse figures on it, this puzzle is essentially the same as the Skewb but with some added orientations of faces).

There seems to be no end to the diversity of such puzzles. 'Christoph's Jewel', designed by the German mathematician Christoph Bandelow, based on an invention of Josef Trajber, is essentially a truncated 'Rubik octahedron' and may be solved using 'super-Rubik's Cube moves' (see [H]). [For fun, you can take a Rubik's Cube, strip off all the stickers (using soap and water), and replace them with new stickers modeling a Rubik octahedron. This is because the octahedron is the dual solid of the cube, as described in the chapter on symmetry and the Platonic solids, below.]

There are even battery-run puzzles. For example the 'Orbix' puzzle, that has 12 buttons, one at each vertex of an icosahedron, which light up. This is a permutation puzzle if you think of a move (which switches certain of the buttons on/off) as permuting the elements of the set of all subsets of the 12 buttons (the subset of buttons which are lit) amongst themselves. We'll see this puzzle and other related ones later on in the chapter on Merlin's Machines.

If this isn't enough, you can find mention of other 'Rubik's Cube-like puzzles' in many other works, such as [Ru], [Si], [H], [B], [GT] and [Jwww].

# Chapter 5

# What's commutative and purple?

Q: 'What's commutative and purple?'
A: 'An abelian grape'.
*Ancient Math Joke*

A story told by Freeman Dyson (1923-), one of the great mathematical physicists of our times, goes as follows: in the early part of the last century mathematician Oswald Veblen and the physicist James Jeans were discussing the reform of the mathematical curriculum at Princeton University. Jeans argued that group theory should be omitted, claiming that group theory was a subject that will never be of any use to physics. Veblen must have won the argument because group theory continued to be taught. It is ironic indeed that group theory not only grew into one of the central themes of physics, but much of the ground-breaking research actually took place at Princeton!

In a rough sense, group theory is the mathematical study of symmetry. The Rubik's Cube displays a remarkable amount of symmetry. This chapter is an introduction to group theory - another useful tool for the Rubik's Cube mechanic wishing to fix his cube, so to speak.

When we studied permutation puzzles in Chapter 4, one of the criteria was that each move was 'invertible'. This is, in fact, one of the conditions for the set of all legal moves of a permutation puzzle to form a group.

A group can be defined very generally as a set having a small number of properties. One of these properties is that you need to be able to combine two elements of the set in some particular way to get another element. In the case of the Rubik's Cube, if you combine two moves, you get another move. The precise abstract way to state this type of property is using the notion of a 'binary operation'. A **binary operation** $*$ on a set $G$ is a function that associates to each pair of elements $(g_1, g_2)$ of $G$ a single element $g_3$, also denoted $g_3 = g_1 * g_2$, in $G$: $* : G \times G \to G$. A group $G$ has a binary operation $*$ (called the 'group operation') satisfying certain properties to be given later. For example, one property is that any element has an 'inverse element' associated to it. In the case of the Rubik's Cube, if you make any move (or sequence of moves), you can always undo the effect of that move by simply reversing every step. This 'reverse move' is the 'inverse' of the original move, as we will see. Another point to keep in mind is that the notation $*$ for the binary operation is *not* standard. Some people use $+$, some use $*$, some use $\cdot$, some omit it and simply use juxtaposition of elements. Which notation is used depends on the group in question but hopefully, this will be unambiguous from the context.

Before defining a group, we must decide, just as for sets, on what notation to use to describe or write down a group $G$. If $G$ is finite then one way is to list all the elements in $G$ and list (or tabulate) all the values of the function $*$. Another method is to describe $G$ in terms of some properties and then define a binary operation $*$ on $G$. A third method is to give a 'presentation' of $G$ (more on this later). Each method has its advantages and disadvantages. We shall eventually use all three approaches.

First, though we start with some examples.

## 5.1   The unit quaternions

In the fall of 1843, William R. Hamilton (1805-1865) was walking along the Royal Canal in Ireland with his wife. It was then and there that Hamilton found a generalization of the complex numbers - quaternions. Quaternions are 'numbers' of the form $a + bi + cj + dk$, where $a, b, c, d \in \mathbb{R}$ are real numbers, $i$ is the usual $\sqrt{-1}$ and $j$ and $k$ satisfy

$$i^2 = j^2 = k^2 = ijk = -1.$$

In fact, Hamilton could not resist the impulse to carve the formulae for the quaternions in the stone of Brougham Bridge as he and his wife passed it.

Quaternions are not only useful for physics but also for computer graphics animation. They can be added or subtracted easily:

$$(a_1+b_1i+c_1j+d_1k)+(a_2+b_2i+c_2j+d_2k) = (a_1+a_2)+(b_1+b_2)i+(c_1+c_2)j+(d_1+d_2)k.$$

To multiply two quaternions, you use the distributive law and the multiplication rules for $i, j, k$ given above. One important difference between multiplying two quaternions $q_1, q_2$ together and multiplying two real numbers $r_1, r_2$ together is that, in general, $q_1q_2 \neq q_2q_1$ (for example, $ij \neq ji$) whereas we always have $r_1r_2 = r_2r_1$ (for example, $\sqrt{2} \cdot 3 = 3 \cdot \sqrt{2}$).

Let $Q$ denote the **quaternion group**:

$$Q = \{1, -1, i, -i, j, -j, k, -k\}.$$

The elements of $Q$ are known as **unit quaternions**. As above, the elements satisfy the rules: $i^2 = j^2 = k^2 = -1, ij = k, jk = i, ki = j$, and in general,

$xy = -yx$ for $x, y$ belonging to $\{i, j, k\}$. Multiplication for this group can be visualized by the following picture.

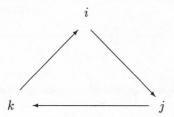

This handy mnemonic device indicates that $ij = k$, since the ordering is consistent with the arrows, and $ji = -k$, since the ordering is not consistent with the arrows.

Though Hamilton discovered the quaternions, it apparently was Cayley who first noticed that the unit quaternions form a group.

## 5.2 Finite cyclic groups

Consider the set of Rubik's Cube moves $G = \{1, R, R^2, R^3\}$. We make several more-or-less obvious observations:

- If you make any other rotations of the right face of the cube you won't get any new moves, i.e. ones not included in this set. In particular, if you compose any two moves in this set you get another move in this set.

- The move that 'undoes' the effect of $R$ is in this set (in fact, $R$ followed by $R^3$ is the identity move 1, so $R^3$ 'undoes' $R$).

- If $* : G \times G \to G$ is simply the map defined by sending a pair $(R^i, R^j)$, with $0 \le i, j \le 3$, to the composition of these two moves, $R^i * R^j = R^{i+j}$, then $*$ is a binary operation.

It turns out that this set $G$ with this operation $*$ is an example of a group.

**Example 5.2.1.** *Let $C_{12}$ be the set whose elements are $\{0, 1, ..., 11\}$ and for which the group operation $*$ is simply 'addition mod 12'. This is how one adds time on a clock (except that we call '12 o'clock' '0 o'clock'). Thus $5 + 8 = 1$, $1 + 11 = 0$, and so on.*

*Questions: What is the ('inverse') element that, when added to 5, gives you 0? What is the inverse element of 1?*

*This group is called the **cyclic group of order 12**.*

**Definition 5.2.1.** *Let $n > 1$ be an integer and let $C_n$ be the group whose elements are $\{0, 1, ..., n-1\}$ and for which the group operation is simply 'addition modulo $n$'. This group is called the **cyclic group of order $n$**. It is often times also denoted by $\mathbb{Z}/n\mathbb{Z}$.*

## 5.3 The dihedral group

Pick an integer $n > 2$ and let $R$ be a regular $n$-gon centered about the origin in the plane. If $n = 3$ then $R$ is an equilateral triangle, if $n = 4$ then $R$ is a square, if $n = 5$ then $R$ is a pentagon, and so on. Let $G$ denote the set of all linear transformations of the plane to itself that preserve the figure $R$. The binary operation $\circ : G \times G \to G$ given by composition of functions gives $G$ the structure of a group. This group is called the **group of symmetries of $R$**.

Incidentally, if we regard $R$ as a figure in 3-space centered above the origin and let $G$ denote the set of all linear transformations *of 3-space* then we obtain a slightly larger group in some cases (see Example 9.3.4 below and [NST] for more details).

Label the vertices of the $n$-gon as 1, 2, ..., $n$. The group $G$ permutes these vertices amongst themselves, hence each $g \in G$ may be regarded as a permutation of the set of vertices $V = \{1, 2, ..., n\}$. In this way, we may regard $G$ as a permutation group since it is the subgroup of $S_n$ generated by the elements of $G$.

The fact that this group has $2n$ elements follows from a simple counting argument: Let $r \in G$ denote the element that rotates $R$ by $2\pi/n$ radians counterclockwise about the center. Let $L$ be a line of symmetry of $R$ that bisects the figure into two halves. Let $s$ denote the element of $G$ that is reflection about $L$. There are $n$ rotations by a multiple of $2\pi/n$ radians about the center in $G$: $1, r, r^2, ..., r^{n-1}$. There are $n$ elements of $G$ that are composed of a reflection about $L$ and a rotations by a multiple of $2\pi/n$ radians about the center: $s, s \circ r, s \circ r^2, ..., s \circ r^{n-1}$. These comprise all the elements of $G$.

The symmetry group of $R$ is known as the **dihedral group of order $2n$**, denoted $D_{2n}$.

**Example 5.3.1.** *Let $G$ be the symmetry group of the square: i.e., the group of symmetries of the square generated by the rigid motions*

$$g_0 = 90 \text{ degrees clockwise rotation about O,}$$
$$g_1 = \text{reflection about } \ell_1,$$
$$g_2 = \text{reflection about } \ell_2,$$
$$g_3 = \text{reflection about } \ell_3,$$
$$g_4 = \text{reflection about } \ell_4,$$

*where $\ell_1, \ell_2, \ell_3$ denote the lines of symmetry in the picture below:*

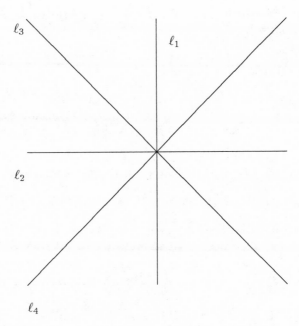

*The elements of G are*

$$1, g_0, g_0^2, g_0^3, g_1, g_2, g_3, g_4.$$

*Let X be the set of vertices of the square. Then G acts on X.*

## 5.4 The symmetric group

> 'The art of doing mathematics consists in finding that special case which contains all the germs of generality.'
> *David Hilbert*

Before defining anything, we shall provide a little motivation for some general notions that will arise later. Each move of the Rubik's Cube may be regarded as a permutation of the set of 54 facets of the cube. It is natural from our perspective to study in general permutations of a set. As Hilbert advised us, we are looking for a really good special case to focus our attention on and the Rubik's Cube group is the example we will use.

First, some basic ideas to get us started.

Let $X$ be any finite set and let $S_X$ denote the set of all permutations of $X$ onto itself:

$$S_X = \{f : X \to X \mid f \text{ is a bijection}\}.$$

This set has the following properties:

1. if $f, g$ belong to $S_X$ then $fg$ (the composition of these permutations) also belongs to $S_X$ ('closed under compositions'),

69

2. if $f, g, h$ all belong to $S_X$ then $(fg)h = f(gh)$, ('associativity'),

3. the identity permutation $I : X \to X$ belongs to $S_X$ ('existence of the identity'),

4. if $f$ belongs to $S_X$ then the inverse permutation $f^{-1}$ also belongs to $S_X$ ('existence of the inverse').

The set $S_X$ is called the **symmetric group of** $X$. We shall usually take for the set $X$ a set of the form $\{1, 2, ..., n\}$, in which case we shall denote the symmetric group by $S_n$. This group is also called the **symmetric group on** $n$ **letters**.

**Example 5.4.1.** *Suppose* $X = \{1, 2, 3\}$. *We can describe* $S_X$ *as*

$$S_X = \{I, s_1 = (1, 2), s_2 = (2, 3), s_3 = (1, 3, 2), s_4 = (1, 2, 3), s_5 = (1, 3)\}.$$

*We can compute all possible products of two elements of the group and tabulate them in a multiplication table as*

|       | $I$   | $s_1$ | $s_2$ | $s_3$ | $s_4$ | $s_5$ |
|-------|-------|-------|-------|-------|-------|-------|
| $I$   | $I$   | $s_1$ | $s_2$ | $s_3$ | $s_4$ | $s_5$ |
| $s_1$ | $s_1$ | $I$   | $s_3$ | $s_2$ | $s_5$ | $s_4$ |
| $s_2$ | $s_2$ | $s_4$ | $I$   | $s_5$ | $s_1$ | $s_3$ |
| $s_3$ | $s_3$ | $s_5$ | $s_1$ | $s_4$ | $I$   | $s_2$ |
| $s_4$ | $s_4$ | $s_2$ | $s_5$ | $I$   | $s_3$ | $s_1$ |
| $s_5$ | $s_5$ | $s_3$ | $s_4$ | $s_1$ | $s_2$ | $I$   |

**Ponderable 5.4.1.** *Verify the four properties of* $S_X$ *mentioned above. (Note that the verification of associativity follows from the associative property of the composition of functions - see the Ponderable 3.1.4).*

## 5.5  General definitions

We take the above four properties of the symmetric group as the four defining properties of a group. The definition of a group below (or 'abstract group', to be precise) was first formulated by Cayley in 1854. (We met Cayley already in §2.2.1 in connection with his work on matrices.)

**Definition 5.5.1.** *Let* $G$ *be a set and suppose there is a mapping*

$$* : G \times G \times G$$
$$(g_1, g_2) \longmapsto g_1 * g_2$$

*(called the group's* **operation***) satisfying*

*(G1) if* $g_1, g_2$ *belong to* $G$ *then* $g_1 * g_2$ *belongs to* $G$ *('G is closed under* $*$*'),*

*(G2) if* $g_1, g_2, g_3$ *belong to* $G$ *then* $(g_1 * g_2) * g_3 = g_1 * (g_2 * g_3)$ *('associativity'),*

*(G3) there is an element $1 \in G$ such that $1 * g = g * 1 = g$ for all $g \in G$ ('existence of an identity'),*

*(G4) if $g$ belongs to $G$ then there is an element $g^{-1} \in G$, called the **inverse of** $g$ such that $g * g^{-1} = g^{-1} * g = 1$ ('existence of inverse').*

*Then $G$ (along with the operation $*$) is a **group**.*

For example, if $R$ is the Rubik's Cube move associated to rotating the right face one quarter turn then $R^{-1} = R^3$.

**Remark 5.5.1.** *In general, when a set $G$ and a binary operation $*$ are explicitly given and when $(G, *)$ is then expected to be a group, usually the hardest condition to verify is associativity. In the special case when $G$ is a set of permutations of a set $X$ and $*$ is simply the usual multiplication of permutations then associativity is easy to verify since all the elements of $G$ are functions (from $X$ to $X$) and the binary operation is function composition. In this case, associativity is more-or-less 'obvious'.*

**Example 5.5.1.** *Actually, this is a 'non-example'. Let $S$ be the set of all legal moves (one can eventually make from a given legally obtained position) of the 15 Puzzle (as described in Chapter 4). In a given position, for example the solved position, there aren't that many possibilities: there are only 2 moves in the solved position and there are never any more than 4 moves possible from any position.*

*From the solved position one can move $(15, 16)$ and $(12, 16)$ (where 16 denotes the blank square) but not for example $(1, 16)$. Since $(15, 16), (12, 16) \in S$ and since $(12, 16)(15, 16)$ is not a legal move, it follows that composition of legal moves is not always legal. This shows that composition is not a binary operation, so property number (G1) fails to hold.*

**Lemma 5.5.1.** *(**Cancellation Law**): Let $G$ be a group and $a, b, c \in G$. If $a * c = b * c$, where $a, b, c \in G$, then $a = b$.*

The proof is left to the reader.

In the above definition, we have not assumed that there was exactly one identity element 1 of $G$ because, in fact, one can show that if there is one then it is unique. (Suppose there are two identity elements, denoted $e_1, e_2$ say. By definition, $e_1 = e_1 * e_2 = e_2$.) Likewise, if $G$ is a group and $g \in G$ then the inverse element of $g$ is unique. This is left to the reader. There are other properties of a group that can be derived from (G1)-(G4). We shall prove them as needed.

The **multiplication table** (also called 'Cayley tables' after Cayley who first introduced them) of a finite group $G$ is a tabulation of the values of the binary operation $*$, as in Example 5.4.1. Let $G = \{g_1, ..., g_n\}$. The multiplication table of $G$ is:

| * | $g_1$ | $g_2$ | $\cdots$ | $g_j$ | $\cdots$ | $g_n$ |
|---|---|---|---|---|---|---|
| $g_1$ | | | | | | |
| $g_2$ | | | | | | |
| $\vdots$ | | | | | | |
| $g_i$ | | | | $g_i * g_j$ | | |
| $\vdots$ | | | | | | |
| $g_n$ | | | | | | |

**Ponderable 5.5.1.** *Compute the multiplication table for $C_3$.*

Some properties:

**Lemma 5.5.2.** *(a) Each element $g_k \in G$ occurs exactly once in each row of the table.*

*(b) Each element $g_k \in G$ occurs exactly once in each column of the table.*

*(c) If the $(i, j)^{th}$ entry of the table is equal to the $(j, i)^{th}$ entry then $g_i * g_j = g_j * g_i$.*

*(d) If the table is symmetric about the diagonal then $g * h = h * g$ for all $g, h \in G$. (In this case, we call $G$ **abelian**.)*

Abelian groups are named after the Norwegian mathematician Niels Abel (1802-1829). Abel's short life was a tragic one dominated by poverty and the death of his father. However he had great mathematical talent and showed before Galois that the roots of the general $5^{th}$ degree polynomial,

$$x^5 + a_1 x^4 + a_2 x^3 + a_3 x^2 + a_4 x + a_5 = 0,$$

cannot be expressed in terms of radicals alone. (Galois proved a more general result about 6 years later.) Unfortunately, his health was not good (sometimes he literally could not afford enough food to eat to keep him healthy) and during a journey to visit his fiancée over Christmas he became seriously ill and died. In his short life, Abel made major contributions to other areas of mathematics as well and was awarded the Grand Prix in mathematics by the Paris Academy after his death.

I can't resist the corny pun: What didn't Isaac Newton work on group theory? He wasn't Abel.

**Definition 5.5.2.** *Let $g$ and $h$ be two elements of a group $G$. We say that $g$ **commutes** with $h$ (or that $g, h$ **commute**) if $g * h = h * g$. We call a group **commutative** (or **abelian**) if every pair of elements $g, h$ belonging to $G$ commute. If $G$ is a group that is not necessarily commutative then we call $G$ **noncommutative** (or **nonabelian**).*

**Example 5.5.2.**   • *The elements in $G = \{1, R, R^2, R^3\}$ all commute with each other, so $G$ is an abelian group.*

72

- *The integers, with ordinary addition as the group operation, is an abelian group.*

**Ponderable 5.5.2.** *Show that any group having exactly 2 elements is abelian.*

**Convention**: When dealing with groups in general we often drop the $*$ and denote multiplication simply by juxtaposition (that is, sometimes we write $gh$ in place of $g*h$), with one exception. If the group $G$ is abelian then one often replaces $*$ by $+$ and then $+$ is *not* dropped.

Now that we know the definition of a group, the question arises: how might they be described? The simplest answer is that we describe a group much as we might describe a set: we could list all its elements and give the multiplication table or we could describe all its elements and their multiplication in terms of some property from which we can verify the four properties of group. Though the first way has the distinct advantage of being explicit, it is this second alternative that is the most common since it is usually more concise.

Our objective is to introduce terminology and techniques that enable us to analyze mathematically permutation puzzles. The type of groups that arise in this context are defined next.

**Definition 5.5.3.** *Let $X$ be a finite set. Let $S = \{g_1, g_2, ..., g_n\}$ be a finite set of elements of permutations of $X$ (so that they all belong to $S_X$). Let $G$ be the set of all possible products of the form*

$$g = x_1 * x_2 ... * x_m, \qquad m > 0,$$

*where each of the $x_1, ..., x_m$ is taken from the set $S$. The set $G$, together with the group operation given by composition of permutations, is called a **permutation group** with **generators** $g_1, ..., g_n$ (or the **permutation group generated by** $S$). We sometimes write*

$$G = \langle g_1, ..., g_n \rangle \subset S_X.$$

**Example 5.5.3.**
- *If $X = \{1, 2, ..., 54\}$ and $S = \{(1, 2, 3, ..., 54)\} \subset S_{54}$ then the permutation group generated by $S$ is a cyclic group $G$ with elements $\{1, g, g^2, ..., g^{53}\}$.*

- *Let $X$ be the set of 54 facets of the Rubik's Cube and let $R, L, U, D, F, B \in S_X$ denote the basic moves of the Rubik's Cube, in the notation introduced in the previous chapter. The permutation group*

$$G = \langle R, L, U, D, F, B \rangle \subset S_X$$

*is called the **Rubik's Cube group**. We shall determine the 'structure' (i.e., its relationship with 'known groups') of this group later in this book.*

It is not too hard to justify our terminology.

**Lemma 5.5.3.** *A permutation group is a group.*

73

**Proof:** Let $G$ be a permutation group as in the above definition.

We shall only prove that each $g \in G$ has an inverse, leaving the remainder of the properties for the reader to verify. The set $\{g^n \mid n \geq 1\} \subset S_X$ is finite. There are $n_1 > 0$, $n_2 > n_1$ such that $g^{n_1} = g^{n_2}$. Then $g^{-1} = g^{n_2 - n_1 - 1}$ since $g \cdot g^{n_2 - n_1 - 1} = 1$. $\square$

Given generators $g_1, ..., g_n \in S_X$, how do you construct the permutation group $G = \langle g_1, ..., g_n \rangle$ generated by them? The algorithm below explains this.

**Algorithm:**

Input: The generators $g_1, ..., g_n$ (as permutations in $S_X$),

Output: The elements of $G$,

$S = \{g_1, ..., g_n, g_1^{-1}, ..., g_n^{-1}\}$,

$L = S \cup \{1\}$,

```
for g in S do
  for h in L do
    if g*h not in L then L = L union {g*h} endif
  endfor
endfor
```

Note that the size of the list L in the for loop changes after each iteration of the loop. The meaning of this is that the if-then command is to be executed exactly once for each element of the list.

**Remark 5.5.2.** *To create a subgroup $G$ of $S_n$ generated by elements $g_1, ..., g_k$ using GAP [Gap] is extremely easy: first, write the generators in disjoint cycle notation. Then type*

```
G:=Group(g1,g2,...,gk);
```

*where* g1,g2,...,gk *is your list of generators. For example,* Group((1,2),(1,2,3) *defines in GAP the group generated by all possible products of the permutations $(1,2)$ and $(1,2,3)$. (Which happens to be $S_3$ itself.)*

**Definition 5.5.4.** *If $G$ is a group then the **order** of $G$, denoted $|G|$, is the number of elements of $G$, if $G$ is a finite set, and $|G| = \infty$, otherwise. If $g$ is an element of the group $G$ then the **order** of $g$, denoted $\mathrm{ord}(g)$, is the smallest positive integer $m$ such that $g^m = 1$, if it exists. If such an integer $m$ does not exist then we say that $g$ has **infinite order**.*

**Example 5.5.4.**
- *We shall see later that the Rubik's cube group $G$ has the property that $|G| = 2^{27} 3^{14} 5^3 7^2 11$, which is roughly $4.3 \times 10^{19}$.*

- *There is an even permutation of order 42 in $S_{12}$, for example $(1,2)(3,4,5)(6,7,8,9,10,11,12)$. There is an odd permutation of order 15 in $S_8$, for example $(1,2,3)(4,5,6,7,8)$.*

- *There is an element in the Rubik's Cube group whose order is 1260 and no element of higher order. J. Butler found the following move of this order: $m = RU^2 D^{-1} B D^{-1}$ (see Bandelow's book [B], page 51, for another simple move of order 1260).*

74

We shall be able to make use of the following fact frequently.

**Theorem 5.5.1.** *(a) (Cauchy) Let p be a prime dividing $|G|$. There is a $g \in G$ of order p.*

*(b) (Lagrange) Let n be an integer not dividing $|G|$. There does not exist a $g \in G$ of order n.*

Augustin Louis Cauchy (1789-1857) made significant contributions to many branches of mathematics and was one of the world's greatest mathematicians of the time.

Part (a) will be proven in §5.8 below (see Corollary 5.11.2) and part (b) is a corollary of Theorem 5.6.1 below.

**Application.** As the Rubik's Cube group $G$ has the property that $|G| = 2^{27}3^{14}5^{3}7^{2}11$, it follows from this and Lagrange's theorem that there is no move of the Rubik's Cube of order 13 (since 13 does not divide $|G|$) but there is one of order 11. (That said, just because we know there exists a move of order 11 doesn't mean we know how to find one!) This is equivalent to saying there exists a cyclic subgroup of the Rubik's Cube group of order 11. There is no cyclic subgroup of the Rubik's Cube group of order 13.

**Ponderable 5.5.3.** *Let $X = \{1,2,3\}$. We use the notation of Example 5.4.1 above.*

*(a) Let G be the permutation group with generator $s_1, G = \langle s_1 \rangle$. Verify that there are only two elements in G.*

*(b) What is the order of $s_5$?*

*(c) Let G be the permutation group with generator $s_3$: $G = \langle s_3 \rangle$. Verify that there are only three elements in G.*

*(d) Find the order of $s_3$.*

*(e) Show that $S_X = \langle s_1, s_2 \rangle$.*

**Definition 5.5.5.** *If G is a group G with only one generator then we say that G is* **cyclic***.*

**Lemma 5.5.4.** *If $G = \langle g \rangle$ is a finite cyclic group with generator g then $|G| = ord(g)$.*

**Proof:** Let $m = ord(g)$, so $g^m = 1$. We can list all the elements of $G$ as follows:

$$1, g, g^2, ..., g^{m-1}.$$

There are m elements in this list. $\square$

## 5.5.1 The Gordon game

Let $(G, *)$ be a finite group, written

$$G = \{g_0 = \text{the identity}, g_1, ..., g_n\},$$

so $g_1 = 0$ if $*$ is addition and $G_1 = 1$ if $*$ is multiplication or composition. You and your opponent share a set of **move tokens**, denoted

$$M = \{g_1, ..., g_n\},$$

and **place tokens**, denoted

$$P = \{g_1, ..., g_n\}.$$

**Rules to play**:

- Players alternate turns. Each turn consists of removing one *move* token and one *place* token according to the conditions listed below. The first person who cannot make a legal play loses.

  Let $m_0 = p_0 = 1$ (or $m_0 = p_0 = 0$, if $G$ is written additively) and let $i = 1$.

- First player picks any move token $m_1 \in M$ and the place token $p_1 = m_1 \in P$. These tokens $m_1$ and $p_1$ are then removed from $M$ and $P$, respectively.

- The next player picks any move token $m_{i+1}$ such that $p_{i+1} = m_{i+1}p_i \in P$. These tokens $m_{i+1}$ and $p_{i+1}$ are then removed from $M$ and $P$, respectively.

- Increment $i$ and go to the previous step.

**Example 5.5.5.** *Let*

$$G = \mathbb{Z}/7\mathbb{Z} = \{0, 1, 2, 3, 4, 5, 6\},$$

*in the notation of Definition 5.2.1. The moves of a game are determined by recording the move tokens. One possible game is*

$$
\begin{array}{ccccccc}
\bullet & 4 & 1 & & 3 & 2 & \\
0 & 1 & 2 & 3 & 4 & 5 & 6
\end{array}
$$

*where the $\bullet$ over the identity element $0$ of the group indicates that it isn't moved and the numbers above a group element indicates when it was moved:*

| player | $m$ | $p$ | $P$ | $M$ |
|--------|-----|-----|-----|-----|
| $1^{st}$ | $m_1 = 2$ | $p_1 = 2$ | $\{1, 3, 4, 5, 6\}$ | $\{1, 3, 4, 5, 6\}$ |
| $2^{nd}$ | $m_2 = 4$ | $p_2 = 2 + 4 = 6$ | $\{1, 3, 4, 5\}$ | $\{1, 3, 5, 6\}$ |
| $1^{st}$ | $m_3 = 6$ | $p_3 = 6 + 6 = 5$ | $\{1, 3, 4\}$ | $\{1, 3, 5\}$ |
| $2^{nd}$ | $m_4 = 3$ | $p_4 = 3 + 5 = 1$ | $\{3, 4\}$ | $\{1, 5\}$ |

$2^{nd}$ player wins.

**Ponderable 5.5.4.** *Play a game!*

**Remark 5.5.3.** *If $G = \mathbb{Z}/p\mathbb{Z}$ (the cyclic group with $p$ elements) there is a conjecture that the $2^{nd}$ player has a winning strategy when $p > 5$ (see Isbell's note [I] for more details). For more general groups, strategies are not known. In fact, they haven't even been conjectured.*

**Remark 5.5.4.** *If you and your opponent both try to drag the game on as long as possible, can you exhaust the set of move tokens and the set of place tokens? The answer is known for abelian groups, dihedral groups and groups of order $< 32$. The general answer seems to be unknown.*

## 5.6 Subgroups

As already noted, the set of moves of the Rubik's Cube forms a group $G$ under the operation of composition. Suppose now you consider a subset of $G$ that also is a group under composition. Such a subset is called a 'subgroup'. For example, $\{1, R, R^2, R^3\}$ is a subgroup of $G$.

**Definition 5.6.1.** *Let $G$ be a group. A **subgroup** of $G$ is a subset $H$ of $G$ such that $H$, together with the operation $*$ inherited as a subset of $G$, satisfies the group operations (G1)-(G4) (with $G$ replaced by $H$ everywhere).*

**Notation**: If $G$ is a group then we will denote the statement '$H$ is a subgroup of $G$' by $H \subset G$.

**Example 5.6.1.**
- $2\mathbb{Z}$ *is a subgroup of* $\mathbb{Z}$.

- $2\mathbb{Z}/10\mathbb{Z} = \{0, 2, 4, 6, 8\}$ *(with addition mod 10) is a subgroup of* $\mathbb{Z}/10\mathbb{Z}$.

- $H = \{1, (1, 2, 3), (1, 3, 2)\}$ *is a subgroup of* $S_3$.

- *A permutation group $G$ generated by elements $g_1, ..., g_n$ belonging to $S_X$ is a subgroup of $S_X$, i.e., $G \subset S_X$.*

One might ask: what are all the subgroups of the Rubik's Cube group? Unfortunately, it turns out that this question is too impractical. There are too many subgroups to list, so a simple answer is not possible. In fact, no one (as far as I am aware) knows exactly how many subgroups the Rubik's Cube group has. However, later, when we have a more useful way of describing a group (using generators and relations - see §10.3), we will explicitly determine some of the subgroups of 'small' order.

We'd like an easy-to-use criteria to determine when a given subgroup $H$ of $S_{54}$ is a subgroup of the Rubik's Cube group $G$. This type of condition doesn't seem to exist, at least not in a practical form, but the following criterion is very useful.

**Theorem 5.6.1.** *(Lagrange) Let $H$ be a subgroup of a finite group $G$. Then $|H|$ divides $|G|$.*

**Proof:** For $x, y \in G$, define $x \sim y$ if $xH = yH$, where

$$xH = \{x * h \mid h \in H\}.$$

This is an equivalence relation. (The curious reader can easily check that the reflexive, symmetry, and transitivity properties hold.) Moreover, the equivalence class of $x$ consists of all elements in $G$ of the form $x * h$, for some $h \in H$, i.e., $[x] = xH$. Let $g_1, ..., g_m \in G$ denote a complete set of representatives for the equivalence classes of $G$. Because of the cancellation law for groups, $|xH| = |H|$ for each $x \in G$. Furthermore, we know that the equivalence classes partition $G$, so

$$G = \cup_{i=1}^{m}[g_i] = \cup_{i=1}^{m} g_i H.$$

Comparing cardinalities of both sides, we obtain $|G| = |g_1 H| + ... + |g_m H| = m|H|$. This proves the theorem. $\square$

**Ponderable 5.6.1.** *Show, as a corollary to the previous Theorem 5.6.1, that Theorem 5.5.1(b) is true.*

**Definition 5.6.2.** *If $H$ and $G$ are finite groups and $H \subset G$ then $|G|/|H|$ (which is an integer by Lagrange's theorem above) is called the **index** of $H$ in $G$, denoted $[G : H] = |G|/|H|$.*

**Example 5.6.2.** *Let*

$$A_n = \{g \in S_n \mid g \text{ is even}\}.$$

*This is a subgroup of $S_n$ called the **alternating subgroup of degree** $n$ . It is known (and not hard to prove using cosets) that $|A_n| = |S_n|/2$.*

**Definition 5.6.3.** *The **center** of a group $G$ is the subgroup $Z(G)$ of all elements that commute with every element of $G$:*

$$Z(G) = \{z \in G \mid z * g = g * z, \text{ for all } g \in G\}.$$

Observe that:

- The identity element always belongs to $G$. (If the identity element is the only element of $Z(G)$ then we say $G$ has **trivial center**.)

- $G$ is commutative if and only if $G = Z(G)$. (This is follows from the definitions.)

**Ponderable 5.6.2.** *Let $G = S_3$. Determine $Z(G)$ using explicit computations. If $n \geq 3$, show that $Z(S_n)$ has trivial center. (Hint: Consider the elements that commute with all the n-cycles.)*

For the center of the Rubik's Cube group, see §5.7.1 and Corollary 11.2.2 below.

## 5.7 Puzzling examples

There are many puzzles on the market that give rise to interesting groups in mathematics. A few simple examples are given in this section. More complicated examples will be studied later.

**Example 5.7.1.** *Consider an infinite chessboard, which we imagine being placed on the Cartesian plane.*

|   | X |   | X |   | X |   | X |   |
|---|---|---|---|---|---|---|---|---|
| X |   | X |   | X |   | X |   |   |
|   | X |   | X |   | X |   | X |   |
| X |   | X |   | X |   | X |   |   |
|   | X |   | X |   | X |   | X |   |
| X |   | X |   | X |   | X |   |   |
|   | X |   | X |   | X |   | X |   |
| X |   | X |   | X |   | X |   |   |

*Label one square as $(0,0)$ and call it* **the origin**. *Label the others $(m,n)$, as one would label the vertices in a lattice in the plane. Place only one chess piece, a king, at $(0,0)$. Label the move one square to the right $x$, one square to the left $x^{-1}$, the move one square forward $y$, one square backwards $y^{-1}$, and label the other moves $xy$, $x^{-1}y$, $x^{-1}y^{-1}$, and $xy^{-1}$, in the obvious way. The set of all possible kings moves may be identifies with the set*

$$\{x^m y^n \mid m,n \in \mathbb{Z}\}.$$

*This is an infinite abelian group under multiplication. The number of ways that the king can reach the square $(m,n)$ in $N$ moves is the coefficient of $x^m y^n$ in the expansion of*

$$(x + x^{-1} + y + y^{-1} + xy + x^{-1}y + x^{-1}y^{-1} + xy^{-1})^N.$$

*For example, using MAPLE one sees that there are exactly 19246920 ways to reach $(1,1)$ from the origin in 10 moves. The corner squares are of course the most interesting to reach.*

*Further details on this construction can be found in the chapter 'Wanderungen von Schachfiguren' by K. Fabel in [BFR].*

**Ponderable 5.7.1.** *Suppose that one day the chess king in the above example comes alive. He starts from $(0,0)$ and begins walking around on his chessboard taking completely random moves at each step. What is the likelyhood he ends up on the square he started on after 3 moves?*

**Ponderable 5.7.2.** *Work out a similar result as in the example above using a knight in place of a king.*

**Example 5.7.2.** *Let $M_R$ be the middle right slice rotation by 90 degrees (viewed from the right face). Define $M_F$, for the front face, and $M_U$ similarly. Consider the subgroup $H$ of the Rubik's Cube group generated by the 'square slice moves',*

$$H = \langle M_R^2, M_F^2, M_U^2 \rangle.$$

*Then $H = \langle M_R^2 \rangle \times \langle M_F^2 \rangle \times \langle M_U^2 \rangle \cong C_2 \times C_2 \times C_2 = C_2^3$, where $C_n$ denotes the cyclic group of order $n$.*

## 5.7.1  The superflip

The collection of all moves of the Rubik's Cube may be viewed as a subgroup $G$ of $S_{48}$. The center of $G$ consists of exactly two elements, the identity and the **superflip** move that has the effect of flipping over every edge, leaving all the corners alone and leaving all the subcubes in their original position ([B], page 48). (We shall prove this later, as a consequence of the 'second fundamental theorem of cube theory' in §11.2.) One move for the superflip is

$$\begin{aligned}
\text{superflip} &= R * L * F * B * U * D * R * L * F * B * U * F^2 * M_R* \\
&\quad *F^2 * U^{-1} * M_R^2 * B^2 * M_R^{-1} * B^2 * U * M_R^2 * D \\
&= R * L * F * B * U * D * R * L * F * B * U * F^2 * R^{-1}* \\
&\quad *L * D^2 * F^{-1} * R^2 * L^2 * D^2 * R * L^{-1}* \\
&\quad *F^2 * D * R^2 * L^2 * D \quad \text{(34 quarterturns)},
\end{aligned}$$

where $M_R$ is middle right slice rotation by 90 degrees (viewed from the right face). Other expressions for this move are Dik T. Winter's move

$$\begin{aligned}
\text{superflip} &= F * B * U^2 * R * F^2 * R^2 * B^2 * U^{-1} * D * F * U^2 * R^{-1} * L^{-1} * U* \\
&\quad *B^2 * D * R^2 * U * B^2 * U \quad \text{(28 quarterturns)},
\end{aligned}$$

and Mike Reid's move (found with a computer search)

$$\begin{aligned}
\text{superflip} &= R^{-1} * U^2 * B * L^{-1} * F * U^{-1} * B * D * F * U * D^{-1} * L * D^2 * F^{-1}* \\
&\quad *R * B^{-1} * D * F^{-1} * U^{-1} * B^{-1} * U * D^{-1} \quad \text{(24 quarterturns)}.
\end{aligned}$$

$$(5.1)$$

Jerry Bryan (in a Feb 19, 1995 posting to the cube-lover's email list, [CL]) showed that there is no fewer number of quarter turn moves taken from

$$\{R, R^{-1}, L, L^{-1}, U, U^{-1}, D, D^{-1}, B, B^{-1}, F, F^{-1}\},$$

that will also give this move. In the jargon, this move is 'minimal in the quarter-turn metric'.

It is known (and we will prove later) that

$$Z(G) = \{1, \text{superflip}\}.$$

**Remark 5.7.1.** *By the way, there is a 'longer' element of the Rubik's Cube group. The superflip composed with the four-spot, that we shall call* **Reid's move** *after Mike Reid, is given by*

$$\text{superflip4spot} = U^2 * D^2 * L * F^2 * U^{-1} * D * R^2 * B * U^{-1} * D^{-1} * R*$$
$$*L * F^2 * R * U * D^{-1} * R^{-1} * L * U * F^{-1} * B^{-1}. \quad (26 \text{ quarterturns}).$$
$$(5.2)$$

*This was also proven by Mike Reid (who discovered this move and the above expression for it with the help of his own computer program) to be minimal in the quarter-turn metric (see the August 2, 1998 posting in [CL]). It is possible that there is no longer move in the Rubik's Cube group in the quarter turn metric.*

## 5.7.2   Example: The two squares group

Let $H = \langle R^2, U^2 \rangle$ denote the group generated by the two square moves, $R^2$ and $U^2$ or the Rubik's Cube. (The reader with a cube in hand may want to try the **Singmaster magic grip** : the thumb and forefinger of the right hand are placed on the front and back face of the $fr$, $br$ edge, the thumb and forefinger of the left hand are placed on the front and back face of the $uf$, $ub$ edge; all moves in this group can be made without taking your fingers off the cube.) This group contains the useful 2-pair edge swap move $(R^2 * U^2)^3$.

We can find all the elements in this group fairly easily:

$$H = \{1, R^2, R^2 * U^2, R^2 * U^2 * R^2, (R^2 * U^2)^2, (R^2 * U^2)^2 * R^2, (R^2 * U^2)^3,$$
$$(R^2 * U^2)^3 * R^2, (R^2 * U^2)^4, (R^2 * U^2)^4 * R^2, (R^2 * U^2)^5, (R^2 * U^2)^5 * R^2\},$$

Therefore, $|H| = 12$. Note that $1 = (R^2 * U^2)^6$, $U^2 = (R^2 * U^2)^5 * R^2$, and $U^2 * R^2 = (R^2 * U^2)^5$.

To discover more about this group, we label the vertices of the cube as follows:

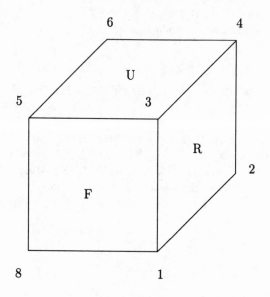

The move $R^2$ acts on the set of vertices by the permutation $(1,4)(2,3)$ and the move $U^2$ acts on the set of vertices by the permutation $(4,5)(3,6)$. We label the vertices of a hexagon as follows:

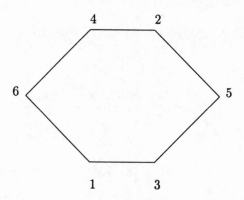

The permutation $(1,4)(2,3)$ is simply the reflection about the line of symmetry containing both 5 and 6. The permutation $(4,5)(3,6)$ is simply the reflection about the line of symmetry containing both 1 and 2. By a fact stated in section 5.3, these two reflections generate the symmetry group of the hexagon.

## 5.8 Commutators

When 'playing' with the Rubik's Cube, certain operations can occur more often than others. Besides combining two moves together, another operation that takes place frequently, is the operation: 'move 1 then move2 then inverse of move 1 then inverse of move 2'. This type of move is called a 'commutator'.

**Definition 5.8.1.** *If $g, h$ are two elements of a group $G$ then we call the element*

$$[g, h] = g * h * g^{-1} * h^{-1}$$

*the* **commutator** *of $g, h$.*

Note that $[g, h] = 1$ if and only if $g, h$ commute. Thus the commutator may be regarded as a rough measurement of the lack of commutativity. To put it simply: putting your underwear on before the pants is not the same as putting on pants before the underwear.

**Ponderable 5.8.1.** *Let $G = S_3$, the symmetric group on 3 letters. Compute the commutators*

$$[s_1, s_2], \quad [s_2, s_1].$$

**Ponderable 5.8.2.** *Let $R, U$ be as in the notation for the Rubik's Cube moves introduced in the previous chapter. Determine the order of the move $[R, U]$. (Ans: 6)*

**Definition 5.8.2.** *(Singmaster [Si]) Let $G$ be the permutation group generated by the permutations $R, L, U, D, F, B$ regarded as permutations in $S_{54}$. The* **Y commutator** *is the element*

$$[F, R^{-1}] = F * R^{-1} * F^{-1} * R.$$

*The* **Z commutator** *is the element*

$$[F, R] = F * R * F^{-1} * R^{-1}.$$

**Ponderable 5.8.3.** *(a) Guess why the Y commutator and the Z commutator have their names.*
*(b) Find the orders of the Y commutator and the Z commutator.*
*(c) Find the order of $[R, [F, U]]$.*

**Example 5.8.1.** *If $x, y$ are basic moves of the Rubik's Cube associated to faces that share an edge then*
*(a) $[x, y]^2$ permutes exactly 3 edges and does not permute any corners,*
*(b) $[x, y]^3$ permutes exactly 2 pairs of corners and does not permute any edges.*

**Definition 5.8.3.** *Let G be any group. The group G' generated by all the commutators*

$$\{[g,h] \mid g,h \text{ belong to } G\}$$

*This is called the* **commutator subgroup** *of G.*

This group may be regarded as a rough measurement of the lack of commutativity of the group G.

**Remark 5.8.1.** *We will see later that the group generated by the basic moves of the Rubik's Cube - R, L, U, D, F, B - has a relatively large commutator subgroup. In other words, roughly speaking 'most' moves of the Rubik's Cube can be generated by commutators such as the Y commutator or the Z commutator.*

**Definition 5.8.4.** *If we repeatedly take commutator subgroups we get a series of groups G, G', G'' = (G')', and so on. The* **derived series** *of a group G is the sequence of subgroups*

$$... \subset (G')' \subset G' \subset G.$$

*A group G is called* **solvable** *if one of the groups in the derived series is the trivial group consisting only of the identity.*

Morally speaking, if abelian groups are the best kind of group then the next best thing is a solvable group.

The idea of a solvable group first arose and its importance emphasized in the work of Galois. However, it was Camille Jordan (1838-1922) who introduced the concept of a composition series (of which the derived series is a special case).

Jordan is also well-known for the 'Jordan normal form' in linear algebra. It is argued in Jordan's entry in [MT] that, in fact, the theory of finite groups began with Jordan, his 1870 text **Traité substitutions et des équations algebraique** being the first book on group theory.

**Ponderable 5.8.4.** *Let G be an abelian group. Show that G is solvable.*

## 5.9 Conjugation

Besides composing two moves together and using the commutator, another operation that takes place frequently on the Rubik's Cube, is the operation: 'move 1 then move 2 then inverse of move 1'. This type of move is called a 'conjugation'.

**Definition 5.9.1.** *: If g, h are two elements of a group G then we call the element*

$$g^h = h^{-1} * g * h$$

*the* **conjugation** *of g by h.*

Note that $g^h = 1$ if and only if $g, h$ commute. Thus the conjugates may be regarded as a rough measurement of the lack of commutativity. The exponential notation is justified by the following facts that are easy to verify.

**Ponderable 5.9.1.** *Show that*
(a) $(g_1 g_2)^h = g_1^h g_2^h$,
(b) $g^{h_1 h_2} = (g^{h_1})^{h_2}$.

**Ponderable 5.9.2.** *Let $G = S_3$, the symmetric group on 3 letters, in the notation of Example 5.4.1. Compute the conjugations*

$$s_1^{s_2}, \qquad s_2^{s_1}.$$

**Ponderable 5.9.3.** *Let $R, U$ be as in the notation for the Rubik's Cube moves introduced in the previous chapter. Determine the order of the move $R^U$.* (Ans: 4)

**Definition 5.9.2.** : *We say two elements $g_1, g_2$ of $G$ are* **conjugate** *if there is an element $h \in G$ such that $g_2 = g_1^h$.*

It turns out that it is easy to see when two permutations $g, h \in S_n$ are conjugate: they are conjugate if and only if the cycles in their respective disjoint cycle decompositions have the same length when arranged from shortest to longest. (This result is due to Cauchy.) For example, the elements

$$g = (6,9)(1,3,4)(2,5,7,8), \qquad h = (1,2)(3,4,5)(6,7,8,9)$$

are conjugate in $S_9$. We shall leave the details and the proof for later - see §9.3.1,

**Ponderable 5.9.4.** *Show that the notion of conjugate defines an equivalence relation. That is, show that*
(a) *any element $g \in G$ is conjugate to itself* (**reflexive**),
(b) *if $g$ is conjugate to $h$ ($g, h$ belonging to $G$) then $h$ is conjugate to $g$* (**symmetry**),
(c) *if $g_1$ is conjugate to $g_2$ and $g_2$ is conjugate to $g_3$ then $g_1$ is conjugate to $g_3$* (**transitivity**).

**Notation**: The set of equivalence classes of $G$ under the equivalence relation given by conjugation, will be denoted $G_*$.

**Ponderable 5.9.5.** *Let $G$ be a finite group. Show that (a) $|G_*| \leq |G|$ and (b) $|G| = |G_*|$ if and only if $G$ is abelian.*

The polynomial

$$p_G(t) = \sum_{g \in G_*} t^{ord(g)},$$

is called the **generating polynomial** of the order function on $G$.

**Ponderable 5.9.6.** *Show two elements that are conjugate must have the same order.*
*(Hint: $(h^{-1}gh)^n = (h^{-1}gh)(h^{-1}gh)...(h^{-1}gh) = h^{-1}g^n h$, for $n = 1, 2, ...$ and $g, h \in G$.)*

85

Therefore, the expression $ord : G \to \mathbb{N}$ can actually be defined on $G_*$: let

$$ord(\{h^{-1}gh \mid h \in G\}) = ord(g).$$

Notice that if $g_1, g_2 \in G_*$ are both of order $d$ then $t^{ord(g_1)} = t^{ord(g_2)} = t^d$. This tells us that

$$p_G(t) = \sum_{g \in G_*} n_G(d)t^d,$$

where $n_G(d)$ denotes the number of elements in $G_*$ of order $d$. In other words, this polynomial encodes, via its coefficients, information on the elements of a given order in $G$. However, since it sums over conjugacy classes (on elements of $G$ themselves), it does not distinguish between two elements of $G$ if they are conjugate.

In [Si], §5.10D, D. Singmaster asks for the possible orders of the elements of the Rubik's Cube group and how many elements of each order there are. (A method for determining this will be described later in this book - see §9.8.2.) This question of Singmaster motivates the following problem.

**Ponderable 5.9.7.** *(hard) Determine $p_G(t)$ for the Rubik's Cube group.*

By hard, I mean you probably won't solve it overnight (at least not unless you have access to a very fast computer).

**Example 5.9.1.** *For $S_8$, the generating polynomial is*

$$t + 4t^2 + 2t^3 + 4t^4 + t^5 + 5t^6 + t^7 + t^8 + t^{10} + t^{12} + t^{15}$$

*and for $S_{12}$ it is*

$$t + 6t^2 + 4t^3 + 9t^4 + 2t^5 + 16t^6 + t^7 + 4t^8 + 2t^9 + 6t^{10} +$$
$$t^{11} + 9t^{12} + 2t^{14} + 2t^{15} + t^{18} + 2t^{20} + t^{21} +$$
$$t^{24} + t^{28} + 3t^{30} + t^{35} + t^{42} + t^{60}.$$

*For example, it follows that there is a permutation of order 42 in $S_{12}$ and a permutation of order 15 in $S_8$.*

Verifying the above examples is a tough exercise in itself. If you are computer-savvy then perhaps one of the packages GAP [Gap], MAGMA, MAPLE, or MATHEMATICA, might come in handy. Have fun!

As was already mentioned, the maximal order (of any element) in the Rubik's Cube group is 1260. So, the degree of the generating polynomial of the Rubik's Cube group is 1260.

**Definition 5.9.3.** *: Fix an element $g$ in a group $G$. The set*

$$Cl(g) = Cl_G(g) = \{h^{-1} * g * h \mid h \in G\}$$

*is called the **conjugacy class of $g$ in $G$**. It is the equivalence class of the element $g$ under the relation given by conjugation.*

Note that if $g_1 \in G$ is conjugate to $g_2 \in G$ then $Cl(g_1) = Cl(g_2)$.

How do you 'compute' conjugacy classes?

**Algorithm**

Input: A set $S$ of generators of a permutation group $G$ and a $x$ belonging to $G$

Output: The conjugacy class of $x$, $Cl_G(x)$

```
class = {x}
for y in class do
  for g in S do
    if g*y*g^{-1} not in class then
      class = class union {g*y*g^{-1}}
    endif
  endfor
endfor
```

Note that the size of the list `class` in the for loop changes after each iteration of the loop. As mentioned before, the meaning of this is that the if-then
command is to be executed exactly once for each element of the list.

**Ponderable 5.9.8.** *Find the elements in $S_4 = \langle (1,2), (2,3), (3,4) \rangle$ that are
conjugate to $(1,2,3,4)$.*

**Theorem 5.9.1.** *Any finite group may be partitioned into its distinct conjugacy
classes,*
$$G = \cup_{g \in G_*} Cl(g).$$

**Ponderable 5.9.9.** *Show that this is a consequence of Theorem 2.3.1.*

If $H$ is a subgroup of $G$ and if $g$ is a fixed element of $G$ then the set
$$H^g = \{ g^{-1} * h * g \mid h \in H \}$$
is a subgroup of $G$. Such a subgroup of $G$ is called a subgroup **conjugate** to
$H$.

**Ponderable 5.9.10.** *Let $S$ be the set of all subgroups of $G$. We define a relation
$R$ on $S$ by*
$$R = \{ (H_1, H_2) \in S \times S \mid H_1 \text{ is conjugate to } H_2 \}.$$

*Show that $R$ is an equivalence relation.*

**Ponderable 5.9.11.** *Let $G = S_n$ and let $H = \langle g \rangle$ be a cyclic subgroup generated
by a permutation $g$ of the set $\{1, 2, ..., n\}$. With respect to the equivalence relation
in the previous problem, show that a subgroup $K$ of $G$ belongs to the equivalence
class $[H]$ of $H$ in $G$ if and only if $K$ is cyclic and is generated by an element $k$
of $G$ conjugate to $g \in G$.*

87

## 5.10 Orbits and actions

Orbits are very important to understand for our purposes. Without them, understanding the mathematics of the Rubik's cube is a little like trying to fix a car without having a socket wrench. The concept of an orbit is an essential tool in the toolbox as it helps us both conceptually and computationally with some of the combinatorics that arise later. Richard Feynmann (1918-1988), the famous physicist, was reported to have said something like: 'My job here is not to make you understand such-and-such but to convince you not to give up. Because no one, not even me, really understands everything about this subject anyway. None-the-less, it is interesting and deserving of our time.' Likewise, orbits are deserving of our time.

Let's start our study of orbits with an example. Let $G$ be the Rubik's Cube group and $X$ the set of facets of the Rubik's Cube. Imagine having two identical cubes, and disassemble one and spread the facets out on a table. Imagine also that when you move the assembled cube, the facets on the table move according to the associated permutation. You can see then that each element of $G$ 'moves' the elements of $X$ around. The mathematical terminology most commonly used to describe this type of situation is to say '$G$ acts on $X$'. The precise general definition is given below.

**Definition 5.10.1.** *Let $X$ be a set and let $G$ be a group. We call $X$ a $G$-set and we say $G$ **acts on** $X$ provided the following conditions hold:*

*1. each $g$ belonging to $G$ gives rise to a function*

$$\phi_g : X \to X,$$

*2. the identity $1$ of the group $G$ defines the identity function on $X$,*

*3. if $g, h$ belong to $G$ then the composite*

$$\phi_{gh} : X \to X$$

*satisfies $\phi_{gh}(x) = \phi_h(\phi_g(x))$.*

*We call this a **left action** since the left-most element (namely, $g$) in the product $gh$ acts first.*

One can, in general, define for all $g \in G$, $\phi_g$ to be the identity map from $X$ to itself. Since this action doesn't do anything, it is called the **trivial action**. We are only interested in non-trivial actions here.

Similarly, we define 'right actions'.

**Definition 5.10.2.** *Let $X$ be a set and let $G$ be a group. We say $G$ **acts on** $X$ **on the right** provided the following conditions hold:*

*1. each $g$ belonging to $G$ gives rise to a function*

$$\phi_g : X \to X,$$

2. *the identity* 1 *of the group* $G$ *defines the identity function on* $X$,

3. *if* $g, h$ *belong to* $G$ *then the composite*

$$\phi_{gh} : X \to X$$

*satisfies* $\phi_{gh}(x) = \phi_g(\phi_h(x))$.

*We call this action a* **right action** *since the right-most element (namely, h) in the product gh acts first.*

**Remark 5.10.1.** *(1) We shall see another interpretation of these definitions later in* §9.2.
*(2) Given a left action* $\phi_g$, *one can create a right action by defining* $\phi'_g = \phi_{g^{-1}}$.

Following the standard convention, the Rubik's Cube will act on the set of facets of the cube on the right.

**Definition 5.10.3.** *Let* $G$ *act on a set* $X$. *We call the action* **transitive** *if for each pair* $x, y$ *belonging to* $X$ *there is a* $g \in G$ *such that* $y = \phi_g(x)$.

In other words, a group $G$ acts transitively on a set $X$ if *any* element $x$ of $X$ can be send to *any* other element $y$ of $X$ by some element $g$ of $G$ (depending on $x, y$).

This notion of transitive is not to be confused with the notion of transitivity of equivalence relations, even though many actions are equivalence relations (though not all are)!

The transitivity of an action turns out to be a strong condition for a group $G$ acting on a set $X$ to satisfy. In other words, if we pick a group acting on a set at 'random' then the action will probably not be transitive. This illustrates how unusual the Rubik's Cube is since it acts transitively on several different sets! For example, it transitively permutes the set of corners subcubes and transitively permutes the set of edge subcubes!

Here are several move examples.

**Example 5.10.1.** *Let* $X$ *be a finite set and let* $G = S_X$ *be the symmetric group of* $X$. *Then* $X$ *is a* $G$-set *and* $G$ *acts transitively on* $X$.

**Ponderable 5.10.1.** *Show that the action in the previous example is transitive.*

**Example 5.10.2.** *Let* $G$ *be the group of all* $2 \times 2$ *invertible matrices with real entries,* $G = GL_2(\mathbb{R})$. *This group acts on the set of column vectors on the left.*

**Ponderable 5.10.2.** *Let* $G$ *be the permutation group generated by the permutations* $R, L, U, D, F, B$, *regarded as elements of* $S_{48}$. *Let* $E$ *denote the set of edges of the cube, that we identify with the set of edge subcubes. Let* $V$ *denote the set of vertices of the cube, which we identify with the set of corner subcubes of the cube. Let* $X$ *be the set of all movable subcubes of the Rubik's Cube (which may identify as the union of* $E$ *and* $V$). *Then* $G$ *acts on* $X$, $E$, *and on* $V$.
  **Question** *(a) Is the action of* $G$ *on* $X$ *transitive?*
  *(a) Is the action of* $G$ *on* $E$ *transitive?*
  *(b) Is the action of* $G$ *on* $V$ *transitive?*

89

**Ponderable 5.10.3.** *Let $G$ be a group and let $X = G$. Define* **left multiplication of $G$ on $X$** *by:*

$$\phi_g : X \to X$$
$$x \longmapsto \phi_g(x) = g * x.$$

*(a) Show that left multiplication defines a left action of $G$ on $X$.*

*(b) Show that this action is transitive.*

*(c) Show that each $\phi_g : G \to G$ is a permutation of the set $G$, so $\phi_g \in S_G$.*

**Ponderable 5.10.4.** *Let $G$ be a group and let $X = G$. Define* **right multiplication of $G$ on $X$** *by:*

$$\phi_g : X \to X$$
$$x \longmapsto \phi_g(x) = x * g.$$

*(a) Show that right multiplication defines a right action of $G$ on $X$.*

*(b) Show that this action is transitive.*

*(c) Show that each $\phi_g : G \to G$ is a permutation of the set $G$, so $\phi_g \in S_G$.*

**Ponderable 5.10.5.** *Let $G$ be a group and let $X = G$. Define* **conjugation on $X$** *by:*

$$\phi_g : X \to X$$
$$x \longmapsto \phi_g(x) = g^{-1} * x * g.$$

*Show that conjugation defines an action of $G$ on $X$ ($X$ and $G$ as above).*

**Ponderable 5.10.6.** *Let $G$ be a group and let $X$ denote the set of all subgroups of $G$. Define* **conjugation on $X$** *by:*

$$\phi_g : X \to X$$
$$x \longmapsto \phi_g(x) = g^{-1} * x * g.$$

*Show that this defines an action of $G$ on $X$.*

**Remark 5.10.2.** *In general, the conjugation actions in the last two problems are not transitive.*

**Definition 5.10.4.** *Let $G$ be a group acting on a set $X$. For each $x$ belonging to $X$, the set*

$$G * x = \{\phi_g(x) \mid g \in G\}$$

*is called the* **orbit** *of $x \in X$ under $G$.*

Given a group acting on a set $X$, how do you 'compute' the orbits? The following algorithm addresses this.

**Algorithm**

Input: A set $S$ of generators of a permutation group $G$ and an $x$ belonging to $X$

Output: The orbit of $x$, $G * x$

```
orbit = {x}
for y in orbit do
   for g in S do
      if g*y not in orbit then orbit = orbit union {g*y} endif
   endfor
endfor
```

Note that the size of the list **orbit** in the for loop changes after each iteration of the loop. As mentioned before, the meaning of this is that the if-then command is to be executed exactly once for each element of the list.

**Ponderable 5.10.7.** *Let $G$ be the Rubik's Cube group and let $x$ be the uf edge facet. Find the orbit of $x$ under the action of $G$ using the above algorithm. Show each step.*

**Ponderable 5.10.8.** *Let $G$ be the group of moves of the Rubik's Cube and let $X$ be the set of vertices of the cube. Let $H$ be the subgroup of $G$ generated by $U * R$. Find:*

*(1) the order of $U * R$,        (Ans: 105)*

*(2) the orbit (in the Singmaster notation) of the ufr vertex in $X$ under $H$.*

**Example 5.10.3.** *Let $X$ be the set of consisting of the 48 facets of the Rubik's Cube that are not center facets - i.e., the 'movable' facets. Let $V$ denote the subset of facets that belong to some corner subcube, $E$ the subset of facets that belong to some edge subcube. Let $G$ denote the Rubik's Cube group. As noted above, $G$ acts on $X$, $V$, $E$. The action of $G$ on $X$ induced an equivalence relation as follows: we say that a facet $f_1$ is 'equivalent' to a facet $f_2$ if there is an element of $G$ (i.e., a move of the Rubik's cube) that sends one facet to the other. By Ponderable 5.10.2, there are exactly two equivalence classes, or orbits, of $G$ in $X$: $V$ and $E$. In particular, the action of $G$ on $V$ is transitive and the action of $G$ on $E$ is transitive.*

Let $G$ be the Rubik's Cube group, $V$ the set of vertices, and let $ruf \in V$ be the right upper front vertex. The left face move $L$ clearly does not affect $ruf$. In this case, we say that $L$ 'stabilizes' or 'fixes' $ruf$. The general definition is given below.

**Definition 5.10.5.** *Let $G$ be a group acting on a set $X$ with the action denoted by $\phi$. For each $x$ belonging to $X$, the subgroup*

$$stab_G(x) = G_x = \{g \in G \mid \phi_g(x) = x\}$$

*is called the **stabilizer** of $x$ in $G$.*

**Ponderable 5.10.9.** *Let $G$ be a group acting on a set $X$, $\phi_g : X \to X$, for all $g \in G$. Show that, for all $x \in X$ and all $g \in G$, we have $stab_G(\phi_g(x)) = g * stab_G(x) * g^{-1}$.*

91

**Example 5.10.4.** *Let $G$ be the group of symmetries of the square (see the example in §5.3 above), let $X$ be the set of vertices of the square, and let $x_0$ be the vertex in the lower right hand corner. Then $stab_G(x_0) = \langle g_3 \rangle$.*

**Ponderable 5.10.10.** *Let $G$ be any group and let $X = G$. Let $G$ act on $X$ by left multiplication:*

$$\phi_g : X \to X$$
$$x \longmapsto \phi_g(x) = g * x.$$

*Show that*

$$stab_G(x) = 1,$$

*for all $x$ belonging to $X = G$.*

**Lemma 5.10.1.** *Let $G$ be any group and let $X = G$. Let $G$ act on $X$ by conjugation:*

$$\phi_g : X \to X$$
$$x \longmapsto \phi_g(x) = g * x * g^{-1}.$$

*Show that*

$$stab_G(x) = \{g \in G \mid g * x = x * g\},$$

*for all $x$ belonging to $X = G$.*

This is relatively easy, so the proof will be omitted.

The 'stabilizer' subgroup for the conjugation action,

$$C_G(x) = \{g \in G \mid g * x = x * g\}$$

is called the **centralizer** of $x$ in $G$.

## 5.11 Cosets

Cosets are certain types of subsets of a group $G$. Like some political parties, they come in two 'flavors': left cosets and right cosets. Like politics, you don't want to get them confused and you rarely (if ever) want to consider using both at the same time. However, unlike politics, they basically look the same, so if you just stick with either left cosets all the time or right cosets all the time, you will be fine. For typographical (this book will avoid politics!) reasons, we shall usually use left cosets in this book.

**Example 5.11.1.** *Let $G$ be the Rubik's Cube group and $H = \{1, D, D^2, D^3\}$ the subgroup of moves of the down face. Let $g \in G$ be a move. The set $gH = \{g, gD, gD^2, gD^3\}$ is the set of moves that have the same effect as $g$, except you might have moved the down face first. This is called a 'left coset of $D$'. The set $Hg = \{g, Dg, D^2g, D^3g\}$ is called a 'right coset of $D$'. It is the set of moves that have the same effect as $g$, except you might have moved the down face afterwards.*

**Example 5.11.2.** *Let* $H = S_2 = \{1, (1,2)\}$ *and*

$$G = S_3 = \{1, (1,2), (2,3), (1,3), (1,2,3), (1,3,2)\}.$$

*The left cosets of* $H$ *in* $G$ *are*

$$H, \quad (1,2)H = H, \quad (2,3)H = \{(2,3),(1,2,3)\}, \quad (1,3)H = \{(1,3),(1,3,2)\},$$
$$(1,2,3)H = \{(1,2,3),(2,3)\} = (2,3)H, \quad (1,3,2)H = \{(1,3,2),(1,3)\} = (1,3)H.$$

*Note that* $G = H \cup (2,3)H \cup (1,3)H$.
*We leave it to the reader to compute the right cosets of* $H$ *in* $G$.

**Ponderable 5.11.1.** *Let* $H$ *be a subgroup of* $G$ *and let* $g, g' \in G$. *Show that either* $gH = g'H$ *or* $gH \cap g'H = \emptyset$.

Understanding cosets, like orbits, not only helps in counting arguments but also helps us understand the 'group structure' of $G$ (i.e., knowing how $G$ is constructed from certain 'well-understood subgroups'). Notice that if $g \in H$ then $gH = Hg = H$. Conversely, if $gH = Hg = H$ then $g \in H$.

**Definition 5.11.1.** *Let* $G$ *be a group, written multiplicatively, and* $H$ *a subgroup of* $G$. *For* $g$ *belonging to* $G$, *the subset* $g * H$ *of* $G$ *is called a* **left coset** *of* $H$ *in* $G$ *and the subset* $H * g$ *of* $G$ *is called a* **right coset** *of* $H$ *in* $G$.
*If* $G$ *is an abelian group, written additively, and* $H$ *is a sbgroup, then left cosets and right cosets are the same:* $H + g = g + H$.

Whether $G$ is abelian or not, we shall sometimes abuse terminology and drop the word 'left' or 'right' in front of coset.

If $G$ is non-abelian then the set of left cosets usually differs from the set of right cosets. However, when $G$ is finite there are always the same number of left cosets as right cosets. It was Galois, in the early 1830's, who first looked closely at groups for which the set of left cosets was the same as the set of right cosets. (Such groups are now called 'normal' and will be studied later.) Galois' work was applied to the study of roots of polynomials.

**Example 5.11.3.** *Let* $G = \mathbb{Z}$, $H = 12\mathbb{Z}$, *under ordinary addition. If* $n \in G$ *is any integer then* $n + H$ *is a coset of* $H$ *in* $G$. *In fact, all the cosets of* $H$ *can be tabulated. Each one occurs as a row exactly once in the following table:*

| ⋮ | ⋮ | | ⋮ | ⋮ |
|---|---|---|---|---|
| -12 | -11 | ... | -2 | -1 |
| 0 | 1 | ... | 10 | 11 |
| 12 | 13 | ... | 22 | 23 |
| ⋮ | ⋮ | | | ⋮ |

**Ponderable 5.11.2.** *If* $H$ *is finite, show* $|H| = |g * H| = |H * g|$.

**Ponderable 5.11.3.** *If* $X$ *is a left coset of* $H$ *in* $G$ *and* $x$ *is an element of* $G$, *show that* $x * X$ *is also a left coset of* $H$ *in* $G$.

**Notation**: The set of all left cosets is written $G/H$ and the set of all right cosets of $H$ in $G$ is denoted $H\backslash G$.

These two sets don't in general inherit a group structure from G but they are useful none-the-less. ($G/H$ is a group with the 'obvious' multiplication $(g_1 * H) * (g_2 * H) = (g_1 g_2) * H$ if and only if $H$ is a 'normal' subgroup of $G$ - we will define 'normal' below.)

As an example of their usefulness, we have the following relationship between the orbits and the cosets of the stabilizers.

**Proposition 5.11.1.** *Let G be a finite group acting on a set X. Then*

$$|G * x| = |G/stab_G(x)|,$$

*for all x belonging to X.*

**Proof:** The map

$$g * stab_G(x) \longmapsto g * x$$

defines a function $f : G/stab_G(x) \to G*x$. The interested reader can easily check that this function is a bijection since it is both and injection and a surjection. $\square$

**Corollary 5.11.1.** *Let G be a finite group acting on itself by conjugation. Let $S \subset G$ denote a complete set of representatives of the conjugacy classes $G_*$ in G and let $S' = S - Z(G)$, i.e., the subset of S of those elements that are not central. Then*

$$G = \cup_{x \in S} Cl(x) = \cup_{x \in S} G/stab_G(x) = Z(G) \cup \cup_{x \in S'} G/stab_G(x),$$

*for all x belonging to X. In particular,*

$$|G| = \sum_{x \in S} |G/stab_G(x)| = |Z(G)| + \sum_{x \in S'} |G/stab_G(x)|,$$

Sometimes this is called the **class equation** or **class formula** for groups.

**Proof:** In the first displayed equation: The first equation is Theorem 5.9.1. The second equality follows from the above proposition.

By taking cardinalities, the second displayed equation is a consequence of the first. $\square$

We are finally ready to make good on our promise to prove Cauchy's theorem.

**Corollary 5.11.2.** *Theorem 5.5.1(a) holds.*

**sketch**: The argument is by induction on $|G|$.

The result is trivial if $|G| = 1$ (since then no prime divides $|G|$).

Suppose $|G| > 1$ and let $p$ be a prime dividing $|G|$. By the induction hypothesis, we assume that the result is true for all subgroups $H$ of $G$ with $|H| < |G|$.

Suppose $x \in G$ is not central. Then its centralizer $C_G(x)$ is a proper subgroup of $G$. If $p||C_G(x)|$ then the result follows from the induction hypothesis. If $p$ does not divide $|C_G(x)|$ (for all non-central $x$) then since $|G| = |C_G(x)||G/stab_G(x)|$, by Proposition 5.11.1, it follows that $p||G/stab_G(x)|$. By Corollary 5.11.1 above, we must have $p||Z(G)|$. If $Z(G)$ is a proper subgroup of $G$ then we are done by the induction hypothesis.

Thus we may assume $G = Z(G)$ is abelian. If $G$ is cyclic then we leave it to the reader to show that $G$ contains an element of order $p$. We shall assume that $G$ is not cyclic. Let $H$ be a proper subgroup of $G$ of maximal order (this exists since $G$ is not cyclic) and let $a \in G - H$. Then $\langle a \rangle \cap H = \{1\}$ (else $a \in H$) and $G = H \cdot \langle a \rangle$ (else $H$ would not be maximal). Thus $p$ either divides $H$ or $|\langle a \rangle|$. In either case, the result follows from the induction hypothesis. $\square$

**Ponderable 5.11.4.** *Let $G$ be the group of symmetries of the square. Using the notation of Example 5.3.1, compute $G/\langle g_3 \rangle$ and $G * x_0$.*

**Theorem 5.11.1.** *(Lagrange): If $G$ is a finite group and $H$ a subgroup then*

$$|G/H| = |G|/|H|.$$

Joseph Louis Lagrange (1736-1813), an Italian who studied at the Univerity of Turin and later taught mathematics there. He later taught at the University of Berlin and then took a research position at the Académie des Sciences in Paris, though he did teach at the Ecole Polytechnique. He made important contributions to mechanics, applications of calculus, and function theory.

**Corollary 5.11.3.** *If $H, G$ are as above then the order of $H$ divides the order of $G$.*

**proof of Theorem**: Let $X$ be the set of left cosets of $H$ in $G$ and let $G$ act on $X$ by left multiplication. Apply the previous lemma with $x = H$. $\square$

**Ponderable 5.11.5.** *Let $G = S_3$, the symmetric group on 3 letters, and let $H = \langle s_1 \rangle$, in the notation of §5.4 above.*
*(a) Compute $|G/H|$ using Lagrange's Theorem.*
*(b) Explicitly write down all the cosets of $H$ in $G$.*

**Definition 5.11.2.** *: Let $H$ be a subgroup of $G$ and let $C$ be a left coset of $H$ in $G$. We call an element $g$ of $G$ a coset representative of $C$ if $C = g * H$. A complete set of coset representatives is a subset of $G$, $x_1, x_2, ..., x_m$, such that*

$$G/H = \{x_1 * H, ..., x_m * H\},$$

*without repetition (i.e., all the $x_i * H$ are disjoint).*

**Ponderable 5.11.6.** *For $g_1, g_2 \in G$, define $g_1 \sim g_2$ if and only if $g_1$ and $g_2$ belong to the same left coset of $H$ in $G$.*
*(a) Show that $\sim$ is an equivalence relation.*
*(b) Show that the left cosets of $H$ in $G$ partition $G$.*

**Ponderable 5.11.7.** *For $G = S_4$ and $H = S_3$, find a complete set of coset representatives of $G/H$ in $G$.*

**Theorem 5.11.2.** *If $S$ is a complete set of coset representatives of $G/H$ then*

$$G = \cup_{s \in S} s * H$$

*and* $|G| = \sum_{s \in S} |s * H|$.

This follows from the above definition and Theorem 2.3.1.

## 5.12 Campanology, revisited

Let us put the material on bell-ringing from §3.5 in the context of group theory. The fact that this can be done so easily is quite remarkable, considering that group theory wasn't developed for at least 100 years later!

Generating the plain lead on four bells is analogous algebraically to generating the dihedral group of order 8, $D_4$. If $a = (1,2)(3,4)$, which swaps the first two and last two bells, and if $b = (2,3)$, which swaps the middle pair, then the plain lead on four bells corresponds to

$$D_4 = \{1, a, ab, aba, (ab)^2, a(ab)^2, (ab)^3, (ab)^3 a\}.$$

Plain Bob Minimus is equivalent algebraically to generating the symmetric group on 4 elements, $S_4$. Let $a$ and $b$ be as before. If we look at the first column of the Plain Bob Minimus composition, we see that it is nothing more than the dihedral group, $D_4$, which is a subgroup of $S_4$. To generate the second column of $S_4$ we introduce $c = (3,4)$ and let $k = (ab)^3 ac$. The second column corresponds to $kD_4$ and the third column to $k^2 D_4$. The generation of the Plain Bob Minimus shows that $S_4$ can be expressed as the disjoint union of cosets of the subgroup $D_4$, that is, the cosets of $D_4$ in $S_4$ partition $S_4$. There is an important generalization of this fact.

**Theorem 5.12.1.** *For any group $G$ and any subgroup $H$, the cosets of $H$ in $G$ partition $G$.*

This is a restatement of Theorem 5.11.2.

As White [Wh] concludes in his paper, he is not suggesting 'that Fabian Stedman was using group theory explicitly, but rather that group theoretical ideas were implicit in (Stedman's) writings and compositions'.

## 5.13 Dimino's Algorithm

We saw in an earlier chapter a simple algorithm for computing the elements of a permutation group $G$. Because time is money and computer time is limited, efficency is an important issue. We shall discuss a much more efficient algorithm in this section. For more details, see [Bu].

**Notation**: Let $S = \{g_1, g_2, ..., g_n\}$ be a set of generators for a permutation group $G$. Let

$$S_0 = \emptyset,$$
$$S_i = \{g_1, ..., g_i\},$$
$$G_0 = \{1\},$$
$$G_i = \langle S_i \rangle = \text{the group generated by the elements in } S_i,$$

for $1 \leq i \leq n$.

**Algorithm (inductive step)**:

Input: The generators $S$ of $G$ and a list $L$ of all the elements of the permutation subgroup $G_{i-1}$.

Output: A list $L$ of elements of $G_i$ and a list $C$ of coset representatives of $G_i/G_{i-1}$.

```
C = {1}
for g in C do
  for s in S_i do
    if s*g not in L then
        C = C union {s*g}
        L = L union s*g*G_{i-1}
    endif
  endfor
endfor
```

**Algorithm (Dimino)**:

Input: The generators S of G

Output: A list of elements of G

```
(Initial case S_1 = < g1 >)
order = 1, element[1] = 1, g = g1
while g not equal to 1 do
  order = order + 1
  element[order] = g
  g = g*g1
endwhile



(General case)
for i from 2 to n do
  <insert inductive step here>
endfor
```

**Example 5.13.1.** *Let $G = S_3 = \langle s_1, s_2 \rangle$. We use Dimino's algorithm to list all the elements of $G$. We have*

$$G_0 = \{1\} \subset G_1 = \langle s_1 \rangle \subset G_2 = G.$$

*First, we list the elements of $G_1 = \langle s_1 \rangle$. Since $s_1 = (1\ 2)$, it is order 2, so*

$$G_1 = \{1, s_1\}.$$

*This is our list $L$ that we will apply the 'inductive step' of Dimino's algorithm to (with $i = 2$). We start with $C = \{1\}$. Now we look at the left cosets of $G_1$ in $G_2 = G$. We have (with $g = 1, s = s_1$)*

$$s_1 * G_1 = G_1,$$

*so we don't increase the size of $C$ or $L$. Next, we have (with $g = 1, s = s_2$)*

$$s_2 * G_1 = \{s_2, s_2 * s_1\} \neq G_1,$$

*so $L = \{1, s_1, s_2, s_2 * s_1\}, C = \{1, s_2\}$. Next, we have (with $g = s_2, s = s_1$)*

$$s_1 * s_2 * G_1 = \{s_1 * s_2, s_1 * s_2 * s_1\} \neq G_1.$$

*(We know $s_1 * s_2 * G_1 \neq G_1$ since neither of the two elements in $s_1 * s_2 * G_1$ is the identity.) Thus, we increase $L, C$:*

$$L = \{1, s_1, s_2, s_2 * s_1, s_1 * s_2, s_1 * s_2 * s_1\},$$

*and $C = \{1, s_2, s_1 * s_2\}$. We know we may stop here since we know $|S_3| = 6$ but the algorithm still has one more statement to execute. Next, we have (with $g = s_2, s = s_2$)*

$$s_2 * s_2 * G_1 = G_1,$$

*so we don't increase the size of $C$ or $L$ (as expected). This step terminates the algorithm and $S_3 = L$.*

**Ponderable 5.13.1.** *Perform Dimino's algorithm on*

$$S_4 = \langle s_1 = (1\ 2), s_2 = (2\ 3), s_3 = (3\ 4) \rangle.$$

# Chapter 6

# Welcome to the machine

'...Organization is of the utmost importance for military affairs, as it is ... for other disciplines where the gathering process of practical knowledge exceeds the strength of any individual. In mathematics, however, organizing talent plays a most subordinate role. Here weight is carried only by the individual. The slightest idea of a Riemann or a Weierstrass is worth more than all organisational endeavours. There is no royal road to mathematics.'

*Georg Frobenius*

This chapter expounds the mathematical structure in some mathematical models of some games belonging to the 'Merlin's Magic family' of games.

In particular, we show that for a randomly choosen initial state, the probability that

- the 3 × 3 Lights Out can be solved is 1,

- the keychain Lights Out can be solved is 1,

- the 5 × 5 Lights Out can be solved is .25,

- the 6 × 6 Deluxe Lights Out can always be solved is 1,

- the Orbix can always be solved.

## 6.1   Some history

Some years ago, Parker Brothers (also owned by Hasbro) marketed an electronic device called **Merlin's Magic** (the rules book was published copyright 1979). This device was a marketed as a 3 × 3 square array of buttons which could turn off or on. Tiger Electronics, a subsidiary of Hasbro, sometime in the late 1990's marketed several generalizations under the name **Lights Out** (or Lights Out Cube or Lights Out Keychain or ...). These are battery-operated puzzles.

The rule for how each button changes the state is very simple: pressing any button toggles on/off the neighboring buttons (what 'neighboring' means will be defined in the next section). Other games in the same family include Rubik's Clock, Lights Out, and Orbix.

# 6.2 Merlin's Machine

Let $\mathbb{Z}/m\mathbb{Z} = \{0, 1, ..., m-1\}$, where you add, subtract, and multiply mod $m$. When these numbers correspond to the colors of a game, we shall let 0 correspond to 'off' and the non-zero numbers to the different lighted colors.

## 6.2.1 The machine

**The graph:** Here is the graph-theoretical description of the situation. Let $\Gamma$ be a finite, simple, connected graph. Let $V$ denote the set of vertices and $E$ the set of edges of $\Gamma$. $V$ represents the set of buttons. Two vertices are called **neighbors'** if they are connected by an edge. Suppose $V = \{v_1, v_2, ..., v_n\}$. The set of all $n$-tuples of 0's, 1's, ..., $m-1$'s (we think of these $n$-tuples as being indexed by the elements of $V$) will be called the **states** of $\Gamma$ and denoted $(\mathbb{Z}/m\mathbb{Z})^n$. Let $s_0 \in (\mathbb{Z}/m\mathbb{Z})^n$ denote an 'initial state' (and shall assume $s_0$ contains at least one non-zero entry). If $v \in V$ is any vertex, let $t(v) \in (\mathbb{Z}/m\mathbb{Z})^n$ denote the state (sometimes called the **toggle vector**) which is 1 at each coordinate which is either indexed by $v$ or by a neighbor of $v$, and which is 0 otherwise. (In the case of the Orbix puzzle, let $t(v) \in (\mathbb{Z}/m\mathbb{Z})^n$ denote the state (sometimes called the **toggle vector**) which is 1 at each coordinate which is indexed by a neighbor of $v$ (but *not v* itself), and which is 0 otherwise.)

**The rules:**

Here's how the game is played.

1. Pick any vertex $v$.

2. Replace the current state of the game, call it $s$ ($s = s_0$ initially), with $s' = s + t(v)$ (where addition is coordinatewise mod m).

3. If all the coordinates of this new state $s'$ are equal to 0 then stop. Otherwise, go to step 1.

Motivated by §6.4, we shall call this game **Merlin's Machine**.

It is possible to slightly generalize the above construction, replacing a graph by a digraph with multiple (parallel) labeled edges. This generalization is sufficiently different from the others that it has been omitted. However, Rubik's Clock falls into this more general framework but with $m = 12$.

In most examples given below, $m = 2$.

## 6.2.2  The rectangular graph

The general setup was described in §2 above. Since the most popular games are rectangular, we shall construct the graph in this special case for easier reference.

Let $M \geq 2, N \geq 2$ be integers. (For example, in the case of Merlin's Magic, we have $M = N = 3$.) Subdivide a square in the plane into $M \cdot N$ equal subsquares whose edges are parallel to the edges of the larger square. Each of these smaller subsquares represents a button in Merlin's Machine. We shall label the $M \cdot N$ buttons on this $M \times N$ array of squares $(S_{i,j})$, as though they were entries of a matrix. Thus, for example, $S_{1,1}$ is in the upper left-hand corner, $S_{1,N}$ is in the upper right-hand corner, $S_{M,1}$ is in the lower left-hand corner, and $S_{M,N}$ is in the lower right-hand corner.

Let $\Gamma$ denote the graph whose vertices are the $M \cdot N$ squares $S_{i,j}$, where two vertices are connected by an edge if and only if the corresponding squares share a common side. There are $N(M-1) + M(N-1)$ edges. This is the graph of the rectangular Merlin's Machine. The rules are as in §6.2.1 above.

# 6.3  Variants

There are several variants of Merlin's Machine which have been patented and marketed. They all fit into the framework of §6.2.1.

## 6.3.1  Merlin's Magic (actually $3 \times 3$ Lights Out)

For simplicity, we only describe the $3 \times 3$ Lights Out (for Merlin's Magic, see [P], [Sto]). Imagine subdividing a square into 9 equal subsquares in a $3 \times 3$ grid. Take $m = 2$. Here the graph $\Gamma$ of §6.2.1 is the graph of a $3 \times 3$ square array with 9 vertices, where two vertices are connected by an edge if and only if the corresponding squares share a boundary. Thus, each interior vertex has degree 4, each edge vertex (not a corner) has degree 3, and each corner vertex has degree 2. The rules are as in §6.2.1.

For the solution strategy, see [P], [Sto], [Sch1].

## 6.3.2  The Orbix

Here $m = 2$ and the graph $\Gamma$ of §6.2.1 is the graph of the icosahedron with 12 vertices, each having degree 5. (Actually, Orbix has four different puzzle types, of which the one described is the first type. The others have different rules for play. Type 2: Same as type 1, but only lit buttons can be pressed. Type 3: Only unlit buttons can be pressed, and then all buttons *except* the adjacent ones change. Type 4: Only lit buttons can be pressed. If the button opposite the pressed one is on then the adjacent 5 lights change. If the button opposite the pressed one is off, then all buttons except the pressed one change. These other types might require analysis and ideas beyond what is presented here.).

For the solution strategy, see §15.10 below.

101

### 6.3.3 Keychain Lights Out

Imagine subdividing a square into equal subsquares in a $4 \times 4$ grid. Take $m = 2$. Here the graph $\Gamma$ of §6.2.1 is the graph of a $4 \times 4$ array with 16 vertices, where each edge vertex is *also* connected to its mirror-opposite vertex (reflecting about the thick central lines in Picture 6.3.3 below) on the opposite edge. (The corner vertices actually have two mirror-opposites.) Thus each vertex has degree 4. This may be visualized as the graph of a $4 \times 4$ grid on a donut obtained by 'gluing' top and bottom edges and then 'gluing' the left and right edges.

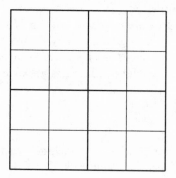

Picture 6.3.3. Keychain Lights Out Grid.

The rules are as in §6.2.1.

(The 'ordinary' $4 \times 4$ game without wrap-around has not been analyzed in detail here. However, see Example 6.5.3.)

### 6.3.4 Lights Out

Imagine subdividing a square into 25 equal subsquares in a $5 \times 5$ grid. Take $m = 2$. Here the graph $\Gamma$ of §6.2.1 is the graph of a $5 \times 5$ square array with 25 vertices, where two vertices are connected by an edge if and only if the corresponding squares share a boundary. Thus, each interior vertex has degree 4, each edge vertex (which is not a corner) has degree 3, and each corner vertex has degree 2. The rules are as in §6.2.1.

### 6.3.5 Deluxe Lights Out

Imagine subdividing a square into equal subsquares in a $6 \times 6$ grid. Take $m = 2$. Here the graph $\Gamma$ of §6.2.1 is the graph of a $6 \times 6$ array with 36 vertices, each interior vertex having degree 4, each edge vertex (which is not a corner) having degree 3, and each corner vertex having degree 2. The rules are as in §6.2.1. Like the Orbix, Deluxe Lights Out has several different types of play, such as 'lit only' (where you can only punch lit buttons) and 'toggle' (where you must alternately punch lit and then unlit buttons). Here we only describe the easiest 'classic' mode of play.

### 6.3.6 Lights Out Cube

Take $m = 2$. For this puzzle, let $\Gamma$ be the graph whose vertices are the 54 facets (i.e., the colored tiles) of a Rubik's Cube. Join two vertices by an edge of the facets share a common boundary. Now play the game following the rules of §6.2.1.

For the solution strategy, see [Ta].

### 6.3.7 Alien Tiles

Take $m = 4$. (This is the only example where $m \neq 2$.) The WWW site www.alientiles.com, maintained by Cliff Pickover and Cam Mckechnie (who have also trademarked the term 'Alien Tiles'), provides one with a game based on an idea similar to Lights Out. Here the game is a $7 \times 7$ array of colored buttons. The graph $\Gamma$ of §6.2.1 is the graph of a $7 \times 7$ array with 49 vertices, where two vertices are connected if their corresponding buttons are either in the same row or in the same column. Now play the game following the rules of §6.2.1.

### 6.3.8 Theoretical generalizations and variants

Besides the Rubik's Clock already mentioned, other related puzzles include the class of 'chip-firing' games investigated, for example, by A. Björner, L. Lovasz, and P. Shor [BLS] and K. Eriksson [E]. Though extremely interesting, their analysis is much more technical and would take us too far afield, so we shall not delve further into them.

## 6.4 Finite state machines

This section introduces a precise language which is helpful for comparing electrical devices.

A **finite state machine** (or **Mealy machine**) is a 5-tuple $(S, I, O, f, g)$ where $S$ is a (finite) set of objects called **states**, $I$ is a finite set of symbols called the **input alphabet**, $O$ is a finite set of symbols called the **output alphabet**, $f : S \times I \rightarrow S$ is the **transition** or **next state** function, $g : S \times I \rightarrow O$ is the **output** function. (One reference for this is, for example, Grimaldi [Gr].)

**Remark 6.4.1.** *The states of the machine can be roughly regarded as the 'memory registers' of the machine. The input alphabet can be roughly regarded as a labeling of the buttons one might press (or moves you make, depending on the constitution of the machine).*

**Remark 6.4.2.** *More precisely, $f$ and/or $g$ need not be defined on all of $S \times I$. One or both may only be defined on some subset of $S \times I$ (in other words, $f$*

and $g$ need only be what is called a **partial function**). When both $f$ and $g$ are functions (each defined on all of $S \times I$) then the machine is called **complete**.

Define $E_{i,j}$ to be the $M \times N$ matrix all of whose entries are 0 *except* for the $(i,j)$-th entry which is equal to 1. An **elementary matrix** is an $M \times N$ matrix which is equal to $E_{i,j}$ for some $1 \leq i \leq M, 1 \leq j \leq N$. Let $M_{M \times N}(\mathbb{Z}/2\mathbb{Z})$ denote the set of all $M \times N$ matrices whose entries are either equal to 0 or equal to 1. Clearly, $E_{i,j} \in M_{M \times N}(\mathbb{Z}/2\mathbb{Z})$. We can, if we want, add two elements of $M_{M \times N}(\mathbb{Z}/2\mathbb{Z})$ (adding each corresponding matrix element mod 2) and get another matrix in $M_{M \times N}(\mathbb{Z}/2\mathbb{Z})$.

**Example 6.4.1.** *(Rubik's Cube) A* **Rubik's Cube** *is a cube, which has been sliced into 27 subcubes of equal sizes, each of which has 6 smaller faces parallel to the 6 faces of the larger cube. In the 'solved state' each face of the larger cube is colored a different (solid) color. To be concrete, suppose we have fixed such a cube in space (say on a table top in front of you, the reader). The 9 facets of the front face shall be colored yellow, the 9 facets of the back face shall be colored green, the 9 facets of the up (top) face shall be colored red, the 9 facets of the down (bottom) face shall be colored orange, the 9 facets of the right face shall be colored blue, the 9 facets of the left face shall be colored white. There are $9 \cdot 6 = 54$ colored facets total. Each of the 6 faces has 3 parallel slices of 9 subcubes, each of which may be independently rotated at angles of $90°$, $180°$, and $270°$.*

*With the yellow-colored center facet facing front, and the red-colored center facet facing up, let $R$ denote the $90°$ clock-wise rotation (as you face that side) of the right slice, let $L$ denote the $90°$ clock-wise rotation of the left slice, let $F$ denote the $90°$ clock-wise rotation of the front slice, let $B$ denote the $90°$ clock-wise rotation of the back slice, let $D$ denote the $90°$ clock-wise rotation of the down slice, and let $U$ denote the $90°$ clock-wise rotation of the up slice. Note that each of these rotations leaves the center facets fixed. There are $9 \cdot 6 - 6 = 48$ colored non-central facets total.*

*Let $S$ denote the set of all possible permutations of the 48 colored facets on the cube (leaving the center facets fixed). Let $I = \{R, L, F, B, U, D\}$. Let $O = S$. Let the transition function $f : S \times I \to S$ be the function sending $(s, i)$ ($s \in S$ and $i \in I$) which takes the state $s$ (the cube with 48 non-central facets scrambled according to some permutation) and the input $i$ (one of the moves $R, L, ..., U$) to the new state $s'$ obtained by performing the move $i$ on the scrambled cube $s$: $f(s, i) = s'$. Let the output function $g$ be the same as $f$.*

*This describes the Rubik's Cube as a finite state machine.*

**Example 6.4.2.** *(rectangular Merlin's Machine) In this example, let $M, N > 2$ be integers. Let*

$$S = M_{M \times N}(\mathbb{Z}/2\mathbb{Z}), \quad I = \{E_{i,j} \mid 1 \leq i \leq M, 1 \leq j \leq N\}, \quad O = M_{M \times N}(\mathbb{Z}/2\mathbb{Z}).$$

*We interpret the input $E_{i,j}$ to mean that the $(i,j)^{th}$ button of the array was pressed. The effect of pressing any button is to toggle the button itself and those* *which are directly north, south, east, or west of the button pressed. There is no wrap-around (at least not in the most basic version) so there are*

- *2 buttons toggled, in case a corner button is pressed,*

- *3 buttons toggled, in case an edge button is pressed,*

- *4 buttons toggled, in case a central button is pressed.*

*To model this as a finite state machine, let the transition function $f : S \times I \to S$ be defined by*

$$f(s, E) = s + t(E), \qquad s \in S, \ E \in I,$$

*where addition is coordinate-wise mod 2 and $t(E)$ is the **toggle matrix**:*

$$t(E_{i,j}) = \begin{cases} E_{i,j} + E_{i-1,j} + E_{i,j-1} + E_{i+1,j} + E_{i,j+1}, & 1 < i < M, \ 1 < j < N, \\ E_{i,j} + E_{i-1,j} + E_{i+1,j} + E_{i,j+1}, & 1 < i < M, \ j = 1, \\ E_{i,j} + E_{i+1,j} + E_{i,j+1}, & i = 1, \ j = 1, \\ E_{i,j} + E_{i-1,j} + E_{i,j+1}, & i = M, \ j = 1, \\ E_{i,j} + E_{i,j-1} + E_{i+1,j} + E_{i,j+1}, & i = 1, \ 1 < j < N, \\ E_{i,j} + E_{i-1,j} + E_{i,j-1} + E_{i,j+1}, & i = M, \ 1 < j < N, \\ E_{i,j} + E_{i-1,j} + E_{i,j-1} + E_{i+1,j}, & 1 < i < M, \ j = N, \\ E_{i,j} + E_{i,j-1} + E_{i+1,j}, & i = 1, \ j = N, \\ E_{i,j} + E_{i-1,j} + E_{i,j-1}, & i = M, \ j = N. \end{cases}$$

*For example, if $M = N = 3$ and $E = E_{1,2}$ then*

$$t(E) = \begin{pmatrix} 1 & 1 & 1 \\ 0 & 1 & 0 \\ 0 & 0 & 0 \end{pmatrix},$$

*and if $E = E_{22}$ then*

$$t(E) = \begin{pmatrix} 0 & 1 & 0 \\ 1 & 1 & 1 \\ 0 & 1 & 0 \end{pmatrix}.$$

*Finally, let the output function $g : S \times I \to O$ be the same as the transition function: $g = f$.*

*This describes Lights Out as a finite state machine.*

## 6.5 The mathematics of the machine

We shall follow Anderson and Feil's [AF] presentation to begin with. The idea is generalized and extended considerably in the papers of Goldwasser, Klostermeyer, et al referenced below. The essential idea presented in this section was posted on the sci.math newsgroup in Feb 1998, months before [AF] appeared, first by Carsten Haese and then by Dave Rusin [Rus]. It appears that Haese thought of the idea independently of Anderson and Feil.

## 6.5.1 The square case

Suppose to begin that $M = N$. Represent each state $s \in M_N(\mathbb{Z}/2\mathbb{Z})$ as an $N^2$-tuple

$$\vec{s} = (s_{1,1}, s_{1,2}, ..., s_{1,N}, s_{2,1}, ..., s_{N,N}).$$

Let $B$ denote the $N \times N$ banded matrix with 1's on the diagonal, subdiagonal, and superdiagonal, and 0's elsewhere:

$$B = \begin{pmatrix} 1 & 1 & 0 & 0 & ... & 0 \\ 1 & 1 & 1 & 0 & ... & 0 \\ 0 & 1 & 1 & 1 & ... & 0 \\ \vdots & & & & & \\ 0 & 0 & ... & 0 & 1 & 1 \end{pmatrix}.$$

Let $I_N$ denote the $N \times N$ identity matrix. Let $A$ denote the $N^2 \times N^2$ matrix

$$A = \begin{pmatrix} B & I_N & 0 & 0 & ... & 0 \\ I_N & B & I_N & 0 & ... & 0 \\ 0 & I_N & B & I_N & ... & 0 \\ \vdots & & & & & \\ 0 & 0 & ... & 0 & I_N & B \end{pmatrix},$$

where (for simplicity of notation) 0 denotes the $N \times N$ matrix of 0's. The $i^{th}$ column of $A$ represents the toggle vector associated to pressing the $i^{th}$ button (in the ordering $(1,1)$, $(1,2)$, ..., $(1,N)$, $(2,1)$, ..., $(N,N)$).

Let $\vec{s}_0$ denote any initial state. The problem of solving the puzzle is that we want to know which buttons to press (i.e., which columns of $A$ to add together) *starting from the solved position* to obtain the given state $\vec{s}_0$. (For once we know this, we can work backwards from $\vec{s}_0$ to solve it.) In other words, we want to add some columns of $A$ together (addition being coordinatewise mod 2) and get $\vec{s}_0$. This fact may be recorded as follows.

**Theorem 6.5.1.** *The puzzle with initial state $\vec{s}_0$ can be solved if and only if the matrix equation $A\vec{x} = \vec{s}_0$ has a solution $\vec{x} \in (\mathbb{Z}/2\mathbb{Z})^{N^2}$.*

With a little knowledge of linear algebra, the following question immediately arises: is $\det(A) = 0$ in $\mathbb{Z}/2\mathbb{Z}$? Using the computer algebra program MAPLE, one finds that the answer is no if $N = 4, 5$ and yes if $N = 3, 6$. This has the following implications.

- No matter what the initial states for the $3 \times 3$ Merlin's Machine and the $6 \times 6$ Merlin's Machine are given, the puzzle can always be solved.

- There are initial states for the $4 \times 4$ Merlin's Machine and the $5 \times 5$ Merlin's Machine from where the puzzle cannot be solved.

What about other values of $N$?

To answer this question, we first introduce a sequence of polynomials. Let

$$f_0(x) = 0, \quad f_1(x) = 1,$$

and for $n > 1$, let

$$f_n(x) = x f_{n-1}(x) + f_{n-2}(x).$$

This defines the polynomials $f_0(x), f_1(x), f_2(x), \ldots$ inductively. The polynomial $f_i(x)$ is called the $i^{th}$ **Fibonacci polynomial**. Let $B_n(x)$ denote the $n \times n$ matrix

$$B_n(x) = \begin{pmatrix} x & 1 & 0 & 0 & \ldots & 0 \\ 1 & x & 1 & 0 & \ldots & 0 \\ 0 & 1 & x & 1 & \ldots & 0 \\ \vdots & & & & & \\ 0 & 0 & \ldots & 0 & 1 & x \end{pmatrix}.$$

**Lemma 6.5.1.** *(Goldwasser, Klostermeyer, Trapp, and Zhang [GKTZ]) For all $n > 1$,*

$$\det(B_n(x)) = f_{n+1}(x).$$

Recall that the determinant was defined in §2.2.5.

The following effectivly answers the question of when the $N \times N$ Merlin's Machine can always be solved, no matter what the initial state is.

**Theorem 6.5.2.** *(Goldwasser, Klostermeyer, and Trapp [GKT]) The $N \times N$ puzzle can be solved with any initial state if and only if the greatest common divisor of $f_{N+1}(x)$ and $f_{N+1}(x+1)$ equals 1.*

## 6.5.2 Downshifting

We need to slow down and intorduce a few more mathematical concepts in linear algebra. These are analogs of standard facts for real matrices which can be found in many linear algebra texts (for example [JN]). However, we shall need them for matrices having coefficients in a finite field.

Let $F$ be either $\mathbb{R}$ or $\mathbb{C}$ or $\mathbb{Z}/p\mathbb{Z}$, where $p$ is a prime.

**Definition 6.5.1.** *Let $V$ be a set with operations $+ : V \times V \to V$ and $\cdot F \times V \to V$. $V$ is called a **vector space over** $F$ (or an $F$-**vector space**) if the following properties hold. For all $\vec{u}, \vec{v}, \vec{w} \in V$ and $a, b \in F$,*

- $\vec{u} + \vec{v} = \vec{v} + \vec{u}$ *(commutativity)*

- $(\vec{u} + \vec{v}) + \vec{w} = \vec{u} + (\vec{v} + \vec{w})$ *(associativity)*

- *the vector $\vec{0} = (0, 0, ..., 0) \in V$ satisfies $\vec{u} + \vec{0} = \vec{u}$ (the zero vector $\vec{0}$ is the additive identity),*

- *for each $\vec{v} \in V$ the element $(-1)\vec{v} = -\vec{v} \in V$ satisfies $\vec{v} + (-\vec{v}) = \vec{0}$ (each element $\vec{v}$ has an additive inverse $-\vec{v}$)*

107

- $(a + b)\vec{v} = a\vec{v} + b\vec{v}$ and $a(\vec{v} + \vec{w}) = a\vec{v} + a\vec{w}$ *(distributive laws)*

- $(ab)\vec{v} = a(b\vec{v})$

- $1 \cdot \vec{v} = \vec{v}.$

In particular, every vector space must contain at least one element (the zero vector). The elements of $V$ are called **vectors** and the elements of $F$ are called **scalars**.

Any subset $W \subset V$ of a vector space which is closed under vector addition (restricting $+$ from $V$ to $W$) and scalar multiplication is called a **subspace of** $V$.

**Example 6.5.1.** *The basic example of a vector space is '3-space', $V = \mathbb{R}^3$, consisting of all vectors $(x, y, z)$ in three dimensions.*

**Example 6.5.2.** *Another example of a vector space is the space of all $m \times n$ matrices having entries in $F$, $M_{m \times n}(F)$ (m rows and n columns). Let $m = n > 1$ and let $1 \leq i, j \leq n$. Define the matrix $E_{ij} \in M_{n \times n}(F)$ to have entry 1 at the $(i, j)$ coordinate and 0 at every other coordinate. This is called the $(i, j)^{th}$* **elementary matrix**.

Suppose $V$ is a vector space over a field $F$. If there are non-zero vectors $\vec{f_1}, \vec{f_2}, ..., \vec{f_k}$ such that every $\vec{v} \in V$ can be written as a 'linear combination' of these $\vec{v_i}$'s, i.e.,

$$\vec{v} = c_1\vec{f_1} + c_2\vec{f_2} + ... + c_k\vec{f_k},$$

for some scalar coefficients $c_i \in F$ $(i = 1, 2, ..., k)$ then we say that $V$ is **spanned** by the elements $\vec{f_1}, \vec{f_2}, ..., \vec{f_k}$. We also say that elements $\vec{f_1}, \vec{f_2}, ..., \vec{f_k}$ form a **spanning set** for $V$.

**Definition 6.5.2.** *If a vector space $V$ has a finite spanning set then we say that $V$ is* **finite dimensional***. The* **dimension** *of $V$ over $F$ is the number of vectors in any minimal spanning set of $V$ (it turns out this number does not depend on the minimal spanning set choosen, so this makes sense). The dimension of $V$ is written $dim(V) = k$. If $V$ is of dimension $k$ then there are $k$ vectors which span $V$ and any such set of $k$ spanning vectors of $V$ is called a* **basis** *of $V$.*

If $W$ is a subspace of $V$ then the number $dim(V) - dim(W)$ is called the **codimension** of $W$ in $V$.

The dimension of a vector space measures how many degrees of freedom - how 'big' - the space is. The codimension of a subspace measures how many degrees of freedom $V$ has relative to $W$.

Let $V = \mathbb{R}^n$ and let $A$ be an $n \times n$ matrix having entries in $F$. Then $A$ defines a function $A : V \to V$. If $v \in V$ is a non-zero vector and $\lambda \in \mathbb{C}$ with the property that

$$Av = \lambda v,$$

then we say $\lambda$ is an **eigenvalue** of $A$ with **eigenvector** $v$.

The intuitive, geometric way to visualize eigenvalues and eigenvectors is to think of the action of $A$ on a vector as being some sort of combination of 'rotations', 'shears' (as in the shear force on an airplane), and 'stretches'. However, if that vector is an eigenvector then the action of $A$ is simply a stretch (by a factor of $\lambda$, at least if $\lambda$ is real). The following is a basic fact about eigenvalues which is found in any textbook on linear algebra (for example, [JN]).

**Lemma 6.5.2.**
*(a) If $A$ is a square matrix then $\lambda$ is an eigenvalue of $A$ if and only if $\det(A - \lambda I) = 0$, where $I$ denotes the identity matrix.*

*(b) If $A$ is an $n \times n$ matrix then $A$ has exactly $n$ eigenvalues (counted according to multiplicity).*

The polynomial

$$p_A(x) = \det(A - xI)$$

is called the **characteristic polynomial** of $A$. A matrix $A$ for which $\lambda$ is a root of $p_A(x) = 0$ if multiplicity $m$ is called an eigenvalue of $A$ of **multiplicity** $m$. This is the multiplicity refered to in part (b) above.

**Remark 6.5.1.** *If you know the fundamental theorem of algebra and the definition of determinant then you can prove (b) follows from (a). The proof of (a) itself is not too hard. Here's a sketch: $Av = \lambda v$ holds if and only if $(A - \lambda I)v = 0$, where here $0$ denotes the $0$ vector in the space of $v$. Since $v$ is non-zero, by assumption, the matrix $A - \lambda I$ cannot be invertible. (And conversely, if $A - \lambda I$ is not one-to-one then $(A - \lambda I)v = 0$ for some non-zero $v$.) By Lemma 2.2.1, $\det(A - \lambda I) = 0$.*

An odd situation arises if $\lambda = 0$. Matrices that have an eigenvalue of $0$ don't behave as 'nicely', in some respects, as matrices for which all the eigenvalues are non-zero. The **rank** of a matrix is the number of non-zero eigenvalues (counted according to multiplicity). The rank is a measurement of how 'bad' a matrix is, in a sense. (Of course, I do not mean to cast immoral dispersions on the character of high ranking matrices!) The following result enables one to compute the rank of a matrix without having to compute the eigenvalues, which can be difficult if $n$ is large.

**Lemma 6.5.3.** *Let $A$ be any matrix as in the lemma above. Let $r_1, ..., r_n$ be the row vectors of $A$. Let $A'$ be the matrix which is the same as $A$ except that either (a) exactly one row, say the $i^{th}$ one, has been replaced by the sum $r_i + cr_j$, where $c$ is any constant and $r_j$ is any other row (so $i \neq j$) if $A$, or (b) the rows of $A'$ are the same as $A$ but with exactly two of them swapped. Then*

$$rank(A) = rank(A').$$

A matrix $A'$ as in the above lemma is said to be obtained from $A$ by **an elementary row operation**. An elementary column operation may be defined similarly. (The definition is left to the reader.) The analogous result also holds for elementary column operations (see for example [JN] for details).

The point to this lemma is that one can often choose the constant $c$ so that $A'$ has more zeros that $A$ does. This is an instance of **Gaussian elimination**. (If it is not possible to find some row $i$ and some constant $c$ for which $A'$ has more zero entries than $A$ then $A$ is said to be **reduced**.)

This result above is due to Carl F. Gauss (1777-1855), perhaps the greatest mathematician of all time. Probably the best story about Gauss concerns the time when he first entered elementary school at the age of 7. His teacher, no doubt tired and wanting a break, gave his class the following problem: 'Find the sum of all the integers from 1 to 100: $1+2+...+100$ =?' He probably expected to get a good bit of peace and quiet but Gauss instantly saw that you could pair 1 and 100, 2 and 99, ..., 50 and 51, and get the answer $50 \cdot 101 = 5050$. For his PhD thesis, he gave the first complete proof of the fundamental theorem of algebra. Gauss made major contributions to number theory, astronomy, geodesy, among others, in his lifetime.

**Gauss' recipe to compute the rank** ( Input: A matrix $A$. Output: the rank of $A$.): Using a sequence of elementary row operations, reduce $A$ to a matrix $U$ that is upper triangular. The number of non-zero entries on the diagonal of $U$ is the rank of $A$.

Gauss was quite a cook!

### 6.5.3 The rectangular case

We reformulate the problem.

Let $V = V_{M,N} = M_{M \times N}(\mathbb{Z}/2\mathbb{Z})$. This is a vector space over the field $\mathbb{Z}/2\mathbb{Z}$ with two elements. Let $W = W_{M,N}$ denote the subspace spanned by the vectors $t(E)$, in the notation of Example 6.4.2 above, where $E$ runs over all $M \times N$ elementary matrices.

The following question was posed by D. Rusin in [Rus]:

**Problem**: What is the codimension of $W$ in $V$?

This question is important for the solution of the rectangular Merlin's Machine due to the following fact. Let $c_{M,N} = dim_{\mathbb{Z}/2\mathbb{Z}}(V_{M,N}/W_{M,N})$ denote the codimension of $W$ in $V$.

**Proposition 6.5.1.** *The $M \times N$ puzzle can be solved with any initial state if and only if $c_{M,N} = 0$.*

**Example 6.5.3.** *(Rusin [Rus]) Several examples were computed by Rusin. The array of values of these numbers for $0 < m, n < 10$ are*

$$
\begin{array}{ccccccccc}
0 & 1 & 0 & 0 & 1 & 0 & 0 & 1 & 0 \\
1 & 0 & 2 & 0 & 1 & 0 & 2 & 0 & 1 \\
0 & 2 & 0 & 0 & 3 & 0 & 0 & 2 & 0 \\
0 & 0 & 0 & 4 & 0 & 0 & 0 & 0 & 4 \\
1 & 1 & 3 & 0 & 2 & 0 & 4 & 1 & 1 \\
0 & 0 & 0 & 0 & 0 & 0 & 0 & 6 & 0 \\
0 & 2 & 0 & 0 & 4 & 0 & 0 & 2 & 0 \\
1 & 0 & 2 & 0 & 1 & 6 & 2 & 0 & 1 \\
0 & 1 & 0 & 4 & 1 & 0 & 0 & 1 & 8 \\
\end{array}
$$

*In particular, not every possible initial state for the the $3 \times 5$ Merlin's Machine puzzle can be solved, since $c_{3,5} \neq 0$.*

Proposition 6.5.1 above is only useful if one can compute the $c_{M,N}$. For low values of $M, N$ this can be done with the help of a computer. For larger values, such calculations become more difficult. The ideas of Goldwasser, Klostermeyer, et al, provide a more computationally effective solution. Theorem 6.5.2 above generalizes to the rectangular case as follows.

**Theorem 6.5.3.** *(Goldwasser, Klostermeyer, and Trapp [GKT]) The $M \times N$ puzzle can be solved with any initial state if and only if the greatest common divisor of $f_{M+1}(x)$ and $f_{N+1}(x+1)$ equals 1.*

## 6.5.4 Alien Tiles again

This is the same as the rectangular case discussed in the subsection above, except for two things:

- we must replace $\mathbb{Z}/2\mathbb{Z}$, representing on/off, by $\mathbb{Z}/4\mathbb{Z}$ (with addition mod 4), representing red, green, blue, and purple, resp.,

- we must change the 'toggle matrix' $t(E)$ in Example 6.4.2 to have a 1 for every matrix entry corresponding to a button in the same row or column as the button pressed (not just the *neighboring* buttons in the same row or column).

The states of this puzzle machine are $M_{M \times N}(\mathbb{Z}/4\mathbb{Z})$. $t'(E)$ is the **toggle matrix** for Alien Tiles:

$$t'(E_{i,j}) = \sum_{i',j'} E_{i',j'},$$

where the sum runs over all $1 \leq i' \leq M$, $1 \leq j' \leq N$, with either $i' = i$ or $j' = j$ or both. If $s$ is the present state then the next state $s'$ is obtained by first choosing an elementary matrix $E$ corrsponding to the button you want to press then computing

$$s' = s + t'(E),$$

where addition is coordinate-wise mod 4.

Let $V = V_{M,N} = M_{M \times N}(\mathbb{Z}/4\mathbb{Z})$. This is a 'module' over $\mathbb{Z}/4\mathbb{Z}$ (mathematicians use the word 'module' in place of 'vector space' when the scalars are taken from a ring such as $\mathbb{Z}/4\mathbb{Z}$). Let $W = W_{M,N}$ denote the submodule (a 'submodule' is analogous to a 'vector subspace') spanned by the vectors $t'(E)$, in the notation above, where $E$ runs over all $M \times N$ elementary matrices.

The following problem naturally arises.

**Ponderable 6.5.1.** *(hard) What is the codimension of $W$ in $V$?*

As far as I know, a general formula for this codimension, as a function of $M$ and $N$, is unknown. However, it is likely that methods of linear algebra apply for small values of $M, N$ as in the reculangular case above. Also, in case $M = N$ the Aien Tiles web site has more information.

111

## 6.5.5 Orbix, revisited

Let us number the lights on the Orbix as follows: with the north pole as 1, the south pole as 12, let the numbering of the remaining buttons, as viewed from above be as in Figure 6.5.5.

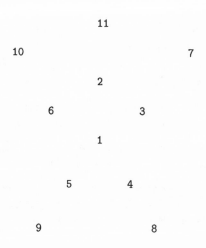

Figure 6.5.5

| Light number | neighbors | antipodes |
|:---:|:---:|:---:|
| 1 | 2,3,4,5,6 | 1,12 |
| 2 | 1,3,6,7,11 | 2,9 |
| 3 | 1,2,4,7,8 | 3,10 |
| 4 | 1,3,5,8,9 | 4,11 |
| 5 | 1,4,6,9,10 | 5,7 |
| 6 | 1,2,5,10,11 | 6,8 |
| 7 | 2,3,8,11,12 | |
| 8 | 3,4,7,9,12 | |
| 9 | 4,5,8,10,12 | |
| 10 | 5,6,9,11,12 | |
| 11 | 2,6,7,10,12 | |
| 12 | 7,8,9,10,11 | |

The adjacency matrix of the graph in §6.2.1 is the $12 \times 12$ matrix $A = (a_{i,j})$, where $a_{i,j} = 1$ if vertex $i$ and vertex $j$ are neighbors (connected by a single edge) and $a_{i,j} = 0$, otherwise. In this case,

$$A = \begin{pmatrix}
0 & 1 & 1 & 1 & 1 & 1 & 0 & 0 & 0 & 0 & 0 & 0 \\
1 & 0 & 1 & 0 & 0 & 1 & 1 & 0 & 0 & 0 & 1 & 0 \\
1 & 1 & 0 & 1 & 0 & 0 & 1 & 1 & 0 & 0 & 0 & 0 \\
1 & 0 & 1 & 0 & 1 & 0 & 0 & 1 & 1 & 0 & 0 & 0 \\
1 & 0 & 0 & 1 & 0 & 1 & 0 & 0 & 1 & 1 & 0 & 0 \\
1 & 1 & 0 & 0 & 1 & 0 & 0 & 0 & 0 & 1 & 1 & 0 \\
0 & 1 & 1 & 0 & 0 & 0 & 0 & 1 & 0 & 0 & 1 & 1 \\
0 & 0 & 1 & 1 & 0 & 0 & 1 & 0 & 1 & 0 & 0 & 1 \\
0 & 0 & 0 & 1 & 1 & 0 & 0 & 1 & 0 & 1 & 0 & 1 \\
0 & 0 & 0 & 0 & 1 & 1 & 0 & 0 & 1 & 0 & 1 & 1 \\
0 & 1 & 0 & 0 & 0 & 1 & 1 & 0 & 0 & 1 & 0 & 1 \\
0 & 0 & 0 & 0 & 0 & 0 & 1 & 1 & 1 & 1 & 1 & 0
\end{pmatrix}$$

For example, the 'toggle vector' for the first button is $(0, 1, 1, 1, 1, 1, 0, 0, 0, 0, 0, 0)$.

**Remark 6.5.2.** • *The matrix $A$ has determinant $625 = 5^4$ and eigenvalues $5, 5^{1/2}, -5^{1/2}, 5^{1/2}, -5^{1/2}, 5^{1/2}, -5^{1/2}, -1, -1, -1, -1, -1$.*

- *The state vector $v_0 = (1, 1, ..., 1)$ is an eigenvector with eigenvalue 5. This state vector has the property that if you punch all the buttons on an unlit Orbix associated to its non-zero entries of $v_0$ then you recover $v_0$.*

- *The state vector $v_1 = (0, 0, 1, 0, 1, 0, 1, 0, 0, 1, 0, 0)$ is an eigenvector with eigenvalue $-1$. This state vector has the property that if you punch all the buttons on an unlit Orbix associated to its non-zero entries of $v_1$ then you recover $v_1$.*

  *The only other vectors with this property are the eigenvectors*

  $$v_2 = (0, 1, 0, 0, 1, 0, 1, 0, 1, 0, 0, 0), \quad v_3 = (0, 0, 0, 1, 1, 0, 1, 0, 0, 0, 1, 0),$$
  $$v_4 = (0, 0, 0, 0, 1, 1, 1, 1, 0, 0, 0, 0), \quad v_5 = (1, 0, 0, 0, 1, 0, 1, 0, 0, 0, 0, 1),$$

  *and their linear combinations.*

- *A move consists mathematically of choosing a vertex $i$ then adding (modulo 2) row $i$ of the adjacency matrix $A$ to the state vector.*

- *If you change the rules of the Orbix and decide to toggle the button pressed as well as its neighbors (as in the Lights Out game) then the $12 \times 12$ matrix* *of toggle vectors is*

$$
\begin{pmatrix}
1 & 1 & 1 & 1 & 1 & 1 & 0 & 0 & 0 & 0 & 0 & 0 \\
1 & 1 & 1 & 0 & 0 & 1 & 1 & 0 & 0 & 0 & 1 & 0 \\
1 & 1 & 1 & 1 & 0 & 0 & 1 & 1 & 0 & 0 & 0 & 0 \\
1 & 0 & 1 & 1 & 1 & 0 & 0 & 1 & 1 & 0 & 0 & 0 \\
1 & 0 & 0 & 1 & 1 & 1 & 0 & 0 & 1 & 1 & 0 & 0 \\
1 & 1 & 0 & 0 & 1 & 1 & 0 & 0 & 0 & 1 & 1 & 0 \\
0 & 1 & 1 & 0 & 0 & 0 & 1 & 1 & 0 & 0 & 1 & 1 \\
0 & 0 & 1 & 1 & 0 & 0 & 1 & 1 & 1 & 0 & 0 & 1 \\
0 & 0 & 0 & 1 & 1 & 0 & 0 & 1 & 1 & 1 & 0 & 1 \\
0 & 0 & 0 & 0 & 1 & 1 & 0 & 0 & 1 & 1 & 1 & 1 \\
0 & 1 & 0 & 0 & 0 & 1 & 1 & 0 & 0 & 1 & 1 & 1 \\
0 & 0 & 0 & 0 & 0 & 0 & 1 & 1 & 1 & 1 & 1 & 1
\end{pmatrix}
$$

*This matrix is singular. It's column span (and row span) is only 6 dimensional over $\mathbb{Z}/2\mathbb{Z}$, and there are only 64 (out of $2^{12} = 4096$ possible) solvable states.*

**Theorem 6.5.4.** *The Orbix puzzle can be always be solved with any initial state.*

## 6.5.6 Return of the Keychain Lights Out

Arranging the state space as in the case of the $4 \times 4$ Merlin's machine, the matrix of toggle vectors is

$$
T := \begin{pmatrix}
1 & 1 & 0 & 1 & 1 & 0 & 0 & 0 & 0 & 0 & 0 & 0 & 1 & 0 & 0 & 0 \\
1 & 1 & 1 & 0 & 0 & 1 & 0 & 0 & 0 & 0 & 0 & 0 & 0 & 1 & 0 & 0 \\
0 & 1 & 1 & 1 & 0 & 0 & 1 & 0 & 0 & 0 & 0 & 0 & 0 & 0 & 1 & 0 \\
1 & 0 & 1 & 1 & 0 & 0 & 0 & 1 & 0 & 0 & 0 & 0 & 0 & 0 & 0 & 1 \\
1 & 0 & 0 & 0 & 1 & 1 & 0 & 1 & 1 & 0 & 0 & 0 & 0 & 0 & 0 & 0 \\
0 & 1 & 0 & 0 & 1 & 1 & 1 & 0 & 0 & 1 & 0 & 0 & 0 & 0 & 0 & 0 \\
0 & 0 & 1 & 0 & 0 & 1 & 1 & 1 & 0 & 0 & 1 & 0 & 0 & 0 & 0 & 0 \\
0 & 0 & 0 & 1 & 1 & 0 & 1 & 1 & 0 & 0 & 0 & 1 & 0 & 0 & 0 & 0 \\
0 & 0 & 0 & 0 & 1 & 0 & 0 & 0 & 1 & 1 & 0 & 1 & 1 & 0 & 0 & 0 \\
0 & 0 & 0 & 0 & 0 & 1 & 0 & 0 & 1 & 1 & 1 & 0 & 0 & 1 & 0 & 0 \\
0 & 0 & 0 & 0 & 0 & 0 & 1 & 0 & 0 & 1 & 1 & 1 & 0 & 0 & 1 & 0 \\
0 & 0 & 0 & 0 & 0 & 0 & 0 & 1 & 1 & 0 & 1 & 1 & 0 & 0 & 0 & 1 \\
1 & 0 & 0 & 0 & 0 & 0 & 0 & 0 & 1 & 0 & 0 & 0 & 1 & 1 & 0 & 1 \\
0 & 1 & 0 & 0 & 0 & 0 & 0 & 0 & 0 & 1 & 0 & 0 & 1 & 1 & 1 & 0 \\
0 & 0 & 1 & 0 & 0 & 0 & 0 & 0 & 0 & 0 & 1 & 0 & 0 & 1 & 1 & 1 \\
0 & 0 & 0 & 1 & 0 & 0 & 0 & 0 & 0 & 0 & 0 & 1 & 1 & 0 & 1 & 1
\end{pmatrix}
$$

Using MAPLE, we can perform Gaussian elimination over $R_2$. It turns out that the rank of $T$ over $R_2$ is 16. This implies the following result.

**Lemma 6.5.4.** *The codimension of the row space of $T$ in the state space is 0. In particular, the keychain puzzle can always be solved.*

# Chapter 7

# 'God's algorithm' and graphs

'Unfortunately what is little recognized is that the most worthwhile scientific books are those in which the author clearly indicates what he does not know; for an author most hurts his readers by concealing difficulties.'

*Evariste Galois*

In this chapter we introduce a graphical interpretation of a permutation group, the Cayley graph. This is then interpreted in the special case of a group arising from a permutation puzzle. Along the lines of Galois' quote above, we shall also discuss the state of our ignorance regarding 'God's algorithm'. See also chapter 16.

## 7.1  In the beginning...

To begin, what's a graph? A **graph** is a pair of countable sets $(V, E)$, where

- $V$ is a countable set of singleton elements called **vertices**,

- $E$ is a subset of the set of all *unordered* pairs $\{\{v_1, v_2\} \mid v_1, v_2 \in V, v_1 \neq v_2\}$. The elements of $E$ are called **edges**.

A graph is drawn by simply connecting points representing vertices together by a line segment if they belong to the same edge.

A **digraph**, or **directed graph**, is a pair of countable sets $(V, E)$, where

- $V$ is a countable set of vertices,

- $E$ is a subset of *ordered* pairs $\{(v_1, v_2) \mid v_1, v_2 \in V, v_1 \neq v_2\}$ called **edges**.

A digraph is drawn by simply connecting points representing vertices together by an arrow if they belong to the same edge $(v_1, v_2)$, the arrow originating at $v_1$ and arrowhead pointing to $v_2$.

If $e = \{v_1, v_2\}$ belongs to $E$ then we say that $e$ is an **edge** from $v_1$ to $v_2$ (or from $v_2$ to $v_1$) and we say $v_1$ and $v_2$ are **neighbors** or **adjacent** in the graph.  If $v$ and $w$ are vertices, a **path** from $v$ to $w$ is a finite sequence of edges beginning at $v$ and ending at $w$:

$$e_0 = \{v, v_1\}, e_1 = \{v_1, v_2\}, ..., e_n = \{v_n, w\}.$$

If there is a path from $v$ to $w$ then we say $v$ is **connected** to $w$.  We say that a graph $(V, E)$ is **connected** if each pair of vertices is connected.  The number of edges eminating from a vertex $v$ is called the **degree** (or **valence**) of $v$, denoted $\text{degree}(v)$.

**Example 7.1.1.** : *If*

$$V = \{a, b, c\}, \quad E = \{\{a, b\}, \{a, c\}, \{b, c\}\},$$

*then we may visualize* $(V, E)$ *as*

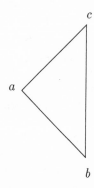

*Each vertex has valence 2.*

**Definition 7.1.1.** : *If $v$ and $w$ are vertices connected to each other in a graph $(V, E)$ then we define the* **distance from** *$v$ to $w$, denoted $d(v, w)$, by*

$$d(v, w) = \min_{v, w \in V \text{ connected}} \#\{\text{edges in a path from } v \text{ to } w\}$$

*By convention, if $v$ and $w$ are not connected then we set $d(v, w) = \infty$. The* **diameter** *of a graph is the largest possible distance:*

$$\text{diam}((V, E)) = \max_{v, w \in V} d(v, w).$$

In the above example, the diameter is 1.

## 7.2 Cayley graphs

Let $G$ be a permutation group,

$$G = \langle g_1, g_2, ..., g_n \rangle \subset S_X.$$

The **Cayley graph** of $G$ with respect to $X = \{g_1, g_2, ..., g_n\}$ is the graph $(V, E)$ whose vertices $V$ are the elements of $G$ and whose edges are determined by the following condition: if $x$ and $y$ belong to $V = G$ then there is an edge from $x$ to $y$ (or from $y$ to $x$) if and only if $y = g_i * x$ or $x = g_i * y$, for some $i = 1, 2, ..., n$.

Cayley graphs are named for Arthur Cayley, who we met already in §2.2.1. Cayley, though starting out as a laywer, eventually published over 900 papers and notes covering nearly every aspect of modern mathematics.

The **Cayley digraph** of $G$ with respect to $X = \{g_1, g_2, ..., g_n\}$ is the digraph $(V, E)$ whose vertices $V$ are the elements of $G$ and whose edges are determined by the following condition: if $x$ and $y$ belong to $V = G$ then there is an edge from $x$ to $y$ if and only if $y = x * g_i$, for some $i = 1, 2, ..., n$.

**Ponderable 7.2.1.** *Show that the Cayley graph of a permutation group is connected.*

**Lemma 7.2.1.** *Let $\Gamma_G = (V, E)$ denote the Cayley graph associated to the permutation group $G = \langle g_1, g_2, ..., g_n \rangle$. Let $N = |\{g_1, g_1^{-1}, g_2, g_2^{-1}, ..., g_n, g_n^{-1}\}|$. Then, for all $v \in V$, degree$(v) = N$.*

**Proof:** Assume not. Then there is a $v \in V = G$ with either

(i) degree$(v) < N$, or

(ii) degree$(v) > N$.

First, we note that, for each $h \in \{g_1, g_1^{-1}, g_2, g_2^{-1}, ..., g_n, g_n^{-1}\}$, the set $\{v, h*v\}$ is an edge of $\Gamma_G$. This follows from the definition of the Cayley graph.

If $r = $ degree$(v) > N$ then, by definition of the Cayley graph, there are distinct $v_1, ..., v_r \in V$ with $v = h_i * v_i$, for all $1 \le i \le r$, where the $h_1, ..., h_r$ are distinct elements of $\{g_1, g_1^{-1}, g_2, g_2^{-1}, ..., g_n, g_n^{-1}\}$. This contradicts the definition of $N$.

If $r = $ degree$(v) < N$ then, by definition of the Cayley graph, there are distinct $h_i, h_j$ in $\{g_1, g_1^{-1}, g_2, g_2^{-1}, ..., g_n, g_n^{-1}\}$ such that $h_i * v = h_j * v$. Since $G$ is a group and $V = G$ (as sets), we may cancel the $v$'s from both sides of the equation $h_i * v = h_j * v$, contradicting the assumption that $h_i$ is distinct from $h_j$. $\square$

**Example 7.2.1.** : *Let*

$$G = \langle s_1, s_2 \rangle = S_3,$$

*where $s_1 = (1,2)$, and $s_2 = (2,3)$. Then the Cayley graph of $G$ with respect to $X = \{s_1, s_2\}$ may be visualized as below.*

117

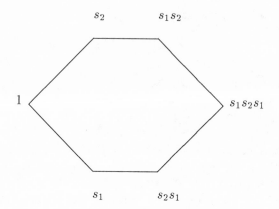

**Ponderable 7.2.2.** *Let*
$$G = \langle s_1, s_2 \rangle = S_3,$$
*where $s_1 = (1\ 2)$, and $s_2 = (2\ 3)$. Find the Cayley digraph of $G$ with respect to $X = \{s_1, s_2\}$.*

**Ponderable 7.2.3.** *Construct the Cayley graph of $C_4$, the cyclic group, with respect to the generator $s = (1, 2, 3, 4)$.*

**Ponderable 7.2.4.** *Construct the Cayley graph of $S_4$, the symmetric group on four letters, with respect to the generators $s_1 = (1\ 2)$, $s_2 = (2\ 3)$ and $s_3 = (3\ 4)$.*

**Ponderable 7.2.5.** *Construct the Cayley digraph of $S_3$ with respect to the generators $f = (1, 3)$, $r = (1, 2, 3)$. (Show, in particular, that $f, r$ do indeed generate $S_3$.)*

**Example 7.2.2.** *: Let*
$$G = \langle R, L, U, D, F, B \rangle \subset S_{54}$$
*be the group of the $3 \times 3$ Rubik's Cube. Each position of the cube corresponds to an element of the group $G$ (i.e., the move you had to make to get to that position). In other words, each position of the cube corresponds to a vertex of the Cayley graph. Each vertex of this graph has valence 12.*

**Ponderable 7.2.6.** *Check this.*

*Moreover, a solution of the Rubik's Cube is simply a path in the graph from the vertex associated to the present position of the cube to the vertex associated to the identity element. The number of moves in the shortest possible solution is simply the distance from the vertex associated to the present position of the cube to the vertex associated to the identity element. The diameter of the Cayley graph of $G$ is the number of moves in the best possible solution in the worst possible case.*

## 7.3 God's algorithm

'...O, cursed spite,
that ever I was to set it right!'
*Shakespeare*, **Hamlet**, *Act 1, scene 5*

**Ponderable 7.3.1.** *(very hard) Let $G$ be the group of a permutation puzzle. Find the diameter of the Cayley graph of $G$.*

Sometimes this is called 'God's algorithm', though here that term is reserved for a harder version of the problem, stated below. This diameter problem is unsolved for must puzzles (including the 3 × 3 Rubik's Cube) and appears to be very difficult computationally in general (see [GJ] for a discussion of similar problems and their complexity). The cases where it is known include (with no attempt at completeness) the following puzzles (the reader is referred to [Lo], [CFS] for more details).

| puzzle | diameter |
|--------|----------|
| Pyraminx | 11 (not including tip moves) |
| Skewb | 11 |
| 2 × 2 Rubik's Cube | 14 (quarter turns) |

These diameters were found with the help of a computer.

Several people have worked on God's algorithm for the Rubik's Cube (in no particular order: Dan Hoey, David Singmaster, Jerry Bryan, Dik Winter, Micheal Reid, Martin Schönert, Mark Longridge, and Richard Korf, Morwen Thistlethwaite, among others). For a while it was guessed that the superflip position is the position which is as far from 'start' (the solved position) as possible. The superflip position is 24 quarter turns from start (as was already mentioned, this fact is due to Jerry Bryan [CL]). However, a longer move was discovered 26 quarter turns away from start (the superflip composed with the four spot, see (5.2)). So, at the moment, one might guess that the diameter of the Rubik's Cube graph is 26.

**Ponderable 7.3.2.** *Let $G$ be the group of a permutation puzzle and let $v$ be a vertex in the Cayley graph of $G$. Find an effective, practical algorithm for determining a path from $v$ to the vertex $v_0$ associated to the identity having length equal to the distance from $v$ to $v_0$.*

This problem is even harder! This algorithm is called **God's algorithm**. A good reference for recent progress on God's algorithm for various Rubik's cube-like puzzles may be found on Mark Longridge's www page [Lo].

**Ponderable 7.3.3.** *Find the Cayley graph of the **sliced squared group***

$$G = \langle M_R^2, M_F^2, M_D^2 \rangle,$$

*where $M_R$ is the middle slice move which turns the middle slice parallel to the right face clockwise 90 degrees (with respect to the right face). Find the diameter of this graph.*

119

Let $\Gamma$ be a graph. A **Hamiltonian circuit** on $\Gamma$ is a sequence of edges forming a path in $\Gamma$ which passes through each vertex exactly once. (If you think of the vertices as cities and the edges as roads then a Hamiltonian circuit is a tour visiting each city exactly once.)

The following unsolved problem was first mentioned in this context by A. Schwenk (in an unpublished paper written while he was at the U.S. Naval Academy).

**Ponderable 7.3.4.** *(very hard) Let $G$ be the group of the $3\times3$ Rubik's Cube puzzle. Does the Cayley graph of $G$ have a Hamiltonian circuit? In other words, can we (in principle) 'visit' each possible position of the Rubik's Cube exactly once, by making one move at a time using only the basic generators $R, L, U, D, F, B$?*

**Remark 7.3.1.** *This is a special case of a more general unsolved **problem**: For an arbitary permutation group with a given generating set, is its Cayley graph Hamiltonian? See [CSW] and [CG] for results in this area.*

An example of a group (with generators) where its Cayley graph is known to have a Hamiltonian circuit is the symmetric group (with generators the set of all transpositions).

**Example 7.3.1.** *Let $G$ be the group $S_n$ with generators given to be the set of all transpositions:*

$$G = S_n, \qquad X = \{(i,j) \mid 1 \le i < j \le n\}.$$

*(There are many more transpositions than necessary to generate $S_n$ since the subset of transpositions of the form $(i, i+1)$, $1 \le i \le n-1$, suffice to generate $S_n$ [R].) The algorithm of Steinhaus (see §3.4) shows that there is a Hamiltonian circuit in the Cayley graph of $S_n$ with respect to $X$.*

The reader interested in more examples is referred to [CG].

William Hamilton, who we met already in §5.1, may have originated the problem of finding Hamiltonian paths (hence the name) by patenting a game called the 'Icosian game' or the 'Hamilton game'. The idea is to find a Hamiltonian path around the vertices of the icosahedron. A picture of the orginal game can be found at the MacTutor History of Mathematics archive [MT]. Hamilton was a prodigy talented in many subjects, who by the age of five, had already learned Latin, Greek, and Hebrew. He eventually turned to mathematics and physics, was knighted in 1835, and was the first foreign member elected to the National Academy of Sciences of the USA.

## 7.4 The graph of the 15 Puzzle

This section discusses the 15 Puzzle from the graph-theoretical point of view following [W].

The 15 Puzzle was introduced in §4.1 above. The object of the puzzle was to order the pieces from one to fifteen from left to right, top to bottom, as shown in the **solved position** given in §4.1.

To solve a mixed up puzzle, one would slide the squares around in the puzzle. In order to do this you must slide a numbered square into the place of the space. We could represent this mathematically by saying that this is a transposition of that numbered square and the blank.

If we label each space in the puzzle as a vertex, and label the vertices numerically, then the resulting graph is represented by the figure below.

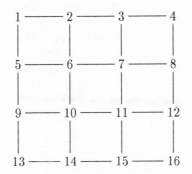

We will let 16 denote the blank and call this graph $\Gamma$. The only legal moves of the puzzle are transpositions of the 16th vertex and a vertex that is adjacent to it. Therefore, any permutation of the vertices produces a labeling on $\Gamma$.

## 7.4.1  General definitions

Now let $\Gamma$ be a simple graph with the vertex set $V(\Gamma)$ of cardinality $N$. (In the above example $N = 16$.) By a **labeling** we mean the placement of the numbers one through $N$ on distinct vertices of $\Gamma$, where $N$ denotes the blank. In other words, a labeling on $\Gamma$ is a bijective mapping $f : V(\Gamma) \to \{1, 2, ..., N\}$. Two labelings $f, g$ on $\Gamma$ are **adjacent** if and only if $g$ is a result of a single transposition on $f$ of vertex $N$ with a vertex adjacent to $N$ on $f$. In other words, $f$ and $g$ are adjacent if they differ by one legal move of the puzzle. From $\Gamma$, we make a new graph $puz(\Gamma)$ as follows [W]: the vertex set $V(puz(\Gamma))$ contains all labelings on $\Gamma$, and two vertices in $puz(\Gamma)$ are joined by an edge if the associated labelings are adjacent.

For example, the labelings in

| 1 | 2 | 3 | 4 |
|----|----|----|----|
| 5 | 6 | 7 | 8 |
| 9 | 10 | 11 |  |
| 13 | 14 | 15 | 12 |

and

| 1 | 2 | 3 | 4 |
|---|---|---|---|
| 5 | 6 | 7 | 8 |
| 9 | 10 | | 11 |
| 13 | 14 | 15 | 12 |

are adjacent.

We can consider a sequence of moves on $\Gamma$ to be a path $p$, such that $p = (x_0, x_1, x_2, ..., x_n)$ where the $x_i$'s are vertices of $\Gamma$, and (if $n \leq 1$) $x_i$ and $x_{i-1}$ are adjacent in $\Gamma$ for $1 \leq i \leq n$. Such a path $p$ is said to be from $x_0$ (its **initial vertex**) to $x_n$ (its **terminal vertex**). The path $p$ is **simple** if $x_0, x_1, x_2, ..., x_n$ are distinct. If $x_0 = x_n$ then (a not necessarily simple path) $p$ is called a **closed path based at** $x_0$. Let $x_0$ be a fixed vertex of $\Gamma$. The **inverse** of a path $p$, denoted $p^{-1}$, is the path which traverses $p$ in the opposite direction. The **identity path** is the path from $x_0$ to itself which contains no edges of $\Gamma$. The set of paths based at $x_0$ forms a group (under composition of paths) called the **homotopy group** of $\Gamma$ based at $x_0$, denoted $\Gamma(x_0)$.

Now suppose we paint the blocks of the 15 Puzzle in a checkerboard pattern:

| | × | | × |
|---|---|---|---|
| × | | × | |
| | × | | × |
| × | | × | |

In this arrangement the blank would start on a white square. If we were to move the blank up, then it is now on a black square. That is one transposition, therefore the movement is odd. If we then move the blank to the left, the blank would be on a white square. This is a total of two transpositions; therefore, the movement is even. After three transpositions the blank would be on a black square, therefore it would be an odd permutation. Therefore if the blank ends on a white square, an even permutation has occurred. If the blank ends on a black square, an odd permutation has occurred.

A **legal position** of the 15 Puzzle is any sequence of legal transpositions starting from the solved position such that the blank ends up in the bottom right-hand corner. Each such position corresponds to a permutation of the 15 numbered vertices and hence to an element of the symmetric group $S_{15}$. The set of all such permutations (arising as a sequence of transpositions) in $S_{15}$ forms a group called the **(homotopy) group of the 15 Puzzle**.

Note that the group of the 15 Puzzle is isomorphic to the homotopy group of the 15 Puzzle graph based at the 'blank vertex'.

If we assign the number 16 to the blank, then we can see that we can arrange the pieces of the puzzle in 16! different ways. However, if we take only legal positions of the 15 Puzzle, then we are fixing one of the pieces. As a result the number of ways to permute the rest of the pieces, with the blank on the white square at the bottom right-hand corner, is at most 15!. All such permutations have to be even, by the checkerboard analysis above. 15!/2 is the number of even permutations of 15 elements (there are an equal number of even and odd permutations). From this we see that the 15 Puzzle has 15!/2 possible legal positions.

**Theorem 7.4.1.** *The positions with the empty space at the bottom right that can be reached from the start position of the 15 Puzzle by shifting tiles are in a bijective correspondence with the $15!/2 = 1,307,674,368,000$ even permutations of the numbers from 1 to 15.*

**Remark 7.4.1.** *The 14-15 Puzzle cannot be solved, because it is an odd permutation. It only has one transposition, 14 interchanged with 15.*

**Proof:** If we label the empty space 16, then every possible position of the puzzle may be regarded as an element of the symmetric group $S_{16}$ and an element of $puz(\Gamma)$. There are 16! elements in $S_{16}$, and 16! vertices of $puz(\Gamma)$. With the argument earlier we can show that all legal positions of the puzzle are obtained by an even number of transpositions. Therefore, all legal moves are even permutations of the puzzle.

Now we must show that there is a certain 3-cycle in the group of the 15 Puzzle. For example, if we shift the three pieces surrounding the empty space around in a circle following the order of moves south-east-north-west then the three-cycle $(11, 12, 15)$ is produced. It can be shown that if you fix the 11 and 12 pieces, then any other piece can take the place of the 15 by following one of the cycles in the figures below.

| 1 | 2 | 3 | 4 |
|---|---|---|---|
| 5 | 6 | 7 | 8 |
| 9 | 10 | blank | 11 |
| 13 | 14 | 15 | 12 |

| | | | |
|---|---|---|---|
| 1 | 2 | 3 | 4 |
| 5 | 6 | 7 | 8 |
| 9 | 10 | blank | 11 |
| 13 | 14 | 15 | 12 |

By Lemma 9.4.4, such 3-cycles generate $A_{15}$. This proves the theorem.
□

**Remark 7.4.2.** *Another proof is possible. Alternatively, with some work we can show that any number can replace the 11 in the three-cycle, and we can show that any other number can replace the 12. From this we can conclude that any three-cycle can be formed. Since every even permutation is a combination of three-cycles (by Proposition 9.4.1), every even permutation of the 15 Puzzle can be reached.*

With this information, we can make a generalization to rectangular puzzles of size $m \times n$ with $m > 1$ and $n > 1$.

**Theorem 7.4.2.** *(Wilson [W]) The homotopy group of an $m \times n$ rectangular puzzle is the alternating group $A_{mn-1}$.*

The proof of this is similar to the proof for the $4 \times 4$ puzzle, if $m > 3$ and $n > 3$. (The special cases when $1 < m < 4$ or $1 < n < 4$ must be treated separately). The size of the alternating group is given by $(mn - 1)!/2$.

## 7.4.2 Remarks on applications

Cayley graphs have been used by computer scientists to model interconnection networks for parallel processors (see [CFS], [CG] for some references).

The problem of finding an efficient algorithm for the shortest solution to the $m \times n$ puzzle is difficult. It amounts to finding the shortest path between two points in a graph which is, in general, a difficult problem computationally [GJ].

# Chapter 8

# Symmetry and the Platonic solids

'Plato said God geometrizes continually.'
*Plutarch, Convivialium disputationum, liber 8,2*

We have encountered regular polygons, such as the hexagon, in the section on the dihedral group. A regular polyhedron is the 3-dimensional analog: a solid figure whose 2-dimensional faces are flat congruent regular polygons and whose 1-dimensional vertices are all congruent. Many puzzles are made by slicing up a regular polyhedron, so it is natural to study them and their symmetries. Such objects also occur in chemistry when the structure of crystals is investigated.

## 8.1 Descriptions

There are exactly 5 different types of regular polyhedra. We refer to Coxeter [Cox] for a proof.

The **Platonic solids** are the 5 regular polyhedrons:

| polyhedron | # faces | # vertices | # edges | group | p,q |
|------------|---------|------------|---------|-------|-----|
| tetrahedron | 4 | 4 | 6 | T | 3,3 |
| hexahedron | 6 | 8 | 12 | O | 4,3 |
| octahedron | 8 | 6 | 12 | O | 3,4 |
| dodecahedron | 12 | 20 | 30 | I | 5,3 |
| icosahedron | 20 | 12 | 30 | I | 3,5 |

Here:
   p, called the **face degree**, denotes the number of edges bounding each face,
   q, called the **vertex degree**, denotes the number of faces meeting each vertex.
   These solids are named after the great Greek philosopher Plato (427 BC-347 BC), who established Plato's Academy, which flourished for 900 years until 529 AD.

A vertex of one of these solids is therefore specified by the q-tuple of faces meeting that vertex. We saw several examples of this already when we specified notation for the movements of the associated Rubik's Cube-like puzzles in chapter 4.

These solids may be drawn in rectangular coordinates using

| polyhedron | coordinates |
|------------|-------------|
| tetrahedron | (1,1,1), (1,-1,-1), (-1,-1,1), (-1,1,-1) |
| hexahedron | (1,1,1), (1,1,-1), (1,-1,1), (-1,1,1),<br>(1,-1,-1), (-1,1,-1), (-1,-1,1), (-1,-1,-1) |
| octahedron | (1,0,0), (0,0,1), (0,1,0),<br>(-1,0,0), (0,-1,0), (0,0,-1) |
| dodecahedron | $(0,\pm\phi^{-1},\pm\phi)$, $(\pm\phi^{-1},\pm\phi,0)$,<br>$(\pm\phi,0,\pm\phi^{-1})$, $(\pm1,\pm1,\pm1)$, |
| icosahedron | $(1,0,\phi)$, $(1,0,-\phi)$, $(-1,0,\phi)$, $(-1,0,-\phi)$,<br>$(0,\phi,1)$, $(0,\phi,-1)$, $(0,-\phi,1)$, $(0,-\phi,-1)$,<br>$(\phi,1,0)$, $(\phi,-1,0)$, $(-\phi,1,0)$, $(-\phi,-1,0)$ |

where $\phi$ denotes the golden ratio.

If $P_1, P_2, P_3$ are three vertices of an icosahedron which form a triangular face then $(P_1 + P_2 + P_3)/3$ forms a vertex of the dual dodecahedron and every vertex of the dual dodecahedron arises in this way.

The three 'Platonic groups' (the group of 'symmetries' of these figures) will be described below. Their names:

- $T$ = symmetric group of the tetrahedron = **tetrahedral group**,

- $O$ = symmetric group of the octahedron (or cube) = **octahedral group**,

- $I$ = symmetric group of the icosahedron (or dodecahedron) = **icosahedral group**.

## 8.2   Background on symmetries in 3-space

This subsection presents, with some proofs, background on isometries in 3 dimensions necessary for understanding the symmetry groups of the Platonic solids. Motivated by studies of the structure of crystals, Camille Jordan was the first to consider the classification of symmetries in 3-space. Jordan (1838-1922) was one of the founders of group theory and used groups to better understand crystals.

We fix once and for all the 'right-hand-rule' orientation in 3-space. We call a distance-preserving transformation in 3-space which fixes the origin a **symmetry of 3-space**. We say that such a symmetry is **orientation preserving** if it preserves the right-hand rule orientation.

**Example 8.2.1.** : *Let $s : \mathbb{R}^3 \to \mathbb{R}^3$ denote the function which takes each vector $v$ belonging to $\mathbb{R}^3$ and returns its reflection $s(v)$ about the yz-plane. This is not orientation preserving since it reverses the direction of a counterclockwise moving circular path in the yz-plane. In terms of rectangular coordinates, $s(x, y, z) = (-x, y, z)$.*

Let
$$\mathbb{R}^3 = \{(x, y, z) \mid x, y, z \text{ real numbers}\}$$

denote 3-space. We also write this, when convenient, as column vectors

$$\mathbb{R}^3 = \{ \begin{pmatrix} x \\ y \\ z \end{pmatrix} \mid x, y, z \text{ real}\}$$

The **distance function** on $\mathbb{R}^3$ is the function

$$d(\vec{v}_1, \vec{v}_2) = \sqrt{(x_1 - x_2)^2 + (y_1 - y_2)^2 + (z_1 - z_2)^2}$$

where $\vec{v}_1 = (x_1, y_1, z_1)$, $\vec{v}_2 = (x_2, y_2, z_2)$. This may be expressed in terms of the **inner product** $\vec{v}_1 \cdot \vec{v}_2 = x_1 x_2 + y_1 y_2 + z_1 z_2$ as $d(\vec{v}_1, \vec{v}_2) = \sqrt{(\vec{v}_1 - \vec{v}_2) \cdot (\vec{v}_1 - \vec{v}_2)}$. Conversely, the **polarization identity**:

$$\vec{v}_1 \cdot \vec{v}_2 = \frac{1}{2}(||\vec{v}_1 + \vec{v}_2||^2 + ||\vec{v}_1 - \vec{v}_2||^2)$$

allows one to recover the value of the inner product from the knowledge of the values of the distance function.

We call a function $f : \mathbb{R}^3 \to \mathbb{R}^3$ an **isometry** if it satisfies

$$d(f(v_1), f(v_2)) = d(v_1, v_2)$$

for all $v_1$ and $v_2$ belonging to $\mathbb{R}^3$.

We want to understand isometries a little better since they will preserve distances (and, in particular, preserve the shapes of solids) and therefore provide us with the kinds of symmetries of 3-space we want to consider. We can construct isometries using certain types of 3 × 3 matrices.

**Lemma 8.2.1.** *If $A$ is a 3 × 3 matrix then the function $A : \mathbb{R}^3 \to \mathbb{R}^3$ is an isometry if and only if $A^t * A = I_3$, where $A^t$ denotes the transpose identity matrix (obtained by flipping the entries of $A$ about the diagonal).*

127

**Remark 8.2.1.** *In particular, if $A$ is an isometry then* $\det(A)^2 = \det(A^t)\det(A) = \det(A^t * A) = \det(I_3) = 1$.

**Proof:** The distance function is preserved if and only if the dot product function is preserved. (This is a consequence of the 'polarization identity' above.) Let $m(\vec{v}, \vec{w}) = \vec{v} \cdot \vec{w}$, where $\cdot$ denotes the vector dot product. Since $m(A\vec{v}, B\vec{w}) = \vec{v} \cdot (A^t * B)\vec{w}$, we have

$$m(A\vec{v}, A\vec{w}) = m(\vec{v}, \vec{w}), \qquad \forall v, w \in \mathbb{R}^3$$

if and only if

$$\vec{v} \cdot (A^t * B)\vec{w} = \vec{v} \cdot \vec{w}, \qquad \forall v, w \in \mathbb{R}^3$$

if an only if $A^t * A = I_3$. $\square$

You may have been wondering how one could construct an isometry. This lemma gives us lots of examples.

**Example 8.2.2.** *A rotation matrix in 3-dimensions may be written in the form*

$$R(\phi, \theta, \psi) = \begin{pmatrix} r_{11} & r_{12} & \sin(\theta)\sin(\psi) \\ r_{21} & r_{22} & \sin(\theta)\cos(\psi) \\ \sin(\phi)\sin(\theta) & -\sin(\theta)\cos(\phi) & \cos(\theta), \end{pmatrix}$$

*where*

$$r_{11} = \cos(\phi)\cos(\psi) - \cos(\theta)\sin(\phi)\sin(\psi),$$
$$r_{12} = \sin(\phi)\cos(\psi) + \cos(\theta)\cos(\phi)\sin(\psi),$$
$$r_{21} = -\cos(\phi)\sin(\psi) - \cos(\theta)\sin(\phi)\cos(\psi),$$
$$r_{22} = -\sin(\phi)\sin(\psi) + \cos(\theta)\cos(\phi)\cos(\psi).$$

*and where the angles $\phi, \theta, \psi$ are the 'Euler angles'. This represents the rotation of 3-space obtained by the following sequence of rotations: rotate by angle $\psi$ about the z-axis, rotate by the angle $\theta$ about the x-axis ($0 \leq \theta \leq \pi$), then rotate by angle $\phi$ about the z-axis again.*

*Although this is an interesting fact due to its explicitness, we shall not use this expression.*

**Question**: Are there any isometries which do not come from matrices as in the above lemma? Yes: any translation gives rise to an isometry but translation is not linear, so cannot be described as a matrix. Moreover, any reflection about a plane containing the origin is also an isometry. A reflection never belongs to the group generated by the translations and rotations.

**Question**: Are there any examples of isometries which do not arise from a composition of a translation and an orthogonal matrix? No: the following theorem classifies all the isometries.

**Theorem 8.2.1.** *A function $f : \mathbb{R}^3 \to \mathbb{R}^3$ is an isometry fixing the origin if and only if $f$ is left multiplication by an orthogonal matrix.*

This will not be proven here (see Artin [Ar], chapter 4, section 5, Proposition 5.16).

As a consequence of this lemma, we see that if the matrix $A$ gives rise to an isometry then $det(A)$ is either equal to 1 or -1 (since $det(A)^2 = det(A^t * A) = det(I_3) = 1$). In particular, the determinant of such a matrix is non-zero, so the matrix is invertible.

**Lemma 8.2.2.** *The set of all $3 \times 3$ matrices $A$ such that the function $A : \mathbb{R}^3 \to \mathbb{R}^3$ is an isometry forms a group under matrix multiplication.*

**Ponderable 8.2.1.** *Verify the group axioms needed to prove this lemma.*

**Notation:** This group will be denoted $O_3(\mathbb{R})$ and called the **orthogonal group** of $\mathbb{R}^3$. We denote by $SO_3(\mathbb{R})$ the following subset

$$SO_3(\mathbb{R}) = \{A \in O_3(\mathbb{R}) \mid det(A) = 1\}.$$

which is called the **special orthogonal group** of $\mathbb{R}^3$.

**Lemma 8.2.3.** $SO_3(\mathbb{R})$ *is a subgroup of* $O_3(\mathbb{R})$.

**Ponderable 8.2.2.** *Verify the group axioms for $SO_3(\mathbb{R})$.*

It is known that the number of cosets in $O_3(\mathbb{R})/SO_3(\mathbb{R})$ is 2. In fact, it is known that

$$O_3(\mathbb{R}) = SO_3(\mathbb{R}) \cup s * SO_3(\mathbb{R}) \quad \text{(disjoint union)} \qquad (8.1)$$

where $s$ is the reflection in the above example (this follows from [Ar], chapter 4, section 5).

**Lemma 8.2.4.** *The isometry $A$ in $O_3(\mathbb{R})$ is orientation preserving if and only if $det(A) = 1$.*

We will not prove this lemma here (see for example, [Ar] or [NST]).

## 8.3   Symmetries of the tetrahedron

Fix a tetrahedron centered at the origin, with one vertex along the z-axis. Each edge has an 'opposite' edge on the tetrahedron (which is actually perpendicular to it if you look at it straight on). Each vertex has an 'opposite' face.

There are orientation preserving symmetries (called **rotations**) of the tetrahedron and orientation reversing symmetries of the tetrahedron. The orientation preserving symmetries of the tetrahedron will be denoted $ST$. They are obtained as follows:

- the 4 axes of symmetry through the centers of the faces yield 2 elements each (120 degree clockwise rotation when viewed from outside and a 240 degree rotation), for a total of 8 elements,

  (This 'tetrahedral symmetry' allows for the mechanical construction of the Pyraminx.)

129

- the 3 pairs of edges (formed by an edge and its opposite) yield one element each (a 180 degree rotation), for a total of 3 elements.

These, plus the identity, give 12 elements in $ST$.

Using the coset decomposition (8.1), we have $T = ST \cup s * ST$ (disjoint), so

$$|T| = |ST| + |s * ST| = 12 + 12 = 24,$$

by the addition principle (Theorem 2.4.1).

**Remark 8.3.1.** *It turns out that $ST$ is essentially the alternating group $A_4$ of even permutations of $S_4$ and $T$ is essentially $S_4$ itself. We shall state the precise result in the next chapter.*

## 8.4 Symmetries of the cube

We fix a cube centered about the origin in 3-space. The set of centers of the faces of a cube forms a set of vertices of an octahedron drawn inside the cube. This octahedron is called the 'dual' polyhedron. These two polyhedra have the same symmetry group, which we denote by $O$. There are orientation preserving symmetries, or rotations, of the cube and orientation reversing symmetries of the cube. The orientation preserving symmetries of the cube will be denoted $SO$. They are obtained as follows:

- the 3 axes of symmetry through the centers of the faces yield 3 elements each (90 degree clockwise rotation when viewed from outside, a 180 degree rotation, and a 270 degree rotation), for a total of 9 elements,

  (This 'hexahedral symmetry' allows for the mechanical construction of the Rubik's Cube.)

- the 4 axes through the opposing vertices yield 2 elements each (all of order 3), for a total of 8 elements,

  (This 'tetrahedral symmetry' allows for the construction of the Skewb [H].)

- the 6 axes through the opposing mid-edge points yield 1 element each (of order 2), for a total of 6 elements.

These elements, plus the identity, yield 24 elements.

**Lemma 8.4.1.** *There are 24 orientation preserving elements in $O$, i.e., $|SO| = 24$.*

The above sketch is one way to see why this is true. Here's another.

**Proof:** Let $V$ be the set of vertices of the cube. The group $SO$ acts on the set $V$. Fix a $v$ belonging to $V$ and let $H = stab_{SO}(v)$. One can check that $|H| = 3$ (since the only symmetry which fixes $v$ is a rotation $g$ about the line through $v$ and its opposite vertex. Since $g$ is order 3, $H = <g>$ is order 3 as well). We have $|V| = 8$, so by a lemma in the previous chapter on orbits and stabilizers, we have $|SO/H| = |V|$. By Lagrange's theorem, $|SO| = |SO/H||H| = 8 \cdot 3 = 24$. $\square$

Now we know $SO$, what is $O$? Note that $s$, the reflection in the example in the previous section, belongs to $O$. Using the coset decomposition of the previous section, we have the coset decomposition

$$O = SO \cup s * SO \quad \text{(disjoint union)}.$$

We know that $|s * SO| = |SO| = 24$, so the addition principle again comes to the rescue, giving us the following result.

**Lemma 8.4.2.** *The order of the octahedral group is* $|O| = 48$.

**Remark 8.4.1.** *It turns out that $SO$ is essentially the symmetric group $S_4$ and $O$ is 'isomorphic to the direct product' $S_4 \times C_2$. We shall state the precise definitions and result in the next chapter.*

## 8.5 Symmetries of the dodecahedron

The set of centers of the faces of a dodecahedron forms a set of vertices of an icosahedron drawn inside. This icosahedron is called the 'dual' polyhedron. We fix a dodecahedron in 3-space so that the vertices of the dual icosahedron are as listed at the beginning of this chapter. Let $SI$ denote the group of orientation preserving symmetries of the dodecahedron. Note $SI$ is a finite subgroup of $SO_3(\mathbb{R})$. Let $I$ denote the group of all symmetries of the dodecahedron. Note $I$ is a finite subgroup of $O_3(\mathbb{R})$ and that $SI$ is a subgroup of $I$. Let $F$ denote the set of faces of the dodecahedron, so $|F| = 12$. $SI$ acts on $F$.

**Lemma 8.5.1.** *$SI$ acts on $F$ transitively.*

We won't prove this. If you look at a dodecahedron it follows 'by inspection'. The reason why this is useful is that it tells us that if $x$ is any face then any other face can be obtained from x by applying some element of $SI$. In other words, the orbit of $x$ is all of $F$: $SI * x = F$.

If $x$ is any face then the only orientation preserving symmetries which don't send $x$ to a different face is a rotation by an integer multiple of 72 degrees about the line passing through the center of $x$ and the center of its opposite face. There are, for each face $x$, exactly 5 distinct rotations of this type. Therefore,

$$|stab_{SI}(x)| = 5.$$

By a lemma in the section on orbits, we have

$$SI/stab_{SI}(x) = SI * x,$$

so $|SI| = |stab_{SI}(x)||SI * x| = 5 \cdot 12 = 60$.

The elements of SI include:

- rotation by $2 * \pi * k/5$, $k = 0, 1, 2, 3, 4$, about the line passing through the center of a face and its opposite,

  (This 'dodecahedral symmetry' allows for the construction of the Megaminx.)

- rotation by $2 * \pi * k/3$, $k = 0, 1, 2$, about the line passing through a vertex and its opposite,

- rotation by $\pi$ about the center of an edge.

Subgroups include:

- stabilizer of a vertex. These are all cyclic of order 3, and they are all conjugate. There are 10 distinct such subgroups since a vertex and its opposite share the same stabilizer.

- stabilizer of a face. These are all cyclic of order 5, and they are all conjugate. There are 6 distinct such subgroups since a face and its opposite share the same stabilizer.

- stabilizer of an edge. These are all cyclic of order 2, and they are all conjugate. There are 15 distinct such subgroups.

**Ponderable 8.5.1.** *Verify all these.*

**Remark 8.5.1.** *It turns out that $SI$ is essentially the alternating group $A_5$ of even permutations of $S_5$ and $I$ is 'isomorphic to the direct product' $A_5 \times S_5$. We shall state the precise result in the next chapter.*

For an excellent discussion of the symmetries of the icosahedron, see [Ba].

## 8.6 Some thoughts on the icosahedron

'Six is a number in itself, and not because God created the world in six days; rather the contrary is true. God created the world in six days because this number is perfect, and would remain perfect even if the work of the six days did not exist.'

*St. Augustine (354-430),* **The City of God**

This section is based on one of John Baez's weekly internet columns [Ba].

A **duad** is a pair of diagonals (a diagonal is a segment from a vertex to its antipodal opposite vertex) of the icosahedron. The top of the icosahedron has 6 vertices and each diagonal must have exactly one of these 6 vertices as an endpoint. There are 12 vertices, hence 6 diagonals, hence

$$\binom{6}{2} = 15$$

different duads. Each duad determines a 'golden rectangle' (i.e., a rectangle whose ratio length/width is either the golden ratio $\phi = (1 + \sqrt{5})/2$ or its inverse.

We may identify a duad with a pair of distinct integers $\{(i, j) \mid 1 \leq i < j \leq 6\}$, i.e., with a 2-cycle in $S_6$.

A duad may be pictured as in Figure 8.6.

**icosahedron with duad**

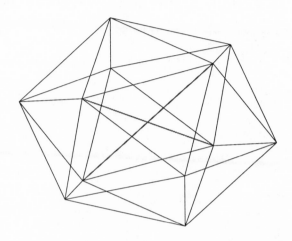

Each element of the rotation group of the icosahedron (i.e., the group of orientation-preserving symmetries of the icosahedron) must send a duad to a duad. Each duad has 4-fold symmetry, i.e., can be sent to itself in 4 ways. There are 15 duads, so there are $4 \cdot 15 = 60$ ways to send a duad to another. This is precisely the number of orientation-preserving symmetries of the icosahedron.

James Sylvester (the same one we met already in §2.2.1) called a partitioning

$$X = X_1 \cup ... \cup X_n \qquad \text{(disjoint)},$$

of a set $X$ a **syntheme** if each of the sets $X_i, 1 \leq i \leq n$, has the same number of elements. If we take

$$X = \{\text{set of diagonals of the icosahedron}\}$$

then a syntheme is a set of three duads, no two having a diagonal in common. There are

$$\binom{6}{2}\binom{4}{2}/3! = 15$$

different synthemes (the 3! since there are 3! ways to permute the duads amongst themselves). A syntheme may be represented by a coloring of the vertices on the top of the icosahedron using three colors, each for exactly two vertices. We may identify a syntheme with a product of 3 distinct 2-cycles in $S_6$. We partition the set of 15 duads into 5 groups of 3 as follows. (Recall each syntheme is a triple of duads.) First, pick a syntheme, $A_1$. Pick another syntheme $A_2$, so that $A_1, A_2$ have no duad in common. Continue on in this way until you pick five synthemes

133

$A_1, .., A_5$, no two of which have a duad in common. Such a choice of 5 synthemes is called a **pentad**. There a 6 pentads, which we label $P_1, ..., P_6$ in any way you like. (A list of the six pentads is given in [R], chapter 7.) Any permutation of the 6 diagonals of the icosahedron gives rise to a permutation of the set of 6 pentads. Hence any permutation of the 6 diagonals of the icosahedron, which may be regarded as an element of $S_6$, gives rise to a permutation of the set of 6 pentads, which may also be regarded as an element of $S_6$. This gives a map

$$f : S_6 \to S_6.$$

This example will be discussed further in §9.3.2. See also [R] and [Ba].

# Chapter 9

# The illegal cube group

On front: *186,000 mps.*
On back: *It's not just a good idea, it's the law!*
  *T-shirt humor*

Groups in mathematics are analogous to molecules in chemistry. We want to know what groups look like, to know how to describe them, how to compare them, how to make more of them, to know if they fall into families with similar properties, and so on.

Given two molecules, a chemist wants to know how to compare them, to understand their similarities and differences. Given two groups $G_1, G_2$, an analogous question is to ask how 'similar' are they? (Exactly what is meant by 'similar' will be explained later.) We shall, in this chapter, introduce notions and techniques useful for comparing two groups. In a later chapter, we will focus on the $3 \times 3$ Rubik's Cube group by comparing it to 'better understood' (or at least more 'usual') groups, such as the cyclic groups and the symmetric groups which we have already studied. In this chapter, we shall focus on understanding a somewhat simpler but closely related group, the group of all 'legal' and 'illegal' moves of the Rubik's Cube. (In an 'illegal' move you can disassemble and reassemble the cube but you cannot peel off and swap individual stickers.) Remarkably, the illegal Rubik's Cube group is simpler to understand than the Rubik's Cube group itself, even though it is much larger.

To 'explicitly construct' the Rubik's Cube group, we need to know how to construct

- quotient groups,

- direct products,

- semi-direct products.

These are studied in this chapter. Once the Rubik's Cube group is constructed, it is a relatively easy matter to find its order, among other things.

## 9.1 Functions between two groups

To compare two groups, we often first talk about functions between them. If $f : G_1 \to G_2$ is a function from a group $G_1$ to a group $G_2$ then, as you might imagine, if $f$ and $G_1$ were well understood then some information about $G_2$ could probably be obtained. It is the purpose of this chapter to find out to what extent this can be made precise.

A homomorphism between two groups is, roughly speaking, a function between them which preserves the (respective) group operations. These special types of functions are the most useful for our purposes.

**Definition 9.1.1.** *Let $G_1, G_2$ be groups, with $*_1$ denoting the group operation for $G_1$ and $*_2$ the group operation for $G_2$. A function $f : G_1 \to G_2$ is a* **homomorphism** *if and only if, for all $a, b \in G_1$, we have*

$$f(a *_1 b) = f(a) *_2 f(b).$$

**Ponderable 9.1.1.** *Prove the following: If $f : G_1 \to G_2$ is a homomorphism of groups then*

$$f(G_1) = \{g \in G_2 \mid g = f(x), \text{ for some } x \in G_1\}$$

*is a subgroup of $G_2$.*

The subgroup $f(G_1) \subset G_2$ is called the **image of** $f$ and is sometimes denoted $im(f)$.

**Example 9.1.1.** *Let $G$ be a group and $h$ a fixed element of $G$. Define $f : G \to G$ by*

$$f(g) = h^{-1} * g * h, \qquad g \in G.$$

*Then the following simple trick*

$$f(a * b) = h^{-1} * (a * b) * h = h^{-1} * a * h * h^{-1} * b * h = f(a) * f(b)$$

*shows that $f$ is a homomorphism. In this case, $im(f) = G$, i.e., $f$ is surjective.*

**Ponderable 9.1.2.** *Let*

$$A = \begin{pmatrix} 0 & 1 & 0 \\ 1 & 0 & 0 \\ 0 & 0 & 1 \end{pmatrix}, \qquad B = \begin{pmatrix} 1 & 0 & 0 \\ 0 & 0 & 1 \\ 0 & 1 & 0 \end{pmatrix}$$

*Now, let $G = \langle A, B \rangle$ denote the group of all matrices which can be written as any arbitrary product of these two matrices (in any order and with as many terms as you want). Show that we have*

$$G = \{I_3 = \begin{pmatrix} 1 & 0 & 0 \\ 0 & 1 & 0 \\ 0 & 0 & 1 \end{pmatrix}, \begin{pmatrix} 1 & 0 & 0 \\ 0 & 0 & 1 \\ 0 & 1 & 0 \end{pmatrix}, \begin{pmatrix} 0 & 1 & 0 \\ 1 & 0 & 0 \\ 0 & 0 & 1 \end{pmatrix},$$
$$\begin{pmatrix} 0 & 1 & 0 \\ 0 & 0 & 1 \\ 1 & 0 & 0 \end{pmatrix}, \begin{pmatrix} 0 & 0 & 1 \\ 1 & 0 & 0 \\ 0 & 1 & 0 \end{pmatrix}, \begin{pmatrix} 0 & 0 & 1 \\ 0 & 1 & 0 \\ 1 & 0 & 0 \end{pmatrix}\}$$

*(The interested reader may want to try to check this by regarding each such matrix as a permutation matrix.) Define* $f : G \rightarrow S_3$ *by*

| $g$ | $f(g)$ |
|-----|--------|
| $I_3$ | $1$ |
| $A$ | $s_1$ |
| $B$ | $s_2$ |
| $A * B$ | $s_1 * s_2$ |
| $B * A$ | $s_2 * s_1$ |
| $A * B * A$ | $s_1 * s_2 * s_1$ |

*Show that this is a homomorphism.*

**Example 9.1.2.** *The function*

$$sign : S_n \rightarrow \{\pm 1\},$$

*which assigns to each permutation its sign, is a homomorphism. This was already proven in §3.1. However, another reason why this is true is because the sign of a permutation g is the determinant of the associated permutation matrix $P(g)$. Since the determinant of the product is the product of the determinants (see §2.2.5), we have*

$$sgn(gh) = \det P(gh) = \det(P(g)P(h)) = \det P(g) \det P(h) = sign(g)sign(h),$$

*for all $g, h \in S_n$. From this it follows that sign is a homomorphism.*

**Lemma 9.1.1.** *If $f : G_1 \rightarrow G_2$ is a homomorphism then*
*(a) $f(e_1) = e_2$, where $e_1$ denotes the identity element of $G_1$ and $e_2$ denotes the identity element of $G_2$,*
*(b) $f(x^{-1}) = f(x)^{-1}$, for all $x$ belonging to $G_1$,*
*(c) $f(y^{-1} *_1 x *_1 y) = f(y)^{-1} *_2 f(x) *_2 f(y)$, for all $x, y$ belonging to $G_1$,*
*where $*_1$ denotes the group operation for $G_1$ and $*_2$ the group operation for $G_2$.*

**Proof:** (a) We have $f(x) = f(x *_1 e_1) = f(x) *_2 f(e_1)$, for any $x \in G_1$. Multiply both sides of this equation on the left by $f(x)^{-1}$.
(b) We have, by part (a), $e_2 = f(e_1) = f(x *_1 x^{-1}) = f(x) *_2 f(x^{-1})$. Multiply both sides of this equation on the left by $f(x)^{-1}$.
(c) Is left to the reader. $\square$

**Ponderable 9.1.3.** *Show part (c) (hint: use the definition and part (b)).*

**Definition 9.1.2.** *Let $G_1, G_2$ be finite groups. We say that $G_1$ **embeds** (or **injects**) into $G_2$ if there exists an injective homomorphism $f : G_1 \rightarrow G_2$. A homomorphism $f : G_1 \rightarrow G_2$ is an **isomorphism** if it is a bijection (as a function between sets). In this case, we call $G_1$ and $G_2$ **isomorphic** and write $G_1 \cong G_2$. An isomorphism from a group $G$ to itself is called an **automorphism**.*

The concept of an isomorphism was first introduced by E. Jordan, who we have already met. Besides group theory, Jordan is famous in mathematical circles for rigorously proving one of the most intuitively obvious, but difficult to rigorously prove, results: Any simple closed curve divides the plane into two regions.

An isomorphism is the notion we will use when we want to say two groups are 'essentially the same group', i.e., one is basically a carbon copy of the other with the elements relabeled and the binary operation modified. For example, if you have any two groups of order 2 then they must be isomorphic. (The interested reader can verify that the map which sends the identity of one group to the identity of the other and the only non-identity of one group to the only non-identity of the other must be a group isomorphism.) In other words, there is only one group of order 2, up to isomorphism.

Let $O(n)$ denote the the number of non-isomorphic groups of order $n$, so $O$ is a function $O : \mathbb{N} \to \mathbb{N}$. This is a rather curiously behaving function, jumping up at integers which have lots of prime factors, such as 16, 24, 32, 48, ... .

**Ponderable 9.1.4.** *Show the following:*

*(a) If $m \geq 1$ then*
$$G = \{e^{2\pi i k/m} \mid 0 \leq k \leq m - 1\}$$
*with group operation given by ordinary multiplication of complex numbers, is isomorphic to the additive group $\mathbb{Z}/m\mathbb{Z}$.*

*(b) If $m > 1, n > 1$ have no prime divisors in common, i.e., if $m, n$ are relatively prime, then $C_m \times C_n \cong C_{mn}$.*

*(Hint: (a) Let $a \in C_m$ be a generator. The map $f = f_{m,a} : C_m \to \mathbb{Z}/m\mathbb{Z}$ which satisfies $f(a) = 1$ extends to a unique homomorphism between these groups. Show that this is an isomorphism. (b) First, show $g(x + mn\mathbb{Z}) = (x + m\mathbb{Z}, x + n\mathbb{Z})$, for all $x \in \mathbb{Z}$ defines a homomorphism $\mathbb{Z}/mn\mathbb{Z} \to \mathbb{Z}/m\mathbb{Z} \times \mathbb{Z}/n\mathbb{Z}$. Next, using the assumption that $m, n$ are relatively prime, show that this function $g$ is an injection. Conclude, since the domain and range of $g$ both have the same cardinality, that $g$ is also a surjection, by Ponderable 2.1.6. Finally, using part (a) show $C_m \times C_n \cong C_{mn}$.)*

## 9.2 Group actions

We need to study actions, begun in chapter 5 above, in more detail.

**Lemma 9.2.1.** *Let $G$ be a group and $X$ a finite set. If $G$ acts on $X$ (on the left, resp. on the right) then there is a homomorphism $G \to S_X$ given by $g \longmapsto \phi_g$. Conversely, if $\phi : G \to S_X$ is a homomorphism then $\phi(g) : X \to X$ defines a (left, resp. right) action of $G$ on $X$.*

This is an immediate consequence of the definition of an action.

**Example 9.2.1.** *Let $G$ be the Rubik's Cube group generated by the basic moves $R, L, U, D, F, B$. For each move $g \in G$, let $\rho(g)$ be the corresponding permutation of the set of vertices $V$ of the cube and let $\sigma(g)$ be the corresponding permutation of the set of edges $E$ of the cube. Then it is possible to show that*
*(a) $\rho : G \to S_V$ is a homomorphism,*
*(b) $\sigma : G \to S_E$ is a homomorphism.*
*To check this, the first fundamental theorem of cube theory in §9.6.1, is useful.*

Let $G$ act on a set $X$, denoted by $\phi_g : X \to X$ for $g \in G$. This action also gives rise to an action on $X \times X$: $\phi_{2,g} : X \times X \to X \times X$ for $g \in G$, where $\phi_{2,g}(x_1, x_2) = (\phi_g(x_1), \phi_g(x_2))$, for each $(x_1, x_2) \in X \times X$. When $G$ acts transitively on $X$ (see Definition 5.10.3 above), it is remarkable enough. Sometimes, $G$ even acts transitively on $X \times X$, a far more special circumstance (in this case, the action is called **doubly transitive**). The definition below generalizes this idea.

**Definition 9.2.1.** *Let a group $G$ act on a set $X$. We call the action $k$-**tuply** **transitive** if for each pair of ordered $k$-tuples $(x_1, x_2, .., x_k), (y_1, y_2, .., y_k)$ of elements belonging to $X$ there is a $g \in G$ such that $y_i = \phi_g(x_i)$ for each $1 \le i \le k$.*

If $k = 2$ then '2-tuply transitive' is the same as 'doubly transitive.'

**Ponderable 9.2.1.** *Is the Rubik's Cube group doubly transitive on the set of edge facets?*

The following result is one illustration of how unique the symmetric group and alternating group are. (Recall that the alternating group $A_n$ was defined in example 5.6.2 above.)

**Theorem 9.2.1.** *If $k > 5$ and $G$ is a group acting $k$-transitively on a finite set $X$ then $G$ is isomorphic to $S_m$ or to $A_n$, for some $m \ge k$ or some $n \ge k + 2$.*
*Conversely, $S_n$ acts $n$-transitively on $\{1, 2, ..., n\}$ and $A_n$ acts $(n-2)$-transitively on $\{1, 2, ..., n\}$.*

This is proven in [R]. Some examples are given in §12.4 below.

**Remark 9.2.1.** *We shall, in conjunction with our group-theoretical determination of the Rubik's Cube group proven later, be able to deduce from Theorem 9.2.1 the following result.*

**Corollary 9.2.1.** *(a) Suppose $c_1, ..., c_6$ are any 6 distinct corner subcubes of the Rubik's cube. Let $c'_1, ..., c'_6$ be any other 6 distinct corner subcubes. There is an element of the Rubik's Cube group $G$ which fixes all edge subcubes and sends $c_i$ to $c'_i$, for all $i = 1, ..., 6$. Loosely speaking, we may express this by saying $G$ acts 6-transitively on the corners, fixing the edges.*
*(b) Suppose $c_1, ..., c_8$ are any 8 distinct corner subcubes of the Rubik's cube. Let $c'_1, ..., c'_8$ be any other 8 distinct corner subcubes. There is an element of the*

Rubik's Cube group $G$ which sends $c_i$ to $c_i'$, for all $i = 1, ..., 8$ (and might not fix all edge subcubes). Loosely speaking, we may express this by saying $G$ acts 8-transitively on the corners (allowing the edges to be moves).

(c) $G$ acts 10-transitively on the edges, fixing the corners.

(d) $G$ acts 12-transitively on the edges but may permute the corners.

## 9.3 When two groups are really the same

**Example 9.3.1.** Let $G$ be the group in Ponderable 9.1.2 and $f : G \to S_3$ the homomorphism. This is in fact an isomorphism.

**Example 9.3.2.** Let $H$ be the subgroup of the Rubik's Cube group generated by the basic move $R$: $H = \langle R \rangle$. Then $H \cong C_4$ (where $C_4$ denotes the cyclic group of order 4).

**Example 9.3.3.** Recall that to each permutation $g$ of the set $\{1, 2, ..., n\}$ we can associate a $n \times n$ permutation matrix $P(g)$ in such a way that

$$P(g) \begin{pmatrix} x_1 \\ x_2 \\ \vdots \\ x_n \end{pmatrix} = \begin{pmatrix} x_{g(1)} \\ x_{g(2)} \\ \vdots \\ x_{g(n)} \end{pmatrix}.$$

Here the image of $i$ under the permutation $g$ is denoted $g(i)$, though in fact one plugs $i$ into $g$ from the left.) We let $P_n$ denote the set of all $n \times n$ permutation matrices. This is a group under matrix multiplication. The function

$$P : S_n \to P_n,$$
$$g \longmapsto P(g)$$

is an isomorphism. The proof that this is a bijection and a homomorphism was given earlier in the chapter on permutations.

**Example 9.3.4.** In this example, we consider the symmetry groups of various solids and planar figures, the symmetries being regarded as 3-dimensional transformations (as opposed to 2-dimensional transformations in subsection 5.3).

From [NST], we have the following table of isomorphisms:

| name | notation | isomorphic to |
|---|---|---|
| symmetry group of tetrahedron | $T$ | $S_4$ |
| rotation group of tetrahedron | $ST$ | $A_4$ |
| symmetry group of octahedron | $O$ | $S_4 \times C_2$ |
| rotation group of octahedron | $SO$ | $S_4$ |
| symmetry group of icosahedron | $I$ | $A_5 \times C_2$ |
| rotation group of icosahedron | $SI$ | $A_5$ |
| symmetry group of regular n-gon | | $D_{2n},\ n$ odd |
| | | $D_n \times C_2,\ n$ even |
| rotation group of regular n-gon | | $D_n$ |

**Example 9.3.5.** *This example may be found in [B].*

*Let $Q$ denote the quaternion group:*

$$Q = \{1, -1, i, -i, j, -j, k, -k\},$$

*where $i^2 = j^2 = k^2 = -1, ij = k, jk = i, ki = j$, and in general, $xy = -yx$ for $x, y$ belonging to $i, j, k$. Then $Q$ is isomorphic to the group*

$$Q^* = \langle a, b \rangle \subset G,$$

*where*

$$a = F^2 * M_R * U^{-1} * M_R^{-1} * U^{-1} * M_R * U * M_R^{-1} * U * F^2,$$
$$b = F * U^2 * F^{-1} * U^{-1} * L^{-1} * B^{-1} * U^2 * B * U * L,$$

*via the map $f : Q^* \to Q$ defined by $\phi(a) = i$, $\phi(b) = k$.*

*The proof of this claim will be discussed in a later chapter. (The easiest way to prove this uses ideas we haven't yet introduced.)*

**Lemma 9.3.1.** *As above, let $G$ be a group and $h$ a fixed element of $G$. Define $c_h : G \to G$ by*

$$c_h(g) = h * g * h^{-1}, \quad g \in G.$$

*Show that this is an automorphism.*

**Proof:** To verify this, we must show that $f = c_h$ is an injective and surjective homomorphism (we drop the subscript for simplicity of notation).

First, we show that $f$ is injective. Suppose $f(g_1) = f(g_2)$. Then $f(g_1 * g_2^{-1}) = 1$, so that $h * g_1 * g_2^{-1} * h^{-1} = 1$. Multiply both sides of this equation on the right by $h$ and on the left by $h^{-1}$. We obtain $g_1 * g_2^{-1} = 1$. This implies $g_1 = g_2$, so $f$ is injective.

Now we show $f$ is surjective. Let $x$ be an arbitrary but fixed element of $G$. Let $y = h^{-1} * x * h$. Then

$$f(y) = f(h^{-1} * x * h) = h * h^{-1} * x * h * h^{-1} = x.$$

Therefore, f is surjective.

It was previously verified that $f$ is a homomorphism. $\square$

**Definition 9.3.1.** *An automorphism as in Lemma 9.3.1 is called **inner**. An automorphism of $G$ which is not of this form, for some $h \in G$, is called **outer**.*

The concept of inner and outer automorphisms were introduced by Otto Hölder (1859-1937). Hölder also classified all the finite groups, up to isomorphism, of order $\leq 200$. The classification of *all* finite groups has only recently been finished and is said to require over 3000 pages of details.

**Notation:** The set of all automorphisms of a group $G$ is denoted $Aut(G)$. The subset of inner automorphisms is denoted

$$Inn(G) := \{f \in Aut(G) \mid f = c_h, \text{ some } h \in G\},$$

in the notation of the above Lemma 9.3.1.

**Ponderable 9.3.1.** *(a) Show Aut(G) is a group with composition as the group operation.*

*(b) Show that Inn(G) is a subgroup of Aut(G).*

## 9.3.1 Conjugation in $S_n$

In the early part of the 1900's a brilliant Indian mathematician Srinivasa Ramanujan (1887-1920) lay in hospital in England very ill. The great British mathematician G. Hardy visited him and remarked that he took taxi number 1729, a 'boring number'. 'On the contrary,' Ramanujan argued, 'it's a very interesting number, being the smallest integer which is the sum of three cubes in two different ways.' One of the most interesting and original mathematicians of all time, Ramanujan knew an astonishing number of unusual facts about integers, such as this one.

What does Ramanujan have to do with conjugation? Read on, and you'll find out!

As promised in §5.9, we will find a simple criteria for determining when two permutations (in $S_n$) are conjugate (in $S_n$).

The following result will be of importance to us in a later chapter.

**Lemma 9.3.2.** *Suppose $f : S_n \to S_n$ is an inner automorphism. If $g \in S_n$ is a disjoint product of cycles of length $k_1, ..., k_r$ then $f(g)$ is a disjoint product of cycles of length $k_1, ..., k_r$.*

In other words, an inner automorphism (i.e., conjugation) must 'preserve the cycle structure'.

**Proof:** Since $f$ is inner, let it be conjugation by some element $h \in S_n$ say, so $f(g) = h^{-1}gh$, for all $g \in S_n$. Let $(i)g \in \{1, ..., n\}$ denote the image of $i \in \{1, ..., n\}$ under $g \in S_n$. The lemma is a consequence of the following simple calculation: if $(i)g = j$ then, for all $1 \leq i \leq n$, we have

$$((i)h)(h^{-1}gh) = (j)h. \tag{9.1}$$

In other words, if $g$ sends $i \longmapsto j$ then $h^{-1}gh$ sends $(i)h \longmapsto (j)h$. It follows that $g$ and $h^{-1}gh$ have the cycle structure. $\square$

**Theorem 9.3.1.** *Two elements $g, g' \in S_n$ are conjugate if and only if they have the same cycle structure.*

**Proof:** The Lemma proves the 'only if' direction of this equivalence. Suppose that $g, g' \in S_n$ have the same cycle structure. Write their disjoint cycle decompositions using the lexicographical ordering imposed on the lengths of the cycles occurring in the decomposition: say

$$g = (i_1, \ ... \ , i_{n_1})(i_1, \ ... \ , i_{n_2})...(i_1, \ ... \ , i_{n_k}),$$
$$g' = (i'_1, \ ... \ , i'_{n_1})(i'_1, \ ... \ , i'_{n_2})...(i'_1, \ ... \ , i'_{n_k}),$$

where $1 \leq n_1 \leq ... \leq n_k \leq n$ and $n = n_1 + ... + n_r$. Pick an $h \in S_n$ such that $h : i_j \longmapsto i'_j$, for all $1 \leq j \leq n$. Then $g' = h^{-1}gh$, by (9.1). $\square$

A **partition** of $n$ is an $r$-tuple, $(n_1, ..., n_r)$ of positive integers such that $n_1 \leq ... \leq n_r \leq n$ and $n = n_1 + ... + n_r$. From the results above, it follows that each conjugacy class of $S_n$ corresponds to one and only one partition of $n$. In particular, the number of partitions of $S_n$ is equal to the number $p(n)$ of distinct partitions of $n$. An asymptotic formula for this number was found by the brilliant Indian mathematician Srinivasa Ramanujan. His formula is too complicated for this book, but a special case of it is the following

$$\lim_{n \to \infty} \frac{\log p(n)}{\sqrt{n}} = \pi \sqrt{\frac{2}{3}}.$$

One of the world's greatest mathematical geniuses, Ramanujan worked on number theory, elliptic functions, continued fractions, and infinite series [MT].

### 9.3.2  ... and a side order of automorphisms, please

Though we shall not need it here, the following fact is interesting since it illustrates what a unique role the symmetric group $S_6$ plays in the family of all symmetric groups.

**Theorem 9.3.2.** *If $n \neq 2, 6$ then the homomorphism $\phi : S_n \to Aut(S_n)$ defined by $\phi(g) = c_g$ (where $c$ is as in Lemma 9.3.1 above) is an isomorphism:*

$$S_n \cong Aut(S_n).$$

*If $n = 6$ then $|Aut(S_6)| = 2 \cdot |S_6|$.*

The following example continues the discussion from §8.6.

**Example 9.3.6.** *Any permutation of the 6 diagonals of the icosahedron, which may be regarded as an element of $S_6$, gives rise to a permutation of the set of 6 pentads, which may also be regarded as an element of $S_6$. This gives a map*

$$f : S_6 \to S_6,$$

*which is in fact a homomorphism. This homomorphism is injective so it is actually an automorphism.*

*However, a 2-cycle on the set of 6 diagonals (i.e., swapping exactly 2 diagonals) does not induce a 2-cycle on the set of these 6 pentads. In fact, a 2-cycle on the set of diagonals gives rise to a product of three disjoint 2-cycles on the set of these 6 pentads. Therefore, by the above theorem (which says that an inner automorphism must preserve the cycle structure) this automorphism $f$ cannot be an inner automorphism.*

## 9.4  Kernels are normal, some subgroups are not

Let $f : G_1 \to G_2$ be a homomorphism between two groups. Let

$$ker(f) = \{g \in G_1 \mid f(g) = e_2\},$$

where $e_2$ is the identity element of $G_2$. This set is called the **kernel** of $f$.

**Lemma 9.4.1.** $ker(f)$ *is a subgroup of* $G_1$.

The interested reader can check this for him/herself by using the definition of a homomorphism.

**Example 9.4.1.** *Let*

$$sgn : S_n \to \{\pm 1\}$$

*denote the homomorphism which associates to a permutation either 1, if it is even, or -1, if it is odd. Then* $A_n = ker(sgn) \subset S_n$.

The following properties of the kernel are useful:

**Lemma 9.4.2.** : *Let* $f : G_1 \to G_2$ *be a homomorphism between two groups.*
*(a)* $f$ *is injective if and only if* $ker(f) = \{e_1\}$.
*(b) if g belongs to the kernel and x is any element of* $G_1$ *then* $x^{-1} * g * x$ *must also belong to the kernel.*

**Proof:**  (a) $f$ is injective if and only if $f(g_1) = f(g_2)$ implies $g_1 = g_2$ $(g_1, g_2 \in G_1)$. Note $f(g_1) = f(g_2)$ is true if and only if $f(g_1 * g_2^{-1}) = e_2$. If $ker(f) = \{e_1\}$ then $f(g_1 * g_2^{-1}) = e_2$ implies $g_1 * g_2^{-1} = e_2$, which implies $g_1 = g_2$, which implies $f$ is injective.

Therefore, if $ker(f) = \{e_2\}$ then $f$ is injective. Conversely, if $f$ is injective then $f(x) = f(e_1)(= e_2)$ implies $x = e_1 (x \in G_1)$. This implies $ker(f) = \{e_1\}$.

(b) Multiply both sides of $e_2 = f(g)$ on the left by $f(x)^{-1}$ and on the right by $f(x)$. We get

$$e_2 = f(x)^{-1} * e_2 * f(x) = f(x^{-1}) * f(g) * f(x) = f(x^{-1} * g * x),$$

as desired. $\square$

**Definition 9.4.1.** *Let* $H$ *be a subgroup of* $G$. *We say that* $H$ *is a* **normal** *subgroup if, for each* $g \in G$, $g^{-1} * H * g = H$ *(i.e., for each* $g \in G$ *and each* $h \in H, g^{-1} * h * g$ *belongs to* $H$).

**Notation**: Sometimes we denote 'H is a normal subgroup of G' by

$$H \triangleleft G$$

**Example 9.4.2.** *(a)* $A_n \triangleleft S_n$ *and* $|A_n| = \frac{1}{2}|S_n|$.
*(b) On the other hand, examples of subgroups which are not normal are easy to come by. If* $n > 5$ *and* $H$ *is any non-trivial proper subgroup of* $A_n$ *(for example, any non-trivial cyclic subgroup) then* $H$ *is not normal in* $A_n$ *(see Theorem 9.4.1 below).*

We have already shown the following

**Lemma 9.4.3.** *If* $f : G_1 \to G_2$ *is a homomorphism between two groups then* $ker(f)$ *is a normal subgroup of* $G_1$.

## 9.4.1 The alternating group

The following remarkable result about the alternating group will not be needed to understand the structure of the Rubik's Cube. However, the theorem below is interesting because of its connection with the fact (due to E. Galois) that you cannot solve the general polynomial of degree 5 or higher using radicals, i.e., that there is no analog of the quadratic formula for polynomials of degree 5 or higher. Explaining this connection would, unfortunately, take us too far from our main topic.

**Theorem 9.4.1.** *If $X$ has 5 elements or greater then $A_X$ has no non-trivial proper normal subgroups. In other words, if $H \triangleleft A_X$ is a normal subgroup then either $H = \{1\}$ or $H = A_X$.*

This will not be proven here. (For a proof, see for example [R] or [Ar], chapter 14.)

However, the next fact about the alternating group *will* be needed later in our determination of the structure of the Rubik's Cube group. This fact also arose in connection with our discussion of the 'legal positions' of the 15 Puzzle in a previous chapter.

**Proposition 9.4.1.** *Let $H$ be the subgroup of $S_n$ generated by all the 3-cycles in $S_n$ then $H = A_n$.*

**Proof:** Since $sgn : S_n \to \{\pm 1\}$ is a homomorphism, and since any 3-cycle is even, any product of 3-cycles must also be even. Therefore, $H \subset A_n$. If $g \in A_n$ then $g$ must swap an even number of the inequalities $1 < 2 < ... < n-1 < n$, by Definition 3.1.1. Therefore, (since any permutation may be written as a product of 2-cycles, Theorem 3.4.1) $g$ must be composed of permutations of the form $(i, j)(k, l)$ or $(i, j)(j, k)$. But $(i, j)(k, l) = (i, j, k)(j, k, l)$ and $(i, j)(j, k) = (i, j, k)$. Therefore, $g \in H$. This implies $A_n \subset H$, so $A_n = H$. $\square$

The following more precise result is very useful for the purposes of the analysis of permutation puzzles.

**Lemma 9.4.4.** *([W]) Let $X$ be a finite set, $|X| \geq 3$ and fix $u, v$ as elements in $X$. Then the 3-cycles $(u, v, x)$, as $x$ runs over all elements of $X - \{u, v\}$, generate $A_X$.*

This lemma gives an even stronger statement than the previous claim. Now, instead of only one element being fixed, there are two elements fixed and the alternating group is still generated.

## 9.5 Quotient groups

> QUOTIENT, n. A number showing how many times a sum of money belonging to one person is contained in the pocket of another – usually about as many times as it can be got there.
> *Ambrose Bierce,* **The devil's dictionary**

One of the most useful facts about normal subgroups is the following result/definition.

**Lemma 9.5.1.** *If $H$ is a normal subgroup of $G$ then the coset space $G/H$ with the binary operation,*

$$aH * bH = (ab)H, \quad (aH)^{-1} = a^{-1}H,$$

*for all $a, b$ belonging to $G$, is a group. The identity element of this group is the trivial coset $H$.*

This group $G/H$ is called the **quotient group** of $G$ by $H$ and is sometimes pronounced '$G$ mod $H$'. Older texts may use the terminology 'factor group' instead. This notion, for an abstract group, was first introduced by Otto Hölder, who we have already met. He originally started out training to become an engineer. Hölder made important contributions not only to group theory but also to Fourier series.

**Example 9.5.1.** *If $f : G_1 \to G_2$ is a homomorphism between two groups then $G_1/ker(f)$ is a quotient group.*

Next we introduce an important idea first emphasized by E. Galois, the French mathematician mentioned in the introduction. Though Galois attended secondary school, he had trouble entering the French university system. Sadly, by the time he entered the École Normale Súperieure in November 1829, he also began to get caught up in the activities of the French revolution which toppled Charles X. His political activities resulted in him being expelled from school in December 1830. Though apparently he had been a rather rebellious teenager, this only frustrated him further. He died in a fight with another Frenchman.

Galois' work was the first to begin to illuminate the 'basic building blocks' of the collection of finite groups (i.e., the analog of the idea that atoms are the 'basic building blocks' of molecules). Galois introduced the ideas of solvable groups and normal groups, in the context of permutation groups. Roughly speaking, one may say that the basic building blocks of finite groups are those groups which have no proper non-trivial normal subgroups. Intuitively, this is because a non-trivial quotient group (by a normal subgroup) is closely related to the original group but smaller in size (and hence perhaps subject to analysis by an inductive argument of some type). These basic building blocks are called 'simple' groups.

**Definition 9.5.1.** *A* **simple group** *is a group with no proper normal subgroups other than the trivial subgroup $\{1\}$.*

**Example 9.5.2.** *If $p$ is a prime then $C_p$ (the cyclic group having $p$ elements) is simple. In fact, if $G$ is any group which is both abelian and simple then there is a prime $p$ such that $G \cong C_p$. If $n > 4$ then $A_n$ is simple (as was stated above in Theorem 9.4.1). These facts are proven in [R].*

*Simple groups are not very abundant. In fact, the first non-abelian simple group is of order 60 (it's $A_5$).*

The following basic result describes the quotient group $G_1/ker(f)$.

**Theorem 9.5.1. (first isomorphism theorem)** *If $f : G_1 \to G_2$ is a homomorphism between two groups then $G_1/ker(f)$ is isomorphic to $f(G_1)$.*

**Proof:** $ker(f)$ is a normal subgroup of $G_1$, so $G_1/ker(f)$ is a group. We must show that this group is isomorphic to the group $f(G_1)$. Define $\phi : G_1/ker(f) \to f(G_2)$ by $\phi(g \cdot ker(f)) = f(g)$, for $g \in G_1$. We must show
(a) $\phi$ is well-defined,
(b) $\phi$ is a homomorphism,
(c) $\phi$ is a bijection.
If $g \cdot ker(f) = g' \cdot ker(f)$ then $g^{-1}g' \in ker(f)$, since $ker(f)$ is a group. This implies $f(g^{-1}g') \in f(ker(f)) = \{1\}$, so $f(g) = f(g')$. This implies $\phi$ is well-defined.
Since $ker(f)$ is normal, $(g \cdot ker(f))(g' \cdot ker(f)) = gg'(g'^{-1}ker(f)g')ker(f) = gg' \cdot ker(f)$. Therefore $\phi((g \cdot ker(f))(g' \cdot ker(f))) = \phi(gg' \cdot ker(f)) = f(gg') = f(g)f(g') = \phi(g \cdot ker(f))\phi(g' \cdot ker(f))$, for all $g, g' \in G$. This implies $\phi$ is a homomorphism.
It is clear that $\phi$ is surjective. To show that $\phi$ is a bijection, it suffices to prove $\phi$ is an injection. Suppose that $\phi(g \cdot ker(f)) = \phi(g' \cdot ker(f))$, for some $g, g' \in G$. Then $f(g) = f(g')$, so $f(g^{-1}g') = 1$. By definition of the kernel, this implies $g^{-1}g' \in ker(f)$, so $g \cdot ker(f) = g' \cdot ker(f)$. This implies $\phi$ is injective.
$\square$

The other isomomorphism theorems will not be needed but will be stated to help to illustrate the usefulness of the notion of normality.

**Theorem 9.5.2. (second isomorphism theorem)** *If $H, N$ are subgroups of a group $G$ and if $N$ is normal then*
*(a) $H \cap N$ is normal in $H$,*
*(b) there is an isomorphism*

$$H/(H \cap N) \cong NH/N.$$

**Theorem 9.5.3. (third isomorphism theorem)** *If $N_1, N_2$ are subgroups of a group $G$, if $N_1 \subset N_2$, and if $N_1$ and $N_2$ are both normal then*
*(a) $N_2/N_1$ is normal in $G/N_1$,*
*(b) there is an isomorphism between*

$$(G/N_1)/(N_2/N_1) \cong G/N_2.$$

We shall not prove these results here - see [G] or [R]. These are due to Emmy Noether (1882-1935), one of the first women to obtain a PhD in mathematics in Germany (or anywhere for that matter). Though extremely talented, the fact that she was Jewish and a woman made her progress in mathematics difficult. D. Hilbert and F. Klein, two of the world's leading mathematicians at the time, supported her and she conducted research and lectured at the University of Göttingen until the Nazis came to power in the 1930's. She left for Bryn Mawr College in the USA but died a few years later. She made many important contributions to modern algebra.

# 9.6 Dabbling in direct products

Given two integers $n_1$ and $n_2$, you can always form another one from them, using multiplication, denoted $n_1 n_2$. Analogously, given two groups $H_1$ and $H_2$, you can always form another group from them using the Cartesian product construction (which we saw already in the context of sets in §2.1), denoted $H_1 \times H_2$.

**Definition 9.6.1.** *Let $H_1, H_2$ be two subgroups. We say that a group $G$ is the* **direct product** *of $H_1$ with $H_2$, written*

$$G = H_1 \times H_2,$$

*if*

*(a) $G = H_1 \times H_2$ (Cartesian product, as sets),*

*(b) the group operation on $G$ is given 'coordinate-wise' (still denoted '*' for simplicity):*

$$(x_1, y_1) * (x_2, y_2) = (x_1 * x_2, y_1 * y_2),$$

*for $x_1, x_2 \in H_1$, $y_1, y_2 \in H_2$ (where $*$ denotes multiplication in $H_1, H_2$, and $G$).*

**Example 9.6.1.** *Let $G$ be (as a set) the Cartesian product $G = C_2 \times C_3$, where $C_n$ denotes the cyclic group of order $n$ (with addition mod $n$ as the operation, $n = 2, 3$). Define addition on $G$ coordinate-wise (mod 2 in the first coordinate, mod 3 in the second coordinate):*

$$(m_1, n_1) + (m_2, n_2) = (m_1 + m_2, n_1 + n_2),$$

*where $0 \le m_i \le 1$, $0 \le n_j \le 2$, for $i = 1, 2$, $j = 1, 2$.*

**Example 9.6.2.** *The symmetry group $O$ of the octahedron is isomorphic to $S_4 \times C_2$. The symmetry group $I$ of the icosahedron is isomorphic to $A_5 \times C_2$. (This is not isomorphic to $S_5$, despite the fact that they both have the same number of elements and they both contain $A_5$ as a normal subgroup.)*

## 9.6.1 First fundamental theorem of cube theory

The following basic fact is implicitly used in some of the examples which follow. We shall frequently need the following fact.

**Theorem 9.6.1.** *Beginning with a solved cube, label the following facets with an invisible '+' (i.e., mark the spatial position of the facet on the cube with a '+'):*

- *U facet of the uf edge subcube*
- *U facet of the ur edge subcube*
- *F facet of the fr edge subcube*
- *all facets which can be obtained from these by a move of the slice group.*

148

*Label the U and D facets of each corner subcube with an invisible '+'. These '+' signs are called the* **standard reference markings**. *Each move g of the Rubik's Cube yields a new collection of '+' labels, called the* **markings relative to** *g. A position of the Rubik's Cube is determined by the following decision process:*

*(a) How are the edge subcubes permuted?*

*(b) How are the center subcubes permuted?*

*(c) How are the corner subcubes permuted?*

*(d) Which of the relative edge markings are flipped (relative to the standard reference markings)?*

*(e) Which of the relative corner markings are rotated from the standard reference markings and, if so, by how much ($2\pi/3$ or $4\pi/3$ radians clockwise, relative to the standard reference markings)?*

This is labeled as a theorem because of its relative importance for us, not because of its difficulty! This is the 'First Fundamental Theorem of Rubik's Cube theory'.

As an exercise, the reader should convince him or herself that this theorem is correct.

## 9.6.2 Example: cube twists and flips

We recall some notation:

- $X$ is the set of 48 facets of the Rubik's Cube which are not center facets,

- $V$ denotes the subset of facets which belong to some corner subcube,

- $E$ is the subset of facets which belong to some edge subcube.

- Let $G$ denote the Rubik's Cube group.

- Let $F$ be the group generated by all the moves of the Rubik's Cube group which do not permute any corner or edge subcubes but may twist or flip them.

- Let $S_X, S_V, S_E$, denote the symmetric group on $X, V, E$, respectively. We may regard $F, G$, as subgroups of $S_X$. We may also regard $S_V, S_E$ as subgroups of $S_X$ (for example, $S_V$ is the subgroup of $S_X$ which leaves all the elements of $E$ fixed).

- Let

$$G_V = S_V \cap G, \quad G_E = S_E \cap G, \quad F_V = S_V \cap F, \quad F_E = S_E \cap F.$$

Note that the action of $G$ on $X$ induces an equivalence relation as follows: we say that a facet $f_1$ is **equivalent** to a facet $f_2$ if there is a move of the Rubik's Cube which sends one facet to the other. There are exactly two equivalence classes, or orbits, of $G$ in $X$: namely, $V$ and $E$. In particular, the action of $G$

on $V$ is transitive and the action of $G$ on $E$ is transitive. On the other hand, $F$ leaves each vertex (resp., edge) fixed, though it may permute the corner facets (resp., edge facets) associated to a vertex (resp., edge).

**Ponderable 9.6.1.** *Show that:*

*(a) The set $\overline{V}$ of equivalence classes of $F$ acting on $V$ is in bijective correspondence with the set of all vertices of the cube.*

*(b) The set $\overline{E}$ of equivalence classes of $F$ acting on $E$ is in bijective correspondence with the set of all edges of the cube.*

The interesting thing is that we have

$$F = F_V \times F_E \quad \text{(direct product)}.$$

**Ponderable 9.6.2.** *Notation as above.*

*(a) Show $F = F_V \times F_E$ (direct product).*

*(b) Is $S_X = S_V \times S_E$ (direct product)?*

A harder question, which we will answer later in the negative: Is $G = G_V \times G_E$ (direct product)?

## 9.6.3 Example: the slice group of the cube

It is worthwhile studying a subgroup of the Rubik's Cube group which is easier to analyze that the Rubik's Cube group itself. This way, some of the basic ideas and setup are introduced in a simpler setting.

Some of the material below can also be found in [BH], [Si].

Let $H$ be the group $\langle M_R, M_F, M_U \rangle$ generated by the middle slice moves. This group is called the 'slice group'. Let $E$ be the set of edges of the cube (which we identify with the set of edge subcubes), let $C$ be the set of center facets of the cube, and let $X = E \cup C$. $H$ acts on $X$. Note that $H$ does not affect the corners (i.e, the vertices of the cube).

**Ponderable 9.6.3.** *(a) Is the action of $H$ on $X$ transitive?*

*(b) Is the action of $H$ on $C$ transitive?*

*(c) Is the action of $H$ on $E$ transitive?*

*solution*: The answer to (a) is no - an edge subcube cannot be sent to a center facet, for example, so there is an element of $X$ which cannot be sent to any other element of $X$ by an element of $H$. The answer to (b) is yes - any center facet can be sent to any other center facet by an element of $H$. The answer to (c) is no - for example, the $uf$ edge subcube cannot be sent to the $ur$ edge subcube by a slice move, so there is an element of $E$ which cannot be sent to any other element of $E$ by an element of $H$. This completes the solution.

The answer of 'no' to (c) brings about the following question.

**Ponderable 9.6.4.** *What are the orbits of $H$ on $E$?*

*solution*: The answer may be phrased in various ways, but let us look at it in the following way: suppose we call two edge subcubes **equivalent** if one can be sent to the other by a slice move (i.e., an element of H). There are 3 disjoint equivalence classes: all the subcubes in the middle RL-slice are equivalent, all the subcubes in the middle FB-slice are equivalent, and all the subcubes in the middle UD-slice are equivalent. The distinct orbits of H acting on E are the following:

- the middle RL-slice, denoted by $E_{RL}$,

- the middle FB-slice, denoted by $E_{FB}$,

- the middle UD-slice, denoted by $E_{UD}$.

Note that
$$E = E_{RL} \cup E_{FB} \cup E_{UD},$$
is a partitioning of $E$ into the distinct equivalence classes defined by the action of $H$ on $E$. This completes the solution.

Each element of $H$ determines an element in $S_X$. We have a homomorphism

$$f : H \to S_X$$

This is another way of saying that $H$ acts on the set $X$, which we already know. Note that each basic slice move $M$ (so $M$ is either $M_R$, $M_F$, or $M_U$) is, as an element of $S_X$, of the following form:

$$M = (4 - \text{cycle in } S_E)(4 - \text{cycle in } S_C).$$

Conversely, does an element of $S_X$ uniquely determine an element of $H$? In other words, is $f$ injective (i.e., an embedding)?

To answer this, fix an $h \in H$ and think about what $f(h)$ tells us: $f(h)$ tells us which subcube moves to which other subcube but it doesn't tell us, for example, how a subcube is flipped or rotated.

The first fundamental theorem 9.6.1 of the cube inspires the following question.

**Ponderable 9.6.5.** *Can an element of H flip, but not permute, an edge subcube (and possibly permuting or flipping other subcubes of the cube)?*

*solution*: The answer is no. The reason why is that the slice moves can only rotate a given edge subcube within the slice it belongs to. This completes the solution.

It follows, therefore, that the permutations of the edge subcube and centers determine a unique element of the slice group. In other words, we have proven the following

**Proposition 9.6.1.** *The homomorphism*

$$f : H \to S_X$$

*is an embedding.*

**Ponderable 9.6.6.** *The analog of this for the Rubik's Cube group is false! Why?*

$H$ acts on the set $E_{RL}$, so we have a homomorphism

$$r_{RL} : H \to S_{E_{RL}}$$

and similarly, $r_{UD} : H \to S_{E_{UD}}$, $r_{FB} : H \to S_{E_{FB}}$.

$H$ acts on each of the sets $E$ and $C$, so we have homomorphisms

$$r = r_{RL} \times r_{UD} \times r_{FB} : H \to S_{E_{RL}} \times S_{E_{UD}} \times S_{E_{FB}} \subset S_E, \qquad s : H \to S_C,$$

which we can put together to obtain an injective homomorphism

$$r \times s : H \to S_{E_{RL}} \times S_{E_{FB}} \times S_{E_{UD}} \times S_C$$

To determine $H$, we determine the image of $H$ in $S_{E_{RL}} \times S_{E_{FB}} \times S_{E_{UD}} \times S_C$.

To do this, we first look at the image of $H$ in each of $S_{E_{RL}}$, $S_{E_{FB}}$, and $S_{E_{UD}}$. This is easy enough:

- the image of $H$ in $S_{E_{RL}}$ is $\langle M_R \rangle \cong C_4$,

- the image of $H$ in $S_{E_{FB}}$ is $\langle M_F \rangle \cong C_4$,

- the image of $H$ in $S_{E_{UD}}$ is $\langle M_U \rangle \cong C_4$.

Later, we shall want to think of $C_4$ as $\{0, 1, 2, 3\}$, with addition mod 4, and the image of an element $h \in H$ under one of the homomorphisms above, $r_{RL} : H \to S_{E_{FB}}$ say, as an integer $0 \leq r_{RL}(h) \leq 3$.

Next, we must determine the image of $H$ in $S_C$. This is easy if it's looked at in the right way. As far as the movements of the center facets is concerned, the slice moves may be replaced by their corresponding rotations of the entire cube about an axis of symmetry. In this case, we see that the image of $H$ in $S_C$ is the same as the image of the orientation-preserving symmetry group of the cube! This we know, by the discussion in Example 9.3.4 above, is isomorphic to $S_4$.

Putting all this together, we see that the image of $H$ in $S_{E_{RL}} \times S_{E_{FB}} \times S_{E_{UD}} \times S_C$ is isomorphic to a subgroup of

$$C_4^3 \times S_4.$$

We may represent the elements of $H$, therefore, as 4-tuples $(h_1, h_2, h_3, h_4)$, with $h_1, h_2, h_3 \in C_4$ and $h_4 \in S_4$. Since each of the generating moves of $H$ (namely, $M_R$, $M_U$, and $M_F$) satisfies

$$sgn(r(h)) = sgn(s(h)),$$

for all $h \in H$, the image of $H$ cannot be all of $C_4^3 \times S_4$.

**Proposition 9.6.2.** *The image of H in $C_4^3 \times S_4$ is isomorphic to the kernel of the map*

$$t : C_4^3 \times S_4 \to \{\pm 1\}$$
$$(h_1, h_2, h_3, h_4) \longmapsto sgn(h_1) \cdot sgn(h_2) \cdot sgn(h_3) \cdot sgn(h_4),$$

*where each sgn is the sign of the permutation, regarded as an element of $S_X$.*

**Ponderable 9.6.7.** *Show that $|ker(t)| = (4^3 \cdot 4!)/2 = 768$.*

*solution*: We have shown that H is isomorphic to a subgroup of $C_4^3 \times S_4$. In fact, we know that the basic slice moves $M_R, M_U, M_F$ (which generate $H$) all belong to the kernel of $t$, so H is isomorphic to a subgroup of $ker(t) \subset C_4^3 \times S_4$.

It remains to show that every element in $ker(t)$ belongs to $H$. To do this, we consider the projection homomorphism

$$p : H \to S_4$$

obtained by composing the homomorphism $r \times s : H \to C_4^3 \times S_4$ constructed above with the projection homomorphism $C_4^3 \times S_4 \to S_4$. We have shown that $p$ is surjective. Our next objective is to compute the kernel of $p$ and use the first isomorphism theorem to determine $H$.

Claim: The kernel of $p$ is

$$ker(p) = \{h \in H \mid s(h) = 1, \ r_{RL}(h) + r_{UD}(h) + r_{FB}(h) \equiv 0 \ (\text{mod } 2)\}.$$

Note that $ker(p)$ is a subgroup of $H$ so the sign of the permutation $s(h)$ is equal to the sign of the permutation $r(h)$:

$$sgn(s(h)) = sgn(r(h)) = sgn(r_{RL}(h)) \cdot sgn(r_{UD}(h)) \cdot sgn(r_{FB}(h)).$$

This implies that

$$ker(p) \subset \{h \in H \mid s(h) = 1, \ r_{RL}(h) + r_{UD}(h) + r_{FB}(h) \equiv 0 \ (\text{mod } 2)\}.$$

Conversely, pick an $h \in H$ such that $s(h) = 1$ and $r_{RL}(h) + r_{UD}(h) + r_{FB}(h) \equiv 0$ (mod 2). We may represent this element $h$ as a 4-tuple $(n_1, n_2, n_3, s)$, with $0 \leq n_1, n_2, n_3 \leq 3$ and $s = 1 \in S_4$.

For example,

- the element $M_1 = M_R * M_F^{-1} * M_D * M_F$ is represented by the 4-tuple $(1, 1, 0, 1)$,

- the element $M_2 = M_R * M_D * M_F * M_D^{-1}$ is represented by the 4-tuple $(0, 1, 1, 1)$,

- the element $M_3 = M_F * M_D * M_R^{-1} * M_D^{-1} * M_F * M_D^{-1} * M_R * M_D$ is represented by the 4-tuple $(0, 0, 2, 1)$.

These elements generate all the elements of the group

$$\{(a, b, c, 1) \mid a, b, c \in C_4, \ a + b + c \equiv 0 \ (\text{mod } 2)\}.$$

Note that the group

$$\{(a, b, c) \mid a, b, c \in C_4, \ a + b + c \equiv 0 \ (\text{mod } 2)\}$$

is, in turn, the kernel of the map $C_4^3 \to C_2$ given by $(a, b, c) \longmapsto a + b + c \equiv 0 \ (\text{mod } 2)$.

**Ponderable 9.6.8.** *Show that $|\{(a, b, c) \mid a, b, c \in C_4, \ a+b+c \equiv 0 \ (\text{mod } 2)\}| = 4^3/2 = 32$.*

Therefore, the element $h$ chosen above must be expressible as a 'word' in these three elements $M_1, M_2, M_3$. This shows that

$$\{h \in H \mid s(h) = 1, \ r_{RL}(h) + r_{UD}(h) + r_{FB}(h) \equiv 0 \ (\text{mod } 2)\} \subset ker(p),$$

which implies the claim.

To summarize what we have so far: we have a surjective homomorphism $p : H \to S_4$ with kernel $\{h \in H \mid s(h) = 1, \ r_{RL}(h) + r_{UD}(h) + r_{FB}(h) \equiv 0 \ (\text{mod } 2)\}$. The kernel has $|ker(p)| = 32$ elements and the image has $|im(p)| = |S_4| = 4! = 24$ elements. Since, by the first isomomorphism theorem,

$$H/ker(p) \cong S_4,$$

we have $|H| = 32 \cdot 24 = 768$. But the kernel of the homomorphism $t : C_4^3 \times S_4 \to \{\pm 1\}$, which we know contains $H$ as a subgroup, also has 768 elements. This forces $h = ker(t)$. This completes the solution.

### 9.6.4 Example: the slice group of the Megaminx

The subgroup $S$ of the Megaminx group generated by all elements of the form $x * y^{-1}$, where $x, y$ correspond to opposite faces is called the **slice group** of the Megaminx.

**Ponderable 9.6.9.** *(Mark Longridge): Determine $S$.*

This group has recently (via an email sent August 31, 2001) been completely determined by Mike Reid. It has $(15!/2) * 4^{14} * (10!/2) * 6^9 * 3 = 9627755206121277812101663948800000$ elements.

## 9.7 A smorgasbord of semi-direct products

To 'explicitly construct' the Rubik's Cube group, we need to know how to construct semi-direct products. We study semi-direct products in this section. A semi-direct product is a specific type of construction, more general than the direct product, of a new group, $G$, from two groups, $H_1$ and $H_2$, having certain properties. It will turn out that $H_1$ and $H_2$ will be subgroups of $G$.

If a group $G$ contains two subgroups $H_1$ and $H_2$, with $H_1 \lhd G$ normal, such that each element of $G$ can be written uniquely as a product $h_1 h_2$, with $h_1 \in H_1$ and $h_2 \in H_2$ then we say that $G$ is the **semi-direct product** of $H_1$ and $H_2$. In this situation, $H_2$ is called a **complement** of $H_1$. In this section, we shall see another way how a semi-direct product can be expressed. Later, we shall see that the Rubik's Cube group can easily be described using semi-direct products.

**Definition 9.7.1.** *Now suppose that $H_1, H_2$ are both subgroups of a group $G$. We say that $G$ is the* **semi-direct product** *of $H_1$ by $H_2$, written*

$$G = H_1 \rtimes H_2$$

*if*

*(a) $G = H_1 * H_2$,*

*(b) $H_1$ and $H_2$ only have 1, the identity of $G$, in common,*

*(c) $H_1$ is normal in $G$.*

*This is the 'internal version' of the semi-direct product.*

Of course, if we define anything using two apparently different definitions, we'd better be sure that they are equivalent! The fact that they are is a theorem (Theorem 7.23 in [R]) which we won't prove here.

Note that the multiplication rule in $G$ doesn't have to be mentioned since we are assuming here that $G$ is given.

The 'external version' is defined by a construction as follows:

**Definition 9.7.2.** *Assume we have a homomorphism*

$$\phi : H_2 \to Aut(H_1).$$

*Define multiplication on the set $H_1 \times H_2$ by*

$$(x_1, y_1) * (x_2, y_2) = (x_1 * \phi(y_1)(x_2), y_1 * y_2).$$

*This defines a group operation. This group, denoted $H_1 \rtimes_\phi H_2$, is the (external)* **semi-direct product**.

This definition will be used with the example of $H_1 = C_m^n$ and $H_2 = S_n$ in the next chapter.

These last two definitions are equivalent by Theorems 7.22-7.23 in [R].

As a set, $H_1 \rtimes H_2$ is simply the Cartesian product $H_1 \times H_2$.

**Example 9.7.1.** *Let $\mathbb{R}^2$ denote the direct product of the additive group of real numbers with itself:*

$$\mathbb{R}^2 = \{(x, y) \mid x, y \text{ real}\},$$

*with the group operation being addition performed componentwise. Let $C_2$ denote the multiplicative cyclic group with 2 elements, whose elements we write (somewhat abstractly) as $C_2 = \{1, s\}$. (We may think of $s$ as being equal to -1 but there is a reason for this notation which shall be made clear soon.) Define an action of $C_2$ on $\mathbb{R}^2$ by*

$$1(x, y) = (x, y), \quad s(x, y) = (y, x), \quad (x, y) \in \mathbb{R}^2.$$

*Let $G$ be the set*

$$G = \mathbb{R}^2 \times C_2.$$

*Define the binary operation $* : G \times G \to G$ by*

$$(g_1, z_1) * (g_2, z_2) = (g_1 + z_1(g_2), z_1 * z_2),$$

*for all $g_1, g_2 \in G$ and all $z_1, z_2 \in C_2$. This is a group - the semi-direct product of $\mathbb{R}^2$ with $C_2$.*

*To see this, we must answer some questions:*

- *closed under the operation? Yes*

- *existence of identity? Yes, $e = ((0,0), 1)$*

- *existence of inverse? Yes, $((x,y), 1)^{-1} = ((-x, -y), 1)$, and $((x,y), s)^{-1} = ((-y, -x), s)$.*

- *associative? This is the hard one:*

$$\begin{aligned}
((g_1, z_1) * (g_2, z_2)) * (g_3, z_3) &= (g_1 + z_1(g_2), z_1 * z_2) * (g_3, z_3) \\
&= (g_1 + z_1(g_2) + (z_1 * z_2)(g_3), (z_1 * z_2) * z_3) \\
(g_1, z_1) * ((g_2, z_2) * (g_3, z_3)) &= (g_1, z_1) * (g_2 + z_2(g_3), z_2 * z_3) \\
&= (g_1 + z_1(g_2) + z_1(z_2(g_3)), z_1 * (z_2 * z_3))
\end{aligned}$$

*This implies associativity.*

**Ponderable 9.7.1.** *Let $G$ be the semi-direct product constructed in the above example. Show that $H = \{((x,y), 1) \mid (x, y) \in \mathbb{R}^2\}$ is a normal subgroup of $G$. Show $H \cong \mathbb{R}^2$.*

**Example 9.7.2.** *Let*

$$S_3 = \{1, s_1, s_2, s_1 * s_2, s_2 * s_1, s_1 * s_2 * s_1\}, \quad H_1 = \{1, s_2, s_1 * s_2 * s_1\}, \quad H_2 = \{1, s_1\}.$$

*Let*

$$\phi : H_2 \to \text{Aut}(H_1)$$

*be defined by*

$$\begin{aligned}
\phi(1) &= 1 \quad \text{(the identity automorphism)} \\
\phi(s_1)(h) &= s_1^{-1} * h * s_1 = s_1 * h * s_1
\end{aligned}$$

*(since $s_1^{-1} = s_1$), $h \in H_1$.*

*Denote the (external) semi-direct product of $H_1, H_2$ by $G = H_1 \rtimes_\phi H_2$.*

**Ponderable 9.7.2.** *Let $G$ be the semi-direct product constructed in the above example. Show that $H_1$ (which we identify with the subgroup $\{(h,1) \mid h \in H_1\}$ of $G$) is a normal subgroup.*

There is of course a close relationship between internally defined semi-direct products and externally defined ones. The following lemma, which is proven in [R], explains this connection:

**Lemma 9.7.1.** *If $G$ is the (internal) semi-direct product of $H_1$ by $H_2$ (so $H_1$ is a normal subgroup of $G$) then there is a homomorphism*

$$\phi : H_2 \to Aut(H_1)$$

*such that $G \cong H_1 \rtimes_\phi (H_2)$.*

**Example 9.7.3.** *Let $C_d$ be the cyclic group of order $d$, which we may regard as a set as $C_d = \{0, 1, ..., d-1\}$, with addition performed mod $d$. Let $N = C_d^n$, which we regard as the group of $n$-vectors with 'coefficients in $C_d$'. Let $H = S_n$ be the symmetric group on $n$ letters, i.e., the group of all permutations*

$$p : \{1, 2, ..., n\} \to \{1, 2, ..., n\}.$$

*The group $H$ acts on $N$ by permuting the indices, i.e., the coordinates of the vectors. For $f \in S_n$, define $f^* : C_d^n \to C_d^n$ by*

$$f^*(v) = (v_{f^{-1}(1)}, ..., v_{f^{-1}(n)}) = P(f^{-1})v^t, \qquad v = (v_1, ..., v_n) \in C_d^n,$$

*where $P(g)$ is the permutation matrix associated to a permutation $g$ and $v^t$ denotes the transpose (a colun vector) of $v$. Now, for $p, q \in S_n$ and $v, w \in C_d^n$, define*

$$(p, v) * (q, w) = (pq, w + q^*(v)).$$

*This defines a semi-direct product $C_d^n \rtimes S_n$, the* **generalized symmetric group**.
*(It is known that $C_d^n \rtimes S_n$ is isomorphic to the group of all '$C_n$-valued $n \times n$ monomial matrices', [R], Exercise 7.33. A* **monomial matrix** *is a matrix which contains exactly one non-zero entry for each row and column.)*

# 9.8   A reification of wreath products

We briefly survey the basic facts about wreath products. Wreath products are generalizations of semi-direct products. They occur naturally in the theory of the Rubik's Cube group (see [Si], for example). We shall be brief because the wreath products which do occur in the theory of the Rubik's Cube may be, if desired, reformulated in terms of semi-direct products. In fact, it suffices to consider generalized symmetric groups (see Theorem 9.8.1 below), which are even easier to deal with and themselves special types of semi-direct products.

Let $G_1$ be a group, let $G_2$ be a group acting on a finite set $X_2$. Fix some labeling of $X_2$ as say $X_2 = \{h_1, h_2, ..., h_m\}$, where $m = |X_2|$ and let $G_1^{X_2}$ denote the direct product of $G_1$ with itself $m$ times, with the coordinates labeled by the elements of $X_2$.

**Definition 9.8.1.** *The* **wreath product** *of* $G_1, G_2$ *is the group*

$$G_1 \operatorname{wr} G_2 = G_1^{X_2} \rtimes G_2$$

*where the action of* $G_2$ *on* $G_1^{X_2}$ *is via its action on* $X_2$.

In particular, *to each* $t \in G_1 \operatorname{wr} G_2$ *there is a* $g_2 \in G_2$. *We denote this* **projection** *by* $g_2 = pr(t)$. *Define the* **base** *of the wreath product by*

$$B = \{t \in G_1 \operatorname{wr} G_2 \mid pr(t) = 1\},$$

*so* $B = G_1^{X_2}$.

**Example 9.8.1.** *Let* $\mathbb{R}^n$ *denote the direct product of the additive group of reals with itself $n$ times. The group operation on* $\mathbb{R}^n$ *is componentwise addition. Let $S_n$ denote the symmetric group. This acts on* $\mathbb{R}^n$ *by permuting coordinates: if $r \in S_n$ is a permutation then define*

$$r(x_1, ..., x_n) = (x_{r(1)}, ..., x_{r(n)}),$$

*for* $(x_1..., x_n) \in \mathbb{R}^n$. *This action respects the addition operation:*

$$r(x_1 + y_1, ..., x_n + y_n) = r(x_1, ..., x_n) + r(y_1, ..., y_n),$$

$(x_1..., x_n), (y_1..., y_n) \in \mathbb{R}^n$. *(Incidentally, it also preserves scalar multiplication:*

$$r(a * (x_1, ..., x_n)) = a * r(x_1, ..., x_n),$$

*for* $(x_1..., x_n) \in \mathbb{R}^n, a \in \mathbb{R}$, *so $r$ defines an invertible linear transformation on* $\mathbb{R}^n$; *in fact, there is a homomorphism $S_n \to Aut(\mathbb{R}^n)$, where $Aut(\mathbb{R}^n)$ denotes the group of all invertible linear transformations on* $\mathbb{R}^n$; *do you recognize this homomorphism? It has occurred previously.)*
   *Let $G$ be the set*

$$G = \mathbb{R}^n \times S_n$$

*and define a binary operation* $* : G \times G \to G$ *by*

$$(g_1, p_1) * (g_2, p_2) = (g_1 + p_1(g_2), p_1 * p_2),$$

*for all $g_1, g_2 \in G$ and all $p_1, p_2 \in S_n$. This is a group. To see this, we must answer some questions:*

- *Is $G$ closed under the operation* $*$? *Yes.*

- *Does there exist an identity? Yes,* $e = ((0, 0), 1)$.

- *Does there exist an inverse? Yes,* $((x, y), 1)^{-1} = ((-x, -y), 1)$, $((x, y), p)^{-1} = (-p^{-1}(x, y), p^{-1})$.

- *Is $G$ associative? This is the hard one:*

$$((g_1, p_1) * (g_2, p_2)) * (g_3, p_3) = (g_1 + p_1(g_2), p_1 * p_2) * (g_3, p_3)$$
$$= (g_1 + p_1(g_2) + (p_1 * p_2)(g_3), (p_1 * p_2) * p_3)$$
$$(g_1, p_1) * ((g_2, p_2) * (g_3, p_3)) = (g_1, p_1) * (g_2 + p_2(g_3), p_2 * p_3)$$
$$= (g_1 + p_1(g_2) + p_1(p_2(g_3)), p_1 * (p_2 * p_3))$$

*This implies associativity.*

*This group is the wreath product of $\mathbb{R}$ with $S_n$:*

$$G = \mathbb{R} \operatorname{wr} S_n,$$

*where $\mathbb{R}^n$ is the base.*

**Lemma 9.8.1.** *(a) The base $B$, which is isomorphic to the direct product $G_1^{|X_2|}$, is a normal subgroup of $G_1 \operatorname{wr} G_2$.*
*(b) $(G_1 \operatorname{wr} G_2)/B$ is isomorphic to $G_2$.*

This is a consequence of the first isomorphism theorem of group theory. The proof is omitted. (We refer to chapter 8 of [NST] or to [R] for further details.)

## 9.8.1 The illegal Rubik's Cube group

Let $H$ be the **illegal Rubik's Cube group** of all legal and illegal moves of the $3 \times 3$ Rubik's Cube. In other words, in addition to the usual basic moves (namely, $R, L, U, D, F, B$), we allow you to take apart the cube and reassemble the corner and edge subcubes (but you cannot remove stickers from the facets). Let $C_3$ denote the group of all rotations of a particular corner subcube by a 120 degree angle. (Actually, this group depends on the corner being rotated, but since these groups are all isomorphic, we drop the dependence from the notation.) Let $C_2$ denote the group of all flips of a particular edge subcube. (Again, this group depends on the edge being flipped but since these groups are all isomorphic, we drop the dependence from the notation.) We shall prove later that

$$H = (C_3 \operatorname{wr} S_V) \times (C_2 \operatorname{wr} S_E),$$

where $V$ is the set of corner subcubes and $E$ is the set of edge subcubes.

## 9.8.2 Elements of order $d$ in $C_m \operatorname{wr} S_n$

In this section, we will apply our understanding of wreath products to increasing our ability to determine all the elements of order $d$ in the Rubik's Cube group. (The case $d = 2$ will be examined in some detail later.) In some cases (e.g., in the cases we deal with here), wreath products turn out to be relatively concrete and familiar groups.

Let $S(n, m)$ denote the group of all $n \times n$ monomial matrices with entries in $C_m$. We begin with the following result, which allows us to identify $S(n, m)$ with the generalized symmetric group in Example 9.7.3.

159

**Theorem 9.8.1.** *There is an isomorphism between $C_m \operatorname{wr} S_n$ and the group $S(n,m)$ which sends an element $(\vec{v}, f) \in C_m \operatorname{wr} S_n$, $\vec{v} = (v_1, v_2, ..., v_n) \in C_m^n$ and $f \in S_n$, to the matrix $P(f)\vec{v}$.*

This follows from a result (see Theorem 10.5.1) which will be proven in the next chapter. It is also a special case of an exercise in [R].

**Ponderable 9.8.1.** *Show that $\phi : C_m \operatorname{wr} S_n \to S(n,m)$, $\phi(v, f) = P(f)v$, is a homomorphism: $P(f)v + P(g)w = P(fg)(v + g^*(w))$, for $v, w \in C_m^n$ and $f, g \in S_n$.*

It is clear from this theorem that an element $(\vec{v}, f) \in C_m \operatorname{wr} S_n$, $\vec{v} = (v_1, v_2, ..., v_n) \in C_m^n$ and $f \in S_n$, is order $d$ only if the permutation matrix $P(f)$ is order $d$. Indeed,

$$(\vec{v}, f)^2 = (\vec{v} + f^*(\vec{v}), f^2),$$
$$(\vec{v}, f)^3 = (\vec{v} + f^*(\vec{v}) + (f^*)^2(\vec{v}), f^3),$$
$$\vdots$$
$$(\vec{v}, f)^k = (\vec{v} + ... + (f^*)^{k-1}(\vec{v}), f^k).$$

We conclude with the following classification of the elements of order $d$ in the wreath product.

**Proposition 9.8.1.** *An element $(\vec{v}, f) \in C_m \operatorname{wr} S_n$, $\vec{v} = (v_1, v_2, ..., v_n) \in C_m^n$ and $f \in S_n$, is order $d$ if and only if $f^d = 1$ and $\vec{v} + ... + (f^*)^{d-1}(\vec{v}) = 0$.*

This result can, in principle, be used in conjunction with the explicit determination of the Rubik's Cube group given later to determine all the elements of a given order. See §11.3.

# Chapter 10

# Words which move

'Everyone knows what a curve is, until he has studied enough mathematics to become confused through the countless number of possible exceptions.'

*Felix Klein*

Klein (1849-1925), one of the most important figures in group theory, was right on the money with his quote about curves. However, as far as this chapter is concerned, we might replace 'curve' by 'word'. The use of 'words' and 'relations' to define group arose from topology, the study of surfaces in particular. This chapter touches on a broad area of group theory called 'combinatorial group theory', a subject having its origins in the late 1800's and early 1900's. Henri Poincaré (1854-1912), one of the greatest mathematical geniuses of all time, introduced the 'fundamental group' of a surface, which turns out to be naturally defined in terms of 'generators' (the 'letters' of our 'words') and 'relations' ('grammatical rules' our 'words' satisfy). Although topology is well outside the scope of this book, it is interesting to note how the study of one area can ultimately lead to applications of something completely different.

Let us begin with an example. $RUBDUBDUB$ is both a word (well, sort of, anyway) and a move of the Rubik's Cube. Is that move different than $DUBRUBRUB$? Just because they look different doesn't mean that they are. ($R^4$ and $D^4$ look different but they both have the same effect as doing nothing at all!) Basically, (one formulation of) the 'word problem' is the problem of determining when two words (i.e., strings of generators of a group) are the same or not.

The word problem is related to the solution of the Rubik's Cube since if we knew which 'words' in the generators $R$, $L$, $U$, $D$, $F$, $B$, described a given position of the Rubik's Cube then we could use one of them to solve the position. The problem is that this 'word problem' is usually quite hard to solve in practice. It is remarkable that there is a package called `abstab` in GAP, due to Philip Osterlund [Os], which can always find a word in the generators $R$, $L$, $U$, $D$, $F$, $B$ which describes a given position of the Rubik's Cube. It is the purpose

161

of this chapter to introduce some of the mathematical ideas which arise when trying to study the Rubik's Cube group from this perspective.

## 10.1 Words in free groups

To be precise and make sure everyone's on the same page, let's record a few definitions and introduce some notation.

**Definition 10.1.1.** *Given a list $L$ of questions, a* **decision algorithm** *for $L$ is a uniform set of unambiguous instructions which, when applied to any question in $L$ gives the correct answer 'yes' or 'no' after a finite number of steps.*

Let $X = \{x_1, ..., x_n\}$ denote a set and $X^{-1}$ a set disjoint from $X$ whose elements we denote by $\{x_1^{-1}, ..., x_n^{-1}\}$. Assume that the map $x \longmapsto x^{-1}$ defines a bijection $X \to X^{-1}$. It will be convenient to let $x^1 = x, x_i^0 = 1$, where 1 is an element not belonging to $X \cup X^{-1}$ which we will call the **identity element**. A **word on** $X$ is a sequence

$$w = (a_1, a_2, ..., a_N),$$

where $N > 0$ is some integer and each $a_i$ belongs to

$$X \cup X^{-1} \cup \{1\}.$$

The sequence of all 1's is called the **empty word**. The **inverse** of the word $w$ is the word

$$w^{-1} = (a_N^{-1}, ..., a_1^{-1}).$$

If $a_i = y_i^{e_i}$, where $e_i$ is in $\{0, 1, -1\}$ and $y_i \in X$, then we shall write the word $w$ as $w = y_1^{e_1}...y_N^{e_N}$.

**Example 10.1.1.** *Let $X = \{R, L, U, D, F, B\}$. The set of words on $X$ are in a bijective correspondence with the set of sequences of basic moves you can make on the Rubik's Cube.*

We call a word $w = y_1^{e_1}...y_N^{e_N}$ on $X$ **reduced** if either $w$ is empty or if the exponents $e_i$ are non-zero and if there are no $x \in X$ with $x, x^{-1}$ adjacent in $w$.

**Definition 10.1.2.** *The* **free group** *$F_n = F_X$ on the generators $x_1, ..., x_n$ is the group of all reduced words on $X$.*

The notion of a free group was first formulated by Walther von Dyck (1856-1934). von Dyck made important contributions to topology, function theory, and group theory.

The proof that $F_n$ is a group is not entirely easy (verifying the associativity property is perhaps the hardest part), see Theorem 11.1 in [R].

## 10.1.1   Length

If, in the notation above, $w = y_1^{e_1}...y_N^{e_N}$ is a reduced word (so $e_i \in \{0,1,-1\}$) then we call $N$ the **length** (or **reduced length**, to be more precise) of $w$.

If $G = \langle g_1,...,g_k \rangle \subset S_n$ is a finite permutation group generated by permutations $g_1, ..., g_k$ then we may still define the notion of length:

**Definition 10.1.3.** *Suppose $g \in G$ is not the identity, where $G$ is a permutation group as above. Then $g$ may be written*

$$g = y_1^{e_1}...y_N^{e_N},$$

*where each $y_i \in \{g_1,...,g_k\}$ and where $e_i \in \{0,1,-1\}$. The number $N$ and the sets $\{y_1,...,y_N\}$, $\{e_1,...,e_N\}$ may not be unique for a given $g$ but among all such possibilities there is at least one such that the value of $N$ is minimum. We call this the **length** of $g$, denoted $\ell(g)$.*

*Let*

$$P_G(t) = \sum_{g \in G} t^{\ell(g)}.$$

*This is called the* **Poincaré polynomial** *of $G$.*

The length of $g$ is the distance in the Cayley graph between the vertex $g$ and the vertex 1. As was mentioned in the chapter on graph theory, the problem of determining the largest possible distance in the Cayley graph of the Rubik's Cube group is known as 'God's algorithm' and is currently unsolved.

**Example 10.1.2.** *Let $G = S_n$ with generators $g_i = (i, i+1)$, $i = 1,...,n-1$. The Poincaré polynomial is known:*

$$\Pi_{k=1}^{n-1} \frac{t^{k+1} - 1}{t - 1},$$

*by [Hum], §§1.11, 3.15.*
*In case $n = 3$, this is*

$$t^3 + 2t^2 + 2t + 1 = (t+1)(t^2 + t + 1)$$

**Ponderable 10.1.1.** *What is the longest possible length of an element of the Rubik's cube group (with respect to the generators $R, L, U, D, F, B$)?*

Note this is the same as asking for longest possible length of an element of the Rubik's cube group with respect to the generators

$$R, L, U, D, F, B, R^{-1}, L^{-1}, U^{-1}, D^{-1}, F^{-1}, B^{-1}.$$

In other words, it is simply a reformulation of 'God's algorithm.'

**Ponderable 10.1.2.** *What is the Poincaré polyomial of the Rubik's Cube group (with respect to $R, L, U, D, F, B$)?*

This is unknown, though we refer the reader interested in pursuing this to see Thompson's paper [Th].

### 10.1.2 Trees

We may represent the free group graphically as follows. We define the Cayley graph of $F_n$ inductively:

- Draw a vertex for each element of $X \cup X^{-1}$ (these are the vertices, $V_1$ say, for the words of length 1).

- Suppose that you have already drawn all the vertices for the words of length $k - 1$, $V_{k-1}$ let's call them. For each $x \in X \cup X^{-1}$ and each $v \in V_{k-1}$, draw a vertex for each word of length $k$ obtained by multiplying $v$ by $x$ on the right, $v * x$, and connect $v$ and $v * x$ by an edge.

There are infinitely many vertices, each of which has degree $|X \cup X^{-1}|$. Moreover, this graph has no circuits or loops (i.e., no path of edges crosses back over onto itself). Such a graph is called a **tree**.

**Example 10.1.3.** *Let $X = \{R, L, U, D, F, B\}$. The elements of the free group $F_X$ correspond to the mechanically different sequences of basic moves you can make on the Rubik's Cube. Of course, different sequences of moves may yield the same position of the Rubik's Cube (e.g., $R^4$ and 1 are the same position but sequence of moves used to attain them are distinct).*

*There are infinitely many vertices of the Cayley graph of $F_X$, each of which corresponds to a mechanically distinct move of the Rubik's Cube.*

Further details on the background material given here may be found in [R], [MKS].

## 10.2 The word problem

There is a way to list all the elements of $F_n$, called the **lexicographic ordering**. We shall describe a way to list all the words in $F_n$ as though they were in a dictionary. We shall give an algorithm for determining if a word $w \in F_n$ occurs before a word $w'$, in which case we write $w < w'$. In the dictionary below, we shall, for example, distingish between the the identity 1 and the 'non-reduced' word $x_1 * x_1^{-1}$.

The first element in this lexicographically ordered list is the word 1, the next $2n$ words are the words

$$x_1 < x_1^{-1} < ... < x_n < x_n^{-1}.$$

In general, we define $y_1...y_M < z_1...z_N$ if either (a) $M < N$, or (b) $M = N$ and $y_1 < z_1$, or (c) $y_1 = z_1$ and $y_2 < z_2$, or (d) $y_1 = z_1$ and $y_2 = z_2$ and $y_3 < z_3$, or ... .

List all the elements of $F_n$ as

$$F_n = \{w_1, w_2, ...\},$$

so $w_1 = 1, w_2 = x_1, \dots$. Let $G$ be a subgroup of $F_n$ or a permutation group $G = \langle g_1, \dots, g_n \rangle$. If $G$ is a permutation group then we regard a word $w_k$ as an element of $G$ by substituting $g_i$ for each $x_i, 1 \leq i \leq n$.

**Definition 10.2.1.** *Fix a $g \in G$. We say that $G$ has a* **solvable word problem** *if there is a decision algorithm for the list $L$ of questions of the form: Is $w_k = g$ in $G$?*

The word problem was first formulated by Max Dehn (1878-1952), a topologist who worked extensively on group theory. He was a student of David Hilbert and wrote several important papers, however, after he emigrated to the USA in 1940 Dehn had a hard time finding a good teaching position. He taught at several places in the States, including St. Johns University in Annapolis Maryland, and then settled down at Black Mountain College, a small school in North Carolina. There was no mathematics department, yet he remained there until his death. The College itself closed several years later.

**Lemma 10.2.1.** *([R], chapter 12) $F_n$ has a solvable word problem.*

**Proof:** (sketch) The following decision algorithm yields a solvable word problem.

1. If $w$ is a word not equal to 1 then underline the first occurance, if any, of the expression $x_i * x_i^{-1}$. If no such expression occurs in w then go to step 3, otherwise go to step 2.

2. delete the expression $x_i * x_i^{-1}$ from $w$ and go to step 1.

3. If $w = 1$ then stop and return 'yes', otherwise, stop and return 'no'.

$\square$

**Remark 10.2.1.** *There are examples of groups $G$ for which there is no decision algorithm to solve the word problem. We shall not discuss them here (see [R], chapter 12, for more details). Sometimes this is referred to in the literature by saying 'the word problem for $G$ is undecidable'. On the other hand, it has been shown that the Continuum Hypothesis (conjectured by Cantor, 'there are no cardinal numbers between the cardiality of $\mathbb{Q}$ and of $\mathbb{R}$') cannot be proven from the axioms of set theory and that its truth or falsity is consistent with the axioms of set theory. This is sometimes referred to in the literature as saying 'The Continuum Hypothesis is undecidable.' These two uses of the word 'undecidable' are different, so be careful!*

**Lemma 10.2.2.** *Each finite permutation group has a solvable word problem.*

**Ponderable 10.2.1.** *Why?*

**Theorem 10.2.1.** *A decision algorithm for the word problem for the Rubik's Cube group with generators $R, L, U, D, F, B$ is the same as an algorithm for solving the Rubik's Cube.*

In fact, there is an algorithm for writing each element in the Rubik's Cube group (or more generally any permutation group $G$) as a word in the generators $R, L, U, D, F, B$ (or more generally the given generators of $G$). This has been implemented in GAP [Gap] by P. Osterlund. Though this is a remarkable piece of work, it should be pointed out that the words one obtains in this way from Osterlund's program are often of length 100 or more.

## 10.3 Presentations and Plutonian robots

Here's a hypothetical situation: You and a friend each have a robot on the planet Pluto with a scrambled Rubik's cube. (We will call the robots $R^2D^2$ and $R^2B^2$ if you don't mind.) These robots have manual dexterity but no preprogramming on how to solve the cube. Furthermore, assume it is very expensive to program different moves, so you want to teach $R^2D^2$ and $R^2B^2$ the smallest number of separate moves that you can. On the other hand, the moves need not be basic moves ($U$, $R$, ..) since we will assume it costs roughly the same to teach the robot the move $R$ as the longer move $R * U^2 * R^{-1}$, for example. Your solution will be a 'word' in these taught moves. Again, to minimize the cost of transmission, you want the 'word' to be absolutely as short as possible. A prize of 1 million dollars has been set up to the first of you who can get their robot to solve its cube.

In other words, we want to solve the word problem for the cube *and* we want to do it as efficiently as possible. Suppose we know we need $n$ generators and we know that this is the smallest number. How do we make a 'word' as short as possible? To make a word in these generators as small as possible, we must know all the 'relations' between these generators so we can, if necessary, substitute them into the word and perform some cancelation. This is what this section is about.

Let $X$ be a finite set, say $n = |X|$. Let $Y$ be a set of reduced words on $X$. Let $R$ be the smallest normal subgroup of $F_n$ containing $Y$. Since $R$ is normal, the quotient $F_n/R$ is a group.

**Definition 10.3.1.** *Let $G$ be a group. We say that $G$ has **generators** $X$ and **relations** $Y$ if $G$ is isomorphic to $F_n/R$. A collection of generators and relations defining a group is called a **presentation** of the group.*

This concept of a group presentation is due to von Dyck, who we met in connection with free groups. As in the situation with free groups, von Dyck was motivated by considerations in topology.

**Example 10.3.1.** *If $G$ is the Rubik's Cube group then $X$ is a set of moves which can solve any cube, no matter how mixed up it is. $R$ is the standard notation used for the smallest normal subgroup of $F_n$ generated by the realtions satisfied by the generators $g_1, ..., g_n$ of a finitely presented group $G$. Don't confuse the set of relations $R$ with the Rubik's Cube right-face move $R$!*

*If $X = \{R, L, D, U, F, B\}$ then $R$ might contain $\{R^4, L^4, D^4, U^4, F^4, B^4\}$, among other things, standing for the identities $R^4 = 1$, $L^4 = 1$, ... .*

**Example 10.3.2.** *The Rubik's Cube group $G$ is generated by 2 elements. In fact, Martin Schönert used GAP to compute the subgroups generated by 300 random pairs of elements of $G$. 151 of those pairs generated the entire group, so the probability 2 randomly choosen elements generate $G$ is about **50%** (see the December 4, 1995 posting to [CL]).*

*In [B], the following two generators discovered by F. Barnes, are given: $g_1 = UBLUL^{-1}U^{-1}B^{-1}$ and $g_2 = R^2FLD^{-1}R^{-1}$. In other words, as a permutation group, $G = \langle g_1, g_2 \rangle$. We know that there is a finite set of relations $R_1, ..., R_k$ such that there is an isomorphism $\phi : G \to \langle x, y | R_1(x, y) = 1, ..., R_k(x, y) = 1 \rangle$, and moreover the isomorphism satisfies $\phi(g_1) = x$ and $\phi(g_2) = y$. What is a collection of relations $R_1(x, y) = 1, ..., R_k(x, y) = 1$ where $k$ is as small as possible and the total word lengths of the $R_1(x, y), ..., R_k(x, y)$ is as small as possible?*

As a matter of notation, an element $r \in R$ is written as an equation $r = 1$ in $G$.

**Remark 10.3.1.** *For those with a background in topology: Serre [Ser], §3.3, gives a topological interpretation of $R$ as 'the fundamental group of the Cayley graph of $G$ with respect to $X$'.*

**Example 10.3.3.** *The cyclic group of order 3, $C_3$, has one generator $x$ and one relation $x^3 = 1$, so*

$$X = \{x\}, \qquad R = \{(x^3)^k \mid k \in \mathbb{Z}\} \subset F_1 = \{x^k \mid k \in \mathbb{Z}\}.$$

*Here the cosets of $F_1/R$ are $R, xR, x^2R$. The set of these three cosets is closed under multiplication. For example, $(xR)(x^2R) = x(Rx^2)R = xx^2RR = x^3R = R$, so the inverse of the element $xR$ is $x^2R$.*

*More generally, $C_n$ has presentation*

$$C_n = \{x \mid x^n = 1\}.$$

**Ponderable 10.3.1.** *By constructing moves of the Rubik's Cube of order 2, 3, 4, show that the Rubik's Cube 'contains' the subgroups $C_2$, $C_3$, $C_4$.*

**Lemma 10.3.1.** *The group $C_m \times C_n$ is presented by*

$$C_m \times C_n = \{a, b \mid a^m = 1, b^n = 1, ab = ba\} = \{a, b \mid a^m = 1, b^n = 1, aba^{-1}b^{-1} = 1\}.$$

**Proposition 10.3.1.** *In general, if $G$ is generated by $g_1, ..., g_m$ with relations $R_i(g_1, ..., g_m) = 1$ and if $H$ is generated by $h_1, ..., h_n$ with relations $S_i(h_1, ..., h_n) = 1$ then $G \times H$ is the group generated by the $g_i$, $h_j$, with relations $R_i(g_1, ..., g_m) = 1$, $S_i(h_1, ..., h_n) = 1$, and $g_ih_j = h_jg_i$.*

We shall not prove this but refer to [MKS], Exercise 13, §4.1.

**Ponderable 10.3.2.** *By constructing moves of the Rubik's Cube of order 2, 3 which commute show that the Rubik's Cube 'contains' the subgroups $C_2 \times C_3 \cong C_6$.*

**Example 10.3.4.** *If $G = \langle R \rangle$ and $H = \langle L^2 \rangle$ then*

$$G \cong \langle x \mid x^4 = 1 \rangle, \qquad H \cong \langle y \mid y^2 = 1 \rangle,$$

*and*

$$G \times H \cong \langle x \mid x^4 = 1, y^2 = 1, xy = yx \rangle.$$

## 10.4 Generators, relations for groups of order < 26

This section gives a table of all the (non-isomorphic) finite groups of order less than or equal to 25 and their generators and relations. First, Figure 10.4 is included to give a feeling for how complicated the table below can get for larger $n$. It is a graph of the function $O(n)$, where $O(n)$ is the number of non-isomorphic groups of order $n$.

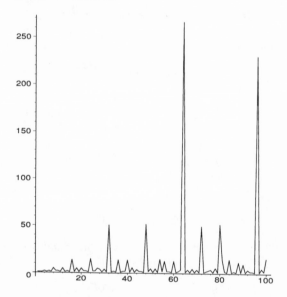

(This was obtained using MAGMA and MAPLE.)

The following table was obtained from the tables in [TW].

**Notation:**

$$C_n = \text{cyclic group of order } n,$$
$$D_n = \text{dihedral group of order } 2n$$
$$Q = \text{quaternion group} = \{-1, 1, -i, i, -j, j, -k, k\},$$
$$Q_{2m} = \text{generalized quaternion group} = \langle a, b \mid aba = b, b^2 = a^m \rangle,$$
$$S_n = \text{symmetric group of permutations of } \{1, 2, ..., n\},$$
$$A_n = \text{alternating group of even permutations of } \{1, 2, ..., n\},$$
$$\mathbb{F}_q = \text{finite field with } q \text{ elements } (q = \text{power of a prime}),$$
$$\mathbb{Z}/n\mathbb{Z} = \text{integers modulo } n.$$

| Order | Group $G$ | generators | relations | notes |
|-------|-----------|------------|-----------|-------|
| 2 | $C_2$ | $a$ | $a^2 = 1$ | |
| 3 | $C_3$ | $a$ | $a^3 = 1$ | $G \cong A_3$ |
| 4 | $C_4$ | $a$ | $a^4 = 1$ | |
| 4 | $C_2 \times C_2$ | $a, b$ | $a^2 = 1, b^2 = 1,$ $ab = ba$ | Klein 4-group $Aut(G) \cong GL(2, \mathbb{F}_2)$ |
| 5 | $C_5$ | $a$ | $a^5 = 1$ | |
| 6 | $C_6 = C_2 \times C_3$ | $a$ | $a^6 = 1$ | |
| 6 | $S_3$ | $a, b$ | $a^3 = 1, b^2 = 1,$ $aba = b$ | $Aut(G) = G$ $G \cong GL(2, \mathbb{F}_2)$ |
| 7 | $C_7$ | $a$ | $a^7 = 1$ | |
| 8 | $C_8$ | $a$ | $a^8 = 1$ | |
| 8 | $C_2 \times C_4$ | $a, b$ | $a^2 = 1, b^4 = 1,$ $ab = ba$ | |
| 8 | $C_2 \times C_2 \times C_2$ | $a, b, c$ | $a^2 = 1, b^2 = 1,$ $c^2 = 1, ab = ba,$ $bc = cb, ac = ca$ | $Aut(G) \cong GL(3, \mathbb{F}_2)$ |
| 8 | $D_4$ | $a, b$ | $a^4 = 1, b^2 = 1,$ $aba = b$ | |
| 8 | $Q$ | $a, b$ | $a^4 = 1, b^2 = a^2,$ $aba = b$ | |
| 9 | $C_9$ | $a$ | $a^9 = 1$ | |
| 9 | $C_3 \times C_3$ | $a, b$ | $a^3 = 1, b^3 = 1,$ $ab = ba$ | $Aut(G) \cong GL(2, \mathbb{F}_3)$ |
| 10 | $C_{10} = C_2 \times C_5$ | $a$ | $a^{10} = 1$ | |
| 10 | $D_5$ | $a, b$ | $a^5 = 1, b^2 = 1,$ $aba = b$ | |
| 11 | $C_{11}$ | $a$ | $a^{11} = 1$ | |
| 12 | $C_{12} = C_3 \times C_4$ | $a$ | $a^{12} = 1$ | |
| 12 | $C_2 \times C_6$ $= C_2 \times C_2 \times C_3$ | $a, b$ | $a^2 = 1, b^6 = 1,$ $ab = ba$ | |
| 12 | $D_6$ | $a, b$ | $a^6 = 1, b^2 = 1,$ $aba = b$ | |
| 12 | $A_4$ | $a, b$ | $a^2 = 1, b^3 = 1,$ $(ba)^3 = 1$ | $Aut(G) = G$ |
| 12 | $Q_6$ | $a, b$ | $a^6 = 1, b^2 = a^3,$ $aba = b$ | 'dicyclic' |

| 13 | $C_{13}$ | $a$ | $a^{13} = 1$ | |
|----|----------|-----|--------------|---|
| 14 | $C_{14} = C_2 \times C_7$ | $a$ | $a^{14} = 1$ | |
| 14 | $D_7$ | $a, b$ | $a^7 = 1, b^2 = 1,$ $aba = b$ | $Aut(G) = G$ |
| 15 | $C_{15} = C_3 \times C_5$ | $a$ | $a^{15} = 1$ | |
| 16 | $C_{16}$ | $a$ | $a^{16} = 1$ | |
| 16 | $C_2 \times C_8$ | $a, b$ | $a^2 = 1, b^8 = 1,$ $ab = ba$ | |
| 16 | $C_4 \times C_4$ | $a, b$ | $a^4 = 1, b^4 = 1,$ $ab = ba$ | $Aut(G)$ $\cong GL(2, \mathbb{Z}/4\mathbb{Z})$ |
| 16 | $C_2^2 \times C_4$ | $a, b, c$ | $a^2 = 1, b^2 = 1,$ $c^2 = 1, ab = ba,$ $ac = ca, bc = cb$ | |
| 16 | $C_2^4$ | $a, b, c, d$ | $a^2 = 1, b^2 = 1,$ $c^2 = 1, d^2 = 1,$ $ab = ba, ac = ca,$ $bc = cb, bd = db,$ $ad = da, cd = dc$ | $Aut(G)$ $\cong GL(4, \mathbb{F}_2)$ |
| 16 | $D_4 \times C_2$ | $a, b, c$ | $a^4 = 1, b^2 = 1,$ $c^2 = 1, aba = b,$ $ac = ca, bc = cb$ | |
| 16 | $Q \times C_2$ | $a, b, c$ | Ponderable | |
| 16 | | $a, b, c$ | $a^2 = 1, b^2 = 1, c^2 = 1,$ $abc = bca = cab$ | |
| 16 | | $a, b$ | $a^2 = 1, b^2 = 1,$ $(ab)^2 = 1, (a^{-1}b)^2 = 1$ | |
| 16 | | $a, b$ | $a^4 = 1, b^4 = 1,$ $aba = b$ | semidirect product of $C_4$ with $C_4$ |
| 16 | | $a, b$ | $a^8 = 1, b^2 = 1,$ $ab = ba^5$ | semidirect product of $C_8$ with $C_2$ ($C_8$ normal) |
| 16 | | $a, b$ | $a^8 = 1, b^2 = 1,$ $ab = ba^3$ | semidirect product of $C_8$ with $C_2$ ($C_8$ normal) |
| 16 | $D_8$ | $a, b$ | $a^8 = 1, b^2 = 1,$ $aba = b$ | semidirect product of $C_8$ with $C_2$ ($C_8$ normal) |
| 16 | $Q_8$ | $a, b$ | $a^8 = 1, b^2 = a^4,$ $aba = b$ | |
| 17 | $C_{17}$ | $a$ | $a^{17} = 1$ | |
| 18 | $C_{18} = C_2 \times C_9$ | $a$ | $a^{18} = 1$ | |

| 18 | $C_3 \times C_6$ $= C_3 \times C_3 \times C_2$ | $a, b$ | $a^3 = 1, b^6 = 1,$ $ab = ba$ | |
|---|---|---|---|---|
| 18 | $S_3 \times C_3$ | $a, b, c$ | $a^3 = 1, b^2 = 1, c^3 = 1$ $aba = b, ac = ca, bc = cb$ | |
| 18 | $D_9$ | $a, b$ | $a^9 = 1, b^2 = 1,$ $aba = b$ | a semidirect product of $C_9$ with $C_2$ ($C_9$ normal), $Aut(G) = G$ |
| 18 | | $a, b, c$ | $a^3 = 1, b^3 = 1, c^2 = 1$ $ab = ba, aca = c, bcb = c$ | |
| 19 | $C_{19}$ | $a$ | $a^{19} = 1$ | |
| 20 | $C_{20} = C_4 \times C_5$ | $a$ | $a^{20} = 1$ | |
| 20 | $C_2 \times C_{10}$ | $a, b$ | $a^2 = 1, b^{10} = 1,$ $ab = ba$ | |
| 20 | $D_{10}$ | $a, b$ | $a^{10} = 1, b^2 = 1,$ $aba = b$ | |
| 20 | $Q_{10}$ | $a, b$ | $a^{10} = 1, b^2 = a^5,$ $aba = b$ | |
| 20 | | $a, b$ | $a^5 = 1, b^4 = 1,$ $ab = ba^3$ | a semidirect product of $C_5$ with $C_4$ ($C_5$ normal), $Aut(G) = G$ |
| 21 | $C_{21} = C_3 \times C_7$ | $a$ | $a^{21} = 1$ | |
| 21 | | $a, b$ | $a^7 = 1, b^3 = 1,$ $ab = ba^4$ | a semidirect product of $C_7$ with $C_3$ ($C_7$ normal), $Aut(G) = G$ |
| 22 | $C_{22} = C_2 \times C_{11}$ | $a$ | $a^{22} = 1$ | |
| 22 | $D_{11}$ | $a, b$ | $a^{11} = 1, b^2 = 1,$ $aba = b$ | $Aut(G) = G$ |
| 23 | $C_{23}$ | $a$ | $a^{23} = 1$ | |
| 24 | $C_{24} = C_3 \times C_8$ | $a$ | $a^{24} = 1$ | |
| 24 | $C_2 \times C_{12}$ $= C_2 \times C_3 \times C_4$ | $a, b$ | $a^2 = 1, b^{12} = 1,$ $ab = ba$ | |
| 24 | $C_2^2 \times C_6$ | $a, b, c$ | $a^2 = 1, b^2 = 1, c^6 = 1,$ $ab = ba, ac = ca, bc = cb$ | |

| 24 | $D_6 \times C_2$ | $a, b, c$ | $a^6 = 1, b^2 = 1, c^2 = 1,$ <br> $aba = b, ac = ca, bc = cb$ | |
|---|---|---|---|---|
| 24 | $A_4 \times C_2$ | $a, b, c$ | Ponderable | |
| 24 | $Q_6 \times C_2$ | $a, b, c$ | Ponderable | |
| 24 | $D_4 \times C_3$ | $a, b, c$ | Ponderable | |
| 24 | $Q \times C_3$ | $a, b, c$ | Ponderable | |
| 24 | $S_3 \times C_4$ | $a, b, c$ | Ponderable | |
| 24 | $D_{12}$ | $a, b$ | $a^{12} = 1, b^2 = 1,$ <br> $aba = b$ | |
| 24 | $Q_{12}$ | $a, b$ | $a^{12} = 1, b^2 = a^6,$ <br> $aba = b$ | |
| 24 | $S_4$ | $a, b$ | $a^4 = 1, b^2 = 1,$ <br> $(ab)^3 = 1$ | $Aut(G) = G$ |
| 24 | $SL(2, \mathbb{F}_3)$ | $a, b, c$ | $a^4 = 1, b^2 = a^2,$ <br> $c^3 = 1, aba = b,$ <br> $ac = cb, bc = cab$ | $Aut(G) \cong Aut(Q)$ <br> $= S_4,$ <br> a semidirect product <br> of $Q$ with $C_3$ <br> ($Q$ normal) |
| 24 | | $a, b$ | $a^3 = 1, b^8 = 1,$ <br> $aba = b$ | a semidirect product <br> of $C_3$ with $C_8$ <br> ($C_3$ normal) |
| 24 | | $a, b, c$ | $a^3 = 1, b^4 = 1,$ <br> $c^2 = 1, bcb = c, aba = b,$ <br> $ac = ca$ | a semidirect product <br> of $C_3$ with $D_4$ <br> ($C_3$ normal) |
| 25 | $C_{25}$ | $a$ | $a^{25} = 1$ | |
| 25 | $C_5^2$ | $a, b$ | Ponderable | |

**Ponderable 10.4.1.** *Find a subgroup of $G$ isomorphic to the Klein 4-group.*

**Example 10.4.1.** *(a) Let $Q$ denote the quaternion group:*

$$Q = \{1, -1, i, -i, j, -j, k, -k\},$$

*where $i^2 = j^2 = k^2 = -1, ij = k, jk = i, ki = j$, and in general, $xy = -yx$ for $x, y$ belonging to $\{i, j, k\}$. Let*

$$Q^* = \langle g_1, g_2 \rangle \subset G$$

*where*

$$g_1 = F^2 * M_R * U^{-1} * M_R^{-1} * U^{-1} * M_R * U * M_R^{-1} * U * F^2,$$
$$g_2 = F * U^2 * F^{-1} * U^{-1} * L^{-1} * B^{-1} * U^2 * B * U * L.$$

*Then $\phi : Q \to Q^*$ is an isomorphism, where $\phi(i) = g_1$ and $\phi(k) = g_2$.*

*This was stated without proof in Example 9.3.5. To prove it, we use the fact that we now know the generators and relations for $Q$. It suffices to check that the above generators $g_i$ satisfy the relations defining $Q$ given in the table above. This is left as an exercise for the interested reader with a Rubik's Cube.*

**Ponderable 10.4.2.** *Which of these groups tabulated above is (isomorphic to) a subgroup of the Rubik's Cube group?*

**Ponderable 10.4.3.** *Of those which are subgroups, find moves associated to the generators given which satisfy the relations given. In other words, find an explicit isomorphism in terms of the generators.*

This is probably much harder but possibly still 'doable', with the help of a program such as GAP or MAGMA.

## 10.5  The presentation problem

The following problem is apparently unsolved at this time.

**Ponderable 10.5.1.** *(Singmaster) Let $G$ be the Rubik's Cube group. Find a set of generators and relations for $G$ of minimal cardinality (i.e., $|X| + |Y|$ is of minimal cardinality).*

**Ponderable 10.5.2.** *Find*

*(a) a set of generators for $G$ of minimal cardinality,*

*(b) a set of relations for $G$ of minimal cardinality,*

*(c) an expression for each such generator as a word in the basic moves $R, L, U, D, F, B$.*

The part (a) is known: there are 2 elements which generate $G$ [Si]. Part (b) is not known, though Dan Hoey's post of Dec 17, 1995 to the cube-lover's list may describe the best known results [CL]. He suggests that $G$ has a set $X$ of 5 generators and a set $Y$ of 44 relations such that the total length of all the reduced words in $Y$ is 605.

### 10.5.1  A presentation for $C_m^n \rtimes S_{n+1}$

Motivated by the problem of D. Singmaster mentioned above, we shall find a presentation for the generalized symmetric group. From this, one can derive a presentation for the illegal Rubik's Cube group.

We can identify the $C_m^n \rtimes S_{n+1}$ with the group of $(n+1) \times (n+1)$ invertible monomial matrices $g$ with coefficients in $C_m$ having the following condition on the determinant: if we write $g = p \cdot d$, where $p$ is a permutation matrix and $d$ is a diagonal matrix then $\det(d) = 1$ (this determinant 1 condition is a condition corresponding to the 'conservation of twists' for the moves of the Rubik's Cube).

We may identify $C_m^n$ with the subgroup

$$\{(x_1, ..., x_{n+1}) \mid x_1 x_2 ... x_{n+1} = 1, \ x_i \in C_m\}$$

and $C_m^n \rtimes S_{n+1}$ as a subgroup of the wreath product $S_{n+1} \ wr \ C_m$.

173

Consider $G = C_m^{n+1} \rtimes S_{n+1}$. The group $S_{n+1}$ has presentation

$$S_{n+1} = \langle a_1, ..., a_n \mid (a_i a_j)^{m_{ij}} = 1, \ \forall 1 \le i, j \le n \rangle,$$

where

$$m_{ij} = \begin{cases} 3, & j = i \pm 1, \\ 2, & |i - j| > 1, \\ 1, & i = j. \end{cases}$$

The following diagram may help to visualize the exponents $m_{ij}$ in the case $n = 4$:

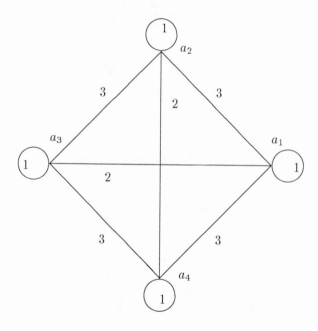

As a group of $(n + 1) \times (n + 1)$ monomial matrices, we identify $a_i$ with the permutation matrix,

$$s_i = \begin{pmatrix} 1 & 0 & & & & \cdots & & 0 \\ & \ddots & & & & & & \\ & & 1 & & & & & \\ & & & 0 & 1 & & & \\ & & & 1 & 0 & & & \\ & & & & & 1 & & \\ & & & & & & \ddots & \\ 0 & \cdots & & & & & 0 & 1 \end{pmatrix}.$$

If $I$ is the $(n+1) \times (n+1)$ identity matrix and if $E_{ij}$ denotes the matrix which is 0 in every entry except the $ij$ entry, which is 1, then

$$s_i = I - E_{ii} - E_{i+1,i+1} + E_{i,i+1} + E_{i+1,i}.$$

The group $C_m$ has presentation

$$C_m = \langle h \mid h^m = 1 \rangle.$$

The group $C_m^n$ has presentation

$$C_m^n = \langle h_1, ..., h_n \mid h_i^m = 1, \ h_i h_j = h_j h_i, \forall 1 \leq i, j \leq n+1 \rangle.$$

We identify $C_m^n$ with the Cartesian product

$$\{(h_1(x_1), h_2(x_2), ..., h_n(x_n)) \mid x_i \in C_m\},$$

where $h_i(t)$ is the diagonal matrix

$$h_i(t) = I - E_{ii} - E_{i+1,i+1} + tE_{i,i} + t^{-1}E_{i+1,i+1}.$$

There are the following identities between the $s_i$ and the $h_j(t)$:

$$\begin{aligned} s_i h_j(t) s_i^{-1} &= h_j(t), \quad |i - j| > 1, \\ s_i h_i(t) s_i^{-1} &= h_i(t)^{-1}, \\ s_{i\pm1} h_j(t) s_{i\pm1}^{-1} &= h_i(t) h_{i\pm1}(t). \end{aligned}$$

This motivates the formulation of the following statement.

**Theorem 10.5.1.** *The group $C_m^n \rtimes S_{n+1}$ is given by generators $a_1, ..., a_n, h_1, ..., h_n$ and relations*

$$\begin{aligned} (a_i a_j)^{m_{ij}} &= 1, \\ \forall 1 \leq i, j &\leq n, \\ h_i^m = 1, \ h_i h_j &= h_j h_i, \forall 1 \leq i, j \leq n \\ a_i h_j a_i^{-1} &= h_j, \quad |i - j| > 1, \\ a_i h_i a_i^{-1} &= h_i^{-1}, \\ a_{i\pm1} h_j a_{i\pm1}^{-1} &= h_i h_{i\pm1}. \end{aligned}$$

**Remark 10.5.1.** *Essentially the same presentation may be found in the paper [DM] by Davies and Morris. We only sketch the proof since it is in [DM].*

## 10.5.2 Idea of the proof

Here is a sketch of the proof (which I thank Dennis Spellman for) of the above theorem. Let $P$ denote the group presented in the above theorem. The claim is, of course, that $C_m^n \rtimes S_{n+1} \cong P$. There is a surjective homomorphism $f : P \to C_m^n \rtimes S_{n+1}$ given by sending the generators to the generators. The problem is to show that this is injective. Let $K = ker(f)$, so

$$|P| = |P/K||K| \geq |P/K| = |C_m^n \rtimes S_{n+1}| = m^n(n+1)!.$$

Note $H = \langle h_1, ..., h_n \mid h_i^m = 1, h_i h_j = h_j h_i \rangle \subset P$ is a normal subgroup of $P$ since each $a_i$ sends a generator of $H$ to a product of them or their inverses. Also, note $H \cong C_m^n$.

We claim that $P/H \cong S_{n+1}$. From this it will follow that $|P| = m^n (n+1)!$, proving that $|K| = 1$, as desired.

To establish $P/H \cong S_{n+1}$, we verify that the presentation on $P/H$ one gets from Theorem 2.1 in [MKS] is the same as that of $S_{n+1}$. Indeed, if $W(a_1, ..., a_n, h_1, ..., h_n) = 1$ is a relation in the presentation, we must determine the word $W(a_1, ..., a_n, h_1, ..., h_n)H$ in $P/H$. Note that every relation 'collapses' and becomes trivial except for the relations $(a_i a_j)^{m_{ij}} = 1$. These relations define the presentation for $S_{n+1}$, as desired. $\square$

**Ponderable 10.5.3.** *How many generators and relations are specified by this theorem when*
   *(a) $n = 7$ and $m = 3$,*
   *(b) $n = 11$ and $m = 2$?*

# Chapter 11

# The (legal) Rubik's Cube group

'The advantage is that mathematics is a field in which one's blunders tend to show very clearly and can be corrected or erased with a stroke of the pencil. It is a field which has often been compared with chess, but differs from the latter in that it is only one's best moments that count and not one's worst.'

*Norbert Wiener,* **Ex-prodigy: my childhood and youth**

In this chapter, we build on the material in the previous chapters to describe mathematically the group of (legal) moves of the $3 \times 3$ Rubik's Cube. For first-timers, it might seem like we've just finished a long steep climb to the top of a hill. However, we are not done! Though the terrain may flatten out, there is still a bit of a hike until we get to our destination.

## 11.1 Mathematical description of the $3 \times 3$ cube moves

In this section, we describe mathematically the moves of the $3 \times 3$ Rubik's Cube. As we will see, this will lead eventually to the description of the Rubik's Cube group as a subgroup of index 12 of a direct product of two wreath products.

### 11.1.1 Notation

First, orient all the corners and edges as in Theorem 9.6.1. These are depicted as follows on the F, R, and U faces:

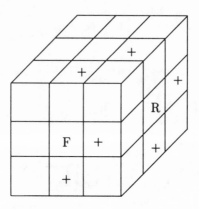

For the other faces: put a '+' on the *dl*, *db*, *lf*, *lu*, *bl*, and *bu* edges. This fixes the edge orientations.

For the corner orientations, put a '+' sign on all the top corner facets and on all the bottom corner facets.

Let $G = < R, L, U, D, F, B >$ be the group of the $3 \times 3$ Rubik's Cube and let $H$ be the 'enlarged' group generated by $R, L, U, D, F, B$ and all the 'illegal' moves (where one is allowed to disassemble and reassemble the cube but not remove any facets). Let $V$ denote the set of vertices of the cube (which we identity with the set of corner subcubes of the Rubik's Cube) and let

$$\rho : H \to S_V$$

denote the homomorphism which associates to each move of the Rubik's Cube the corresponding permutation of the vertices (see Example 9.2.1). Let $E$ denote the set of edges of the cube (which we identity with the set of edge subcubes of the Rubik's Cube) and let

$$\sigma : H \to S_E$$

denote the homomorphism which associates to each move of the Rubik's Cube the corresponding permutation of the edges (see Example 9.2.1).

These homomorphisms $\rho$ and $\sigma$ shall be reformulated in a slightly different way in §11.2.2 below.

## 11.1.2   Corner orientations

Let $v : H \to C_3^8$ be the function which associates to each move $g \in H$ the corresponding corner orientations. More precisely, let $g \in H$ and say $g$ moves corner $i$ to corner $j$. Then $v_i(g) \in C_3$ is the orientation which the $i^{th}$ vertex

178

gets sent to by $g$, where the vertices are labeled as in the diagram shown and where the orientation is the number of $120°$ clockwise twists required to turn the relative reference '+' obtained by moving corner $i$ to $j$ using the move $g$ into the standard reference '+' on corner $j$.

**Example 11.1.1.** *We have*

| $X$ | $\vec{v}(X)$ |
|-----|------|
| $F$ | $(2,0,0,1,1,0,0,2)$ |
| $U$ | $(0,0,0,0,0,0,0,0)$ |
| $F{*}U$ | $(2,0,0,1,1,0,0,2)$ |
| $U{*}F$ | $(2,0,0,1,1,0,0,2)$ |
| $D$ | $(0,0,0,0,0,0,0,0)$ |
| $B$ | $(0,1,2,0,0,2,1,0)$ |
| $R$ | $(1,2,0,0,2,1,0,0)$ |
| $L$ | $(0,0,1,2,0,0,2,1)$ |

**Remark 11.1.1.** *The effect of a move $g \in H$ on the corner orientations may also be regarded as a relabeling of the '+' markings.*

Note that a move $g \in H$ has two effects on the corners:
(a) a permutation $\rho(g) \in S_V$ of the vertices,
(b) a reorientation of the vertices moves in (a).
In particular, for $g, h \in H$, the orientation $\vec{v}(gh)$ can only differ from $v(g)$ in the coordinates corresponding to the vertices permuted by $h$.

We shall now verify that the 'relative' orientation $\vec{v}(gh) - \vec{v}(g)$ is the same as the orientation $\vec{v}(h)$, provided one takes into account the effect of $g$ on the vertices: $\vec{v}(h) = \rho(g)(\vec{v}(gh) - \vec{v}(g))$.

**Lemma 11.1.1.** $\vec{v}(gh) = \vec{v}(g) + \rho(g)^{-1}(\vec{v}(h))$.

**Proof:** The move $gh$ orients the $i^{th}$ corner subcube by $v_i(gh)$ and permutes the vertices by $\rho(gh)$, by definition.

On the other hand, $gh$ will first act by $g$ then $h$. The move $g$ will reorient the $i^{th}$ corner subcube by $v_i(g)$ and send the $i^{th}$ vertex to the $\rho(g)(i)^{th}$ vertex.

To study the subsequent effect of $h$ on this, let us subtract $\vec{v}(g)$ from $\vec{v}(gh)$, so that we are back to our original orientation (we will add $\vec{v}(g)$ back in later). Call this position the **modified cube** for now.

The move $h$ first orients the $j^{th}$ corner subcube of the modified cube by $v_j(h)$ and permutes it to vertex $\rho(h)(j)$. The $i^{th}$ subcube of the modified cube comes from (via $g$) the $\rho(g)^{-1}(i)^{th}$ subcube of the original cube. Thus the $i^{th}$ corner subcube of the modified cube is, by means of $h$, reoriented by $v_{\rho(g)^{-1}(i)}(h)$. To this we must add in $v_i(g)$ to get the total effect of $gh$ on the $i^{th}$ vertex of the original:

$$v_i(gh) = v_i(g) + v_{\rho(g)^{-1}(i)}(h),$$

for each $1 \leq i \leq 8$, which implies Lemma 11.1.1. $\square$

### 11.1.3 Edge orientations

Let $w : H \to C_2^{12}$ be the function which associates to each move $g \in H$ the corresponding edge orientations. More precisely, let $g \in H$ and say $g$ moves edge $i$ to edge $j$. Then $w_i(g) \in C_2$ is the orientation which the $i^{th}$ edge gets sent to by $g$, where the edges are labeled as in the diagram shown and where the orientation is the number of $180°$ flips required to turn the relative reference '+' obtained by moving edge $i$ to $j$ using the move $g$ into the standard reference '+' on edge $j$.

**Example 11.1.2.** *We have*

| $X$ | $\vec{w}(X)$ |
|---|---|
| $F$ | (1,0,0,0,0,0,0,0,1,0,0,0) |
| $U$ | (1,0,1,0,0,0,0,0,0,0,0,0) |
| $F*U$ | (1,0,1,0,1,0,0,0,1,0,0,0) |
| $U*F$ | (1,1,1,0,0,0,0,0,1,0,0,0) |
| $B$ | (0,0,0,0,0,0,1,1,0,0,0,0) |
| $D$ | (0,0,0,0,0,0,0,0,0,1,0,1) |
| $R$ | (0,1,0,0,0,0,0,0,0,1,0,0) |
| $L$ | (0,0,0,0,1,0,0,0,1,0,0,0) |

**Remark 11.1.2.** *The effect of a move $g \in H$ on the edge orientations may also be regarded as a relabeling of the '+' markings.*

Note that a move $g \in H$ has two effects on the edges:
(a) a permutation $\sigma(g) \in S_E$ of the edges,
(b) a reorientation of the edges which were moved in (a).
In particular, for $g, h \in H$, the orientation $\vec{w}(gh)$ can only differ from $\vec{w}(g)$ in the coordinates corresponding to the edges permuted by $h$.

We shall now claim that

$$\vec{w}(gh) = \vec{w}(g) + \sigma(g)^{-1}(\vec{w}(h)), \tag{11.1}$$

i.e., that

$$w_i(gh) = w_i(g) + w_{\sigma(g)^{-1}(i)}(h),$$

for each $1 \le i \le 12$. The proof of this is so similar to the proof of Lemma 11.1.1 that we leave it to the student to modify its proof to verify (11.1).

### 11.1.4 The semi-direct product

Consider the following direct product of two semi-direct products:

$$H' = (C_3^8 \rtimes S_V) \times (C_2^{12} \rtimes S_E).$$

**Remark 11.1.3.** *This may also be written in the notation of wreath products as the following direct product of two wreath products:*

$$H' = (S_V \text{ wr } C_3^8) \times (S_E \text{ wr } C_2^{12}).$$

As a *set*, we think of $H$ as belonging to $C_3^8 \times S_V \times C_3^8 \times S_V$. If we represent elements $h, h'$ of $H$ as $h = (v, r, w, s), h' = (v', r', w', s') \in C_3^8 \times S_V \times C_3^8 \times S_V$ then the group operation will be given by

$$h * h' = (v, r, w, s) * (v', r', w', s') = (v + P(r)(v'), rr', w + P(s)(w'), ss').$$

Consider the function

$$
\begin{aligned}
\iota : H &\to (C_3^8 \rtimes S_V) \times (C_2^{12} \rtimes S_E) \\
g &\longmapsto (v(g), \rho(g), w(g), \sigma(g)).
\end{aligned}
$$

**Proposition 11.1.1.** *$\iota$ is an isomorphism, $H \cong H'$.*

**Proof:** Since

$$
\begin{aligned}
&(\vec{v}(g), \rho(g), \vec{w}(g), \sigma(g)) * (\vec{v}(h), \rho(h), \vec{w}(h), \sigma(h)) \\
&= (\vec{v}(g) + P(\rho(g))(\vec{v}(h)), \rho(g)\rho(h), \vec{w}(g) + P(\sigma(g))(\vec{w}(h)), \sigma(g)\sigma(h)),
\end{aligned}
$$

the map $\iota$ is a homomorphism. Since any reorientation and permutation can be achieved by some illegal move, $\iota$ must be surjective. By theorem 9.6.1, the kernel of $\iota$ is trivial (this is just a fancy way of saying that if no subcube is permuted or reoriented then the cube doesn't change!). $\square$

## 11.2 Structure of the cube group

> 'The scientist does not study nature because it is useful; he studies it because he delights in it, and he delights in it because it is beautiful. If nature were not beautiful, it would not be worth knowing, and if nature were not worth knowing, life would not be worth living.'

> *Henri Poincaré*

The main result of this section is the 'second fundamental theorem of cube theory.' The result of this section is both beautiful and, we hope the reader agrees, is worth knowing.

Some preliminaries. We identify, as in §11.1, each $g \in G$ with a 4-tuple

$$(\vec{v}(g), \rho(g), \vec{w}(g), \sigma(g)),$$

where

- $\rho(g)$ is the corresponding permutation of the set of vertices $V$ of the cube,

- $\sigma(g)$ is the corresponding permutation of the set of edges $E$ of the cube,

- $v(g), w(g)$ are 'orientations' defined in §11.1.

**Remark 11.2.1.** *Let $S_n$ denote the symmetric group on n letters and identify $S_V$ with $S_8$, $S_E$ with $S_{12}$. By example 9.2.1, we know that*
*(a) $\rho : G \to S_8$ is a homomorphism,*
*(b) $\sigma : G \to S_{12}$ is a homomorphism.*

## 11.2.1 Second fundamental theorem of cube theory

**Question**: Given a 4-tuple $(v, r, w, s)$, where $r, s$ are permutations of the corners, resp. edges, as above and

$$v \in C_3^8, \quad w \in C_2^{12},$$

what conditions on $r, s, v, w$ insure that it corresponds to a possible position of the Rubik's Cube?

The following result answers this question. It is, according to [BCG], due to Ann Scott.

**Theorem 11.2.1.** *(Second fundamental theorem of cube theory) A 4-tuple $(\vec{v}, r, \vec{w}, s)$ as above ($r \in S_8$, $s \in S_{12}$, $\vec{v} \in C_3^8$, $\vec{w} \in C_2^{12}$) corresponds to a possible position of the Rubik's Cube if and only if*

*(a) $sgn(r) = sgn(s)$,    ('equal parity as permutations')*

*(b) $v_1 + \ldots + v_8 \equiv 0 \pmod 3$,    ('conservation of total twists')*

*(c) $w_1 + \ldots + w_{12} \equiv 0 \pmod 2$,    ('conservation of total flips').*

**Proof:** For simplicity, we write $v$ for $\vec{v}$.

First we prove the 'only if' part. That is, we assume that $(v, r, w, s) \in S_V \times S_E \times C_3^8 \times C_2^{12}$ represents a (legally obtained!) position of the Rubik's Cube. From this we want to prove (a)-(c).

Let $g \in G$ be the element which moves the Rubik's Cube from the solved position to the position associated to this 4-tuple. Then $r = \rho(g)$ and $s = \sigma(g)$. We know that $g$ may be written as a word in the basic moves $R, L, U, D, F, B$, say $g = X_1 \ldots X_k$, where each $X_i$ is equal to one of the $R, L, U, D, F, B$. Observe that if $X$ is any one of these basic moves then $sgn(\rho(X)) = sgn(\sigma(X))$. Since $sgn$, $\rho$, and $\sigma$ are homomorphisms, it follows that

$$sgn(r) = sgn(\rho(g)) = \prod_{i=1}^{k} sgn(\rho(X_i)) = \prod_{i=1}^{k} sgn(\sigma(X_i)) = sgn(\sigma(X)) = sgn(s).$$

This proves (a).

We have verified (b) for the basic moves in example 11.1.1 above. Note that

(i) the conservation of twists condition in (b) is true for $(v_1, \ldots, v_8)$ if and only if it is true for any permutation $P(p)(v) = (v_{(1)p}, \ldots, v_{(8)p})$,

(ii) if $(v_1, \ldots, v_8)$ and $(v_1', \ldots, v_8')$ each satisfy the conservation of twists condition in (b) then their sum also satisfies it.

As above, write $g$ as a word in the basic moves $R, L, U, D, F, B$, say $g = X_1 \ldots X_k$, where each $X_i$ is equal to one of the $R, L, U, D, F, B$. We assume that this expression is minimal in the sense that we choose the $X_i$ so that $k$ is as small as possible. This $k$ is called the **length** of $g$. (This length is the same as the distance from $g$ to the identity in the Cayley graph of $G$.)

We now prove (b) by induction on the length. We have already checked it for all words of length $k = 1$.

Assume $k > 1$. By the formula giving the orientation of the product of two moves in terms of the two orientations of the moves, we have

$$\vec{v}(X_1...X_{k-1}X_k) = \rho(X_1...X_{k-1})^{-1}(\vec{v}(X_k)) + \vec{v}(X_1...X_{k-1}).$$

The term $\rho(X_1...X_{k-1})^{-1}(\vec{v}(X_k))$ satisfies the conservation of twists condition in (b) by (i) above. The term $\vec{v}(X_1...X_k)$ satisfies the conservation of twists condition in (b) by the induction hypothesis. Their sum satisfies the conservation of twists condition in (b) by (ii) above. This proves (b).

The proof of (c) is very similar to the proof of (b), except that we use example 11.1.2 in place of example 11.1.1.

**Ponderable 11.2.1.** *Provide the details.*

Now, we must prove the theorem in the 'if' direction. In other words, assuming (a), (b), and (c) we must show that there is a corresponding legal position of the Rubik's Cube. This part of the proof is constructive.

First, we prove a special case. Assume that $r$ and $s$ are both the identity and that $(w_1, ..., w_{12}) = (0, ..., 0)$.

There is a move which twists exactly two corners and preserves the orientations and positions of all other subcubes. For example, the move $g = (R^{-1}D^2RB^{-1}U^2B)^2$ twists the ufr corner by $120^o$ clockwise, the bdl corner by $240^o$ clockwise, and preserves the orientations and positions of all other subcubes. This move can be easily modified, by a suitable conjugation, to obtain a move which twists any pair of corners, and preserves the orientations and positions of all other subcubes. These moves generate all possible 8-tuples satisfying the conservation of twists condition in (b). This proves the 'if' part of the theorem in the case that $r$ and $s$ are both the identity and that $(w_1, ..., w_{12}) = (0, ..., 0)$.

Next, we prove another special case. Assume that $r$ and $s$ are both the identity and that $(v_1, ..., v_8) = (0, ..., 0)$.

There is a move which flips exactly two edges and preserves the orientations and positions of all other subcubes. For example, the move

$$g = LFR^{-1}F^{-1}L^{-1}U^2RURU^{-1}R^2U^2R$$

(found in [B], page 112) flips the uf edge, the ur edge, and preserves the orientations and positions of all other subcubes. This move can be easily modified, by a suitable conjugation, to obtain a move which flips any pair of edges, and preserves the orientations and positions of all other subcubes. These moves generate all possible 12-tuples satisfying the conservation of flips condition in (c). This proves the 'if' part of the theorem in the case that $r$ and $s$ are both the identity and that $(v_1, ..., v_8) = (0, ..., 0)$.

As a consequence of these last two special cases, it follows that the 'if' part of the theorem is true in the case that $r$ and $s$ are both the identity.

Finally, we prove our last special case. Assume that $(v_1, ..., v_8) = (0, ..., 0)$ and that $(w_1, ..., w_{12}) = (0, ..., 0)$. Consider the following three claims.

183

- Given any three edges subcubes, there is a move which is a 3-cycle on these edges and preserves the orientations and positions of all other subcubes.

- Given any three corners, there is a move which is a 3-cycle on these corners and preserves the orientations and positions of all other subcubes.

- Given any pair of edges and any pair of corners, there is a move which is a 2-cycle on these edges, a 2-cycle on these corners, and preserves the orientations and positions of all other subcubes.

**Ponderable 11.2.2.** *Verify these three claims.*

By proposition 9.4.1, we know that $A_E$ is generated by the edge 3-cycles above and that $A_V$ is generated by the corner 3-cycles above. In other words, we can construct a position of the Rubik's Cube associated to any 4-tuple $(r, s, 0, 0)$, provided $r \in A_V$ and $s \in A_E$. The subgroup $A_E \times A_V$ is index 4 in $S_E \times S_V$ since $|S_n/A_n| = 2$. The third type of move, the edge-corner 2-cycles above, does not correspond to an element of the subset $A_E \times A_V$ of the Rubik's Cube group because an edge 2-cycle is an odd permutation of the edges. Therefore, if we consider the subgroup of $S_E \times S_V$ generated by all three types of moves we will obtain either all of $S_E \times S_V$ or some subgroup of index 2 which properly contains $A_E \times A_V$. The first possibility can be ruled out since it contradicts the parity condition in (a). The only subgroup of $S_E \times S_V$ of index 2 which properly contains $A_E \times A_V$ is the subgroup of elements satisfying the parity condition in (a).

It follows that the 'if' part of the theorem is true in the case that $v$ and $w$ are both zero.

The theorem is a consequence of these special cases because of the following.

Claim: No matter what position the Rubik's Cube is in, there is always a move which does not permute any subcubes but 'solves' the orientation of the cube so that $v$ and $w$ are both zero.

**Ponderable 11.2.3.** *Prove this claim.*

Finally, the proof of the theorem is finished. $\square$

**Corollary 11.2.1.** *The Rubik's Cube group is given by*

$$G = \{g = (\vec{v}, r, \vec{w}, s) \in H \mid (a), (b), (c) \text{ hold}\},$$

*where (a), (b), (c) are as in the above theorem.*

Now we show that the center of the Rubik's Cube group consists of the identity and the superflip.

**Corollary 11.2.2.** *The center of $G$ consists of two elements: the identity and the 'superflip' element $z = (\vec{v}, r, \vec{w}, s)$ where $\vec{w} = (1, 1, ..., 1) \in C_2^{12}$, $\vec{v} = \vec{0} \in C_3^8$, $s = 1 \in S_{12}$ and $r = 1 \in S_8$.*

**Proof:**  The proof given here is essentially the same as that of Bandelow [B]. Again, write $v$ in place of $\vec{v}$, for simplicity. Recall that the center of $S_n$, $n > 2$, is trivial (see Ponderable 5.6.2 above). Fix $(v, r, w, s) \in G$. We have

$$(v, r, w, s) * (v', r', w', s') = (v + P(r)(v'), rr', w + P(s)(w'), ss')$$
$$= (v', r', w', s') * (v, r, w, s) = (v' + P(r')(v), r'r, w' + P(s')(w), s's),$$

for all $(v', r', w', s') \in G$, only if $r = 1$ and $s = 1$. This implies that $v + v' = v' + P(r')v$, for all $r' \in S_8$. This forces $v$ to be either $(0, 0, ..., 0)$ or $(1, 1, ..., 1)$ or $(2, 2, .., 2)$. Since it must satisfy conservation of twists, it can't be $(1, 1, ..., 1)$ or $(2, 2, .., 2)$. Similarly, $w + w' = w' + P(s')w$, for all $s' \in S_{12}$ forces $w$ to be either $(0, 0, ..., 0)$ or $(1, 1, ..., 1)$. Either of these choices is okay since they both satisfy conservation of flips. $\square$

## 11.2.2  Some consequences

We shall now reformulate the above fact about the Rubik's Cube group from a different point of view. This new perspective allows us to count its elements easier. Let

$$G_0 = \{(\vec{v}, r, \vec{w}, s) \mid r \in S_8,\ s \in S_{12},$$
$$\vec{v} = (v_1, v_2, ..., v_8), v_i \in \{0, 1, 2\}, v_1 + ... + v_8 \equiv 0 \pmod 3,$$
$$\vec{w} = (w_1, w_2, ..., w_{12}), w_i \in \{0, 1\}, w_1 + ... + w_{12} \equiv 0 \pmod 2\}.$$

In other words, $G_0$ contains the permutations of the corners, permutations of the edges, the orientations of the corners, and the orientations of the edges. There is a conservation of flips condition and a conservation of twists condition but no parity condition on the permutations. Define a binary operation $*$ : $G_0 \times G_0 \to G_0$ by

$$(\vec{v}, r, \vec{w}, s) * (\vec{v}', r', \vec{w}', s') = (\vec{v} + P(r)(\vec{v}'), r * r', \vec{w} + P(s)(\vec{w}'), s * s'),$$

where $P$ is the permutation matrix (Definition 3.2.2). This defines a group structure on $G_0$. This is a subgroup of the illegal Rubik's Cube group of index 6.

**Theorem 11.2.2.**  *There is an isomorphism*

$$G_0 \cong (C_3^7 \rtimes S_8) \times (C_2^{11} \rtimes S_{12}),$$

*where $C_n$ is the cyclic group with $n$ elements and $\rtimes$ denotes the semi-direct product and where $C_n^k$ ($n = 2, 3$, $k = 7, 11$) is identified with the subgroup of $C_n^{k+1}$ defined by*

$$\{\vec{v} = (v_1, v_2, ..., v_k) \mid v_i \in \{0, 1, n - 1\}, v_1 + ... + v_k \equiv 0 \pmod n\}.$$

*In particular,*

$$|G_0| = |S_8||S_{12}||C_2^{11}||C_3^7| = 8! \cdot 12! \cdot 2^{11} \cdot 3^7.$$

**Proof:** This follows from the definition of semi-direct product and the Multiplication Principle (Theorem 2.4.2). □

**Corollary 11.2.3.** *The Rubik's Cube group $G$ is the kernel of the homomorphism*

$$\phi : G_0 \to \{1, -1\}$$
$$(\vec{v}, r, \vec{w}, s) \longmapsto sgn(r)sgn(s).$$

*In particular, $G < G_0$ is normal of index 2 and*

$$|G| = 8! \cdot 12! \cdot 2^{10} \cdot 3^7.$$

**Proof:** This follows from the Theorem and the first isomorphism theorem of group theory (Theorem 9.5.1). □

Recall that the commutator subgroup $G_1$ of $G$ is the subgroup consisting of all finite products of commutators, $[g, h] = g * h * g^{-1} * h^{-1}$, where $g, h$ are arbitary elements of $G$.

**Theorem 11.2.3.** $G_1 = \{g \in G \mid sgn(\rho(g)) = sgn(\sigma(g)) = 1\}$.

**Proof:** It is known that the commutator subgroup of $S_n$ is $A_n$, for $n > 4$ (see chapter 3 of [R]; in fact, for $n > 4$, $A_n$ is the only proper non-trivial normal subgroup of $S_n$).

Consider the projection map $\phi : C_k^n \rtimes S_n \to S_n$. The kernel of $\phi$ is $C_k^n$. Moreover,

$$\phi([g, h]) = \phi(g * h * g^{-1} * h^{-1}) = \phi(g) * \phi(h) * \phi(g^{-1}) * \phi(h^{-1}) = [\phi(g), \phi(h)],$$

for any $g, h \in C_k^n \rtimes S_n$. This implies that $\phi$ maps the commutator subgroup $(C_k^n \rtimes S_n)_1$ to the commutator subgroup $(S_n)_1 = A_n$. From this it follows that $(C_k^n \rtimes S_n)_1 = C_k^n \rtimes A_n$. This implies the theorem. □

**Corollary 11.2.4.** $|G_1| = |G|/2$.

**Proof:** The above theorem implies that $|G/G_1| = 2$, by using the first isomomorphism theorem. □

# 11.3 The moves of order 2

In this section we present, as promised, a method for determining the moves of a given order in the Rubik's Cube group. We shall only compute the number of moves of order 2, though the method should, in principle, work more generally. However, perhaps it is possible for the curious and enterprising reader to use a computer find the moves of order $d$, where $d > 2$, in a manner similar to what we do here.

The set of elements of order 2 may be divided into 3 disjoint subsets:
(a) elements moving corner pieces only,
(b) elements moving edge pieces only,

(c) elements moving corner and edge pieces.

Let us look at subset of (a) consisting only of the even permutations of order 2 of the 8 corner pieces. Obviously those elements can only consist of orientation preserving swaps of corner pieces where either 2 pairs (4 corner pieces) or 4 pairs (all 8 corner pieces) are affected. Given a pair of corner pieces, there are always 3 orientation preserving swaps.

(a2), (a3) and (b) (yes, do not forget (b)) has to be calculated analogously.

Recall from §9.8.2 that an element $(\vec{v}, f) \in C_m$ wr $S_n$, $\vec{v} = (v_1, v_2, ..., v_n) \in C_m^n$ and $f \in S_n$, is order $d$ if and only if $f^d = 1$ and $\vec{v} + ... + f^{d-1}(\vec{v}) = 0$.

In particular, elements of order 2 in $C_3^7 \rtimes S_8$ are those $(\vec{v}, r)$ with $r$ of order 2 (hence a product of distinct 2-cycles) and $\vec{v}$ satisfying $v_1 + ... + v_8 \cong 0 \pmod{3}$ and $v_i + v_{(i)r} \cong 0 \pmod{3}$. *Not all such elements correspond to a legal move of the Rubik's Cube.* Using the 'multiplication principle of counting' (Theorem 2.4.2), we compute the number of elements of order 2 in $C_3^7 \rtimes S_8$ to be

$$
\frac{1}{4!} \binom{8}{2} \binom{6}{2} \binom{4}{2} 3^4 +
$$
$$
\frac{1}{3!} \binom{8}{2} \binom{6}{2} \binom{4}{2} 3^3 +
$$
$$
\frac{1}{2!} \binom{8}{2} \binom{6}{2} 3^2 + \binom{8}{2} 3 = 21819. \tag{11.2}
$$

The '$3^3$' term and the '3' term do not count legal Rubik's Cube moves, if taken on their own, since they are not even (hence do not satisfy the 'conservation of parity' condition). The number of elements $(\vec{v}, r)$ of order 2 in $C_3^7 \rtimes S_8$ with $r$ even is

$$
\frac{1}{4!} \binom{8}{2} \binom{6}{2} \binom{4}{2} 3^4 + \frac{1}{2!} \binom{8}{2} \binom{6}{2} 3^2 = 10395. \tag{11.3}
$$

The number of elements $(\vec{v}, r)$ of order 2 in $C_3^7 \rtimes S_8$ with $r$ odd is

$$
\frac{1}{3!} \binom{8}{2} \binom{6}{2} \binom{4}{2} 3^3 + \binom{8}{2} 3 = 11424. \tag{11.4}
$$

Similarly, the 'multiplication principle of counting' (Theorem 2.4.2), we compute the number of elements of order 2 in $C_2^{11} \rtimes S_{12}$ to be

$$
\frac{1}{6!} \binom{12}{2} \binom{10}{2} \cdots \binom{4}{2} 2^6 +
$$
$$
\frac{1}{5!} \binom{12}{2} \binom{10}{2} \cdots \binom{4}{2} 2^7 + \frac{1}{4!} \binom{12}{2} \cdots \binom{6}{2} 2^8
$$
$$
+ \frac{1}{3!} \binom{12}{2} \cdots \binom{8}{2} 2^9 + \frac{1}{2!} \binom{12}{2} \binom{10}{2} 2^{10} + \binom{12}{2} 2^{11} \tag{11.5}
$$
$$
= 30706368.
$$

Again, not all these count legal Rubik's Cube moves, if taken on their own. The number of elements $(\vec{w}, s)$ of order 2 in $C_2^{11} \rtimes S_{12}$ with $s$ even is

$$\frac{1}{6!} \binom{12}{2} \binom{10}{2} \cdots \binom{4}{2} 2^6 + \frac{1}{4!} \binom{12}{2} \cdots \binom{6}{2} 2^8 + \frac{1}{2!} \binom{12}{2} \binom{10}{2} 2^{10}$$
$$= 15491520.$$

(11.6)

The number of elements $(\vec{w}, s)$ of order 2 in $C_2^{11} \rtimes S_{12}$ with $s$ odd is

$$\frac{1}{5!} \binom{12}{2} \binom{10}{2} \cdots \binom{4}{2} 2^7 + \frac{1}{3!} \binom{12}{2} \cdots \binom{8}{2} 2^9 + \binom{12}{2} 2^{11}$$
$$= 15214848.$$

(11.7)

The second fundamental theorem of cube theory 11.2.1 implies that an element $(\vec{v}, r, \vec{w}, s)$ of the Rubik's Cube group is order 2 if and only if it is an element of order 2 in the group $H = (C_3^7 \rtimes S_8) \times (C_2^{11} \rtimes S_{12})$ and $sgn(r) = sgn(s)$. A non-trivial element $(h_1, h_2) \in H$, with $h_1 = (\vec{v}, r)$ and $h_2 = (\vec{w}, s)$, is order 2 if and only if exactly one of the following mutually exclusive cases occurs:

(a) $h_1 \neq 1$, $h_2 = 1$, $h_1^2 = 1$, $sgn(r) = 1$,

(b) $h_1 = 1$, $h_2 \neq 1$, $h_2^2 = 1$, $sgn(s) = 1$,

(c) $h_1 \neq 1$, $h_2 \neq 1$, $h_1^2 = h_2^2 = 1$, $sgn(r) = sgn(s) = 1$,

(c) $h_1 \neq 1$, $h_2 \neq 1$, $h_1^2 = h_2^2 = 1$, $sgn(r) = sgn(s) = -1$.

We count each case. Case (a) is counted in (11.3), case (b) is counted in (11.6), case (c) is counted in (11.3) and (11.6), and case (d) is counted in (11.4) and (11.7). Totaling these up, we find that the number of (non-trivial) elements of order 2 in the Rubik's Cube group is $334864275867 = (3.34..) \times 10^{11}$.

# Chapter 12

# Squares, two-faces, and other subgroups

'In arctic and tropical times,
the integers, addition, and times,
taken mod p will yield
a full finite field,
as p ranges over the primes.'
*Ancient math haiku*

Rather than solve the Rubik's Cube for an arbitrary scrambled position, what if you noticed the scrambler only used certain types of moves closed under composition (such as the slice group)? Could you solve the cube using only those types of moves?

It is remarkable that several 'familiar' groups may be embedding into the Rubik's Cube group, and hence be regarded as a subgroup of the cube group. For example, we have seen in an earlier chapter how to embed the group of quaternions $Q = \{1, -1, i, -i, j, -j, k, -k\}$ inside the Rubik's Cube group. The slice (sub)group is another such example. We shall meet another subgroup - the 'two faces' group - here, which allows us to introduce finite fields.

The subgroup method, discussed in chapter 15, is a method for solving the Rubik's Cube which can be implemented on a computer. One of the groups arising in this method is the group studied in the next section.

## 12.1 The squares subgroup

Let $G$ denote the subgroup of the Rubik's Cube group generated by the *squares* of the basic moves:

$$G = \langle U^2, D^2, R^2, L^2, F^2, B^2 \rangle$$

called the **squares subgroup**. We shall verify below that the order of this group is $2^{13}3^4$. (By the way, as a consequence of this and Burnside's theorem [R], ch. 5, it follows that $G$ is a solvable group. We shall not need this fact.) In this section, we will investigate the group structure of $G$ using the same method which was used to determine the structure of the Rubik's Cube group.

The group $G$ acts on the set of edges and the set of vertices of the cube. There is a choice of orientation of the edges (resp., corners) similar to that in §§10.1.2-3 such that each element of $G$ *preserves the edge (resp., corner) orientations*.

The action $\phi$ of $G$ on the edges $E$ of the cube has exactly 3 orbits: the middle slice parallel to the right face $E_R$, the middle slice parallel to the front face $E_F$, the middle slice parallel to the up face $E_U$. In particular, the group $G$ acts (by restriction) on $E_R$, $E_F$, and $E_U$. The action $\psi$ of $G$ on the set of vertices $V$ has exactly 2 orbits: $V_1 = \{ufr, ubl, dfl, drb\}$, $V_2 = \{ufl, ubr, dfr, dlb\}$. Therfore, the group $G$ acts (by restriction) on $V_1$ and $V_2$. These actions yield associated homomorphisms:

$$\phi : G \to S_E,$$
$$\phi_{E_R} : G \to S_{E_R},$$
$$\phi_{E_F} : G \to S_{E_F},$$
$$\phi_{E_U} : G \to S_{E_U},$$
$$\psi : G \to S_V,$$
$$\psi_1 : G \to S_{V_1},$$
$$\psi_2 : G \to S_{V_2}.$$

**Proposition 12.1.1.** $G = \phi(G) \times \psi(G)$.

The proof of this is left as an exercise (hint: use the second fundamental theorem of Rubik's Cube theory).

**Lemma 12.1.1.** *If $g \in G$ then*

$$sgn(\phi_{E_R}(g))sgn(\phi_{E_F}(g))sgn(\phi_{E_U}(g)) = 1.$$

*Conversely, if $(p_1, p_2, p_3) \in S_{E_R} \times S_{E_F} \times S_{E_U}$ then there is a $g \in G$ such that $p_1 = \phi_{E_R}(g)$, $p_2 = \phi_{E_F}(g)$, $p_3 = \phi_{E_U}(g)$ if and only if $sgn(p_1)sgn(p_2)sgn(p_3) = 1$.*

As a consequence, we find that

$$\phi(G) = ker(sgn \times sgn \times sgn : \phi_{E_R}(G) \times \phi_{E_F}(G) \times \phi_{E_U}(G) \to \{\pm 1\})$$
$$\cong (S_4 \times S_4 \times S_4)/C_2.$$

In particular, $|\phi(G)| = (4!)^3/2 = 2^8 3^3$.

It remains to determine $\psi(G)$. We denote this group by $H$ for notational simplicity. We may label the vertices of the cube $1, 2, ..., 8$ in such a way that $H = \langle u, d, l, r, f, b \rangle$, where $u = (1,3)(2,4)$, $f = (1,8)(4,5)$, $d = (5,7)(6,8)$, $b = (3,6)(2,7)$, $r = (2,5)(1,6)$, $l = (4,7)(3,8)$. The action of $H$ on the set of vertices of the cube has two orbits ($\{1,3,6,8\}$ and $\{2,4,5,7\}$ in our labeling above), which we denote for simplicity by $V_1$ and $V_2$. There are homomorphisms

$$\psi_1 : H \to S_{V_1}, \quad \psi_2 : H \to S_{V_2},$$

but we shall not say much about these. Instead, we determine more about $H$. According to GAP, this group has $|H| = 96$ elements and 10 conjugacy classes (by the way, GAP also says that all the generators $u, ..., b$ are conjugate):

| size | representative |
|------|----------------|
| 1 | 1 |
| 12 | d=(5,7)(6,8) |
| 32 | l*d=(3,6,8)(4,5,7) |
| 3 | d*l*b*l=(2,4)(5,7) |
| 12 | d*b*l=(2,4,7,5)(3,6) |
| 3 | l*b*l*u=(1,3)(6,8) |
| 12 | b*l*u=(1,3,6,8)(4,7) |
| 3 | u*d=(1,3)(2,4)(5,7)(6,8) |
| 12 | l*d*u=(1,3,6,8)(2,4,5,7) |
| 6 | u*r*d*f=(1,3)(2,5)(4,7)(6,8) |

The stabilizer in $H$ of any vertex $v \in V$, written $H_v$, is a subgroup of order 24 isomorphic to the symmetric group $S_4$.

Furthermore, $H$ has a normal subgroup $N$ of order 48 (and index 2), where $N$ is a semidirect product of $C_3$ by $C_2^4$, with $C_2^4$ normal in $N$.

This is all we shall say about $H$.

The order of $G$ is therefore $G = |\phi(G)| \cdot |H| = 96 \cdot (4!)^3/2 = 2^{13}3^4$, as claimed above.

## 12.2 Fast-forwarding though finite fields

In this subsection, we introduce very briefly fields and especially finite fields. There are many excellent books on these subjects, so the treatment here will only give the reader a quick introduction to the basic ideas.

Finite fields are especially useful for communication theory since they are finite, and hence can represent the letters of an alphabet inside a computer, and since they carry a lot of structure, which can aid in making certain types of calculations faster. We shall mostly focus on finite fields in this section.

### 12.2.1 The general definition

A field is a set $F$ with an addition law $+$ and a multiplication law $\cdot$ which obeys a list of properties similar to those for the field of real numbers $\mathbb{R}$. More precisely, we call $(F, +, \cdot)$ a **field** if

(F1) $(F, +)$ is an abelian group, with an identity element denoted 0 ('the additive group of the field'),

(F2) for all $x, y, z \in F$, $(x + y)z = xz + yz$ ('distributive law'),

(F3) $(F - \{0\}, \cdot)$ is an abelian group, with an identity element denoted 1 ('the multiplicative group of the field').

191

**Lemma 12.2.1.** *If $F$ is a finite field then not only is $(F - \{0\}, \cdot)$ an abelian group, it is actually a cyclic group.*

We shall not prove this here (see chapter 13 of Artin [Ar] for a proof).

**Definition 12.2.1.** *Let $F_1, F_2$ be two fields. A function $f : F_1 \to F_2$ is called a* **field isomorphism** *if*

- *when $f$ is restricted to the additive group $(F_1, +)$, call this restriction $f$ again, it yields an isomorphism of groups $f : (F_1, +) \to (F_2, +)$,*

- *when $f$ is restricted to the multiplicative group $(F_1 - \{0\}, \cdot)$, call this restriction $f$ again, it yields an isomorphism of groups $f : (F_1 - \{0\}, \cdot) \to (F_2 - \{0\}, \cdot)$.*

## 12.2.2  A construction of $\mathbb{F}_p$

For each $k \in \mathbb{Z}$, let $\overline{k}$ denote the residue class of $k$ mod $p$. Let $\mathbb{F}_p$ denote the **finite field with $p$ elements**, so $\mathbb{F}_p$ is, as a set,

$$\mathbb{F}_p = \{\overline{0}, \overline{1}, ..., \overline{p-1}\},$$

with addition and multiplication being performed mod $p$.

**Example 12.2.1.** *When $p = 5$, $\mathbb{F}_5$ will denote the finite field with 5 elements, so $\mathbb{F}_5$ is, as a set,*

$$\mathbb{F}_5 = \{\overline{0}, \overline{1}, \overline{2}, \overline{3}, \overline{4}\},$$

*with addition and multiplication being performed mod 5.*

It is a general fact that if $F$ is any finite field then there is a prime number $p$ such that $px = 0$ for all $x \in F$. This prime number is called the **characteristic** of $F$. The easiest example of a finite field with characteristic $p$ is the finite field having $p$ elements, $\mathbb{F}_p$. It is not hard to see that any other finite field $F$ of characteristic $p$ must be isomorphic (as a vector space) to a finite dimensional vector space over $\mathbb{F}_p$. (Even if you've never seen a 'vector space over $\mathbb{F}_p$' defined before, if you know what a 'real vector space' is from Definition 2.1.2 then you've got the right idea.) The dimension of the vector space $F$ is called the **degree** of the field $F$ over $\mathbb{F}_p$, denoted $d = [F : \mathbb{F}_p]$. In this case, we say $F$ is a **field extension** of $\mathbb{F}_p$ (of degree $d$).

**Lemma 12.2.2.** *For fixed $p, d$ there is, up to isomorphism, only one such field extension of degree $d$ and characteristic $p$.*

We shall not prove this. See chapter 13 of Artin [Ar] for a proof.
Next, we shall show how to *construct* such extensions.

## 12.2.3   A construction of finite fields

First, some general remarks on how one might begin to implement finite fields on a computer.

Let $F$ be a finite field. Since $F$ is a finite dimensional vector space containing $\mathbb{F}_p$, it has a vector space basis which we label as

$$e_1 = 1, e_2, ..., e_d.$$

Thus $F$ is, as a set, the collection of elements of the form

$$x_1 e_1 + ... x_n e_n, \qquad x_i \in \mathbb{F}_p.$$

Since $F$ is a field, there are $c_{ij}^k \in \mathbb{F}_p$, which we call **structure constants**, such that

$$e_i e_j = \sum_{k=1}^{n} c_{ij}^k e_k.$$

There are $d_i^k \in \mathbb{F}_p$, which we call **inversion constants**, such that

$$e_i^{-1} = \sum_{k=1}^{n} d_i^k e_k.$$

These $p^3 + p^2$ constants determine how to multiply and divide elements of $F$. All a computer needs to do the arithmetic operations of $F$ can be stored in memory.

**Example 12.2.2.** *Let $p = 5$, so $\mathbb{F}_5 = \{\bar{0}, \bar{1}, \bar{2}, \bar{3}, \bar{4}\}$. The set of squares is given by*

$$\{x^2 \mid x \in \mathbb{F}_5\} = \{\bar{0}, \bar{1}, \bar{4}\}.$$

*In particular, $\bar{2}, \bar{3}$ are not squares in this field. Let $e_2 = \sqrt{\bar{2}}$ be a formal symbol for some element which satisfies $e_2^2 = \bar{2}$. This is a root of the polynomial $x^2 - \bar{2} = 0$.*

*The vector space $F$ over $\mathbb{F}_5$ with basis $\{e_1 = 1, e_2\}$ is 2-dimensional over $\mathbb{F}_5$. Two elements $x_1 e_1 + x_2 e_2 = x_1 + x_2 \sqrt{\bar{2}}$ and $y_1 e_1 + y_2 e_2 = y_1 + y_2 \sqrt{\bar{2}}$ are multiplied by the rule*

$$(x_1 + x_2 \sqrt{\bar{2}}) \cdot (y_1 + y_2 \sqrt{\bar{2}}) = x_1 y_1 + \bar{2} x_2 y_2 + (x_1 y_2 + y_1 x_2) \sqrt{\bar{2}}.$$

*It is a degree 2 field extension since*

$$c_{11}^1 = \bar{1}, \quad c_{11}^2 = \bar{0}, \quad c_{12}^1 = \bar{0}, \quad c_{12}^2 = \bar{1}, \quad c_{21}^1 = \bar{2}, \quad c_{21}^2 = \bar{0},$$

*and*

$$d_1^1 = \bar{1}, \quad d_1^2 = \bar{0}, \quad d_2^1 = \bar{0}, \quad d_2^2 = \bar{3}.$$

The construction used in the above example may be summarized more generally as follows:

1. Pick an element $\overline{m} \in \mathbb{F}_p$ which is not the square of another element, if such an element exists.

2. Let $e_1 = 1$ and $e_2 = \sqrt{\overline{m}}$ be a formal symbol for some element which satisfies $e_2^2 = \overline{m}$.

3. As a set, let $F = \{xe_1 + ye_2 \mid x, y \in \mathbb{F}_p\}$. To define $F$ as a field, let $+$ be 'componentwise addition' mod $p$ and let $\cdot$ be defined by

$$(x_1 + x_2\sqrt{\overline{m}}) \cdot (y_1 + y_2\sqrt{\overline{m}}) = x_1y_1 + \overline{m}x_2y_2 + (x_1y_2 + y_1x_2)\sqrt{\overline{m}}.$$

A finite field $F$ constructed in this way is called a **quadratic extension** of $\mathbb{F}_p$.

**Definition 12.2.2.** *Let $f(x)$ be a polynomial having coefficients in a field $F$. If $f(x)$ does not factor into polynomials of smaller degree having coefficients in $F$ then we call $f(x)$* **an irreducible polynomial over** *$F$. A polynomial that is not irreducible is called* **reducible over** *$F$, or we say it* **factors over** *$F$.*

More generally, let $d > 1$ be an integer.

1. Assume $g(x) = x^{d-1} + x^{d-2} + ... + x + 1$ is irreducible over $\mathbb{F}_p$. Pick an element $\overline{m} \in \mathbb{F}_p$ which is not the $k^{th}$ power of another element in $\mathbb{F}_p$ (for all $1 < k$ dividing $d$). (Since we assumed $g(x)$ was irreducible, such an element must exist.)

2. Let $e_1 = 1$, let $e_2 = \overline{m}^{1/d}$ be a formal symbol that satisfies $e_2^d = \overline{m}$, and (if $d > 2$) let $e_i = e_2^{i-2}$ for $i = 3, ..., d$.

3. As a set, let $F = \{x_1e_1 + ... + x_de_d \mid x_i \in \mathbb{F}_p\}$. To define $F$ as a field, let $+$ be 'componentwise addition' mod $p$ and let $\cdot$ be defined by expanding and collecting $(x_1e_1 + ... + x_de_d) \cdot (y_1e_1 + ... + y_de_d)$.

A finite field $F$ constructed in this way has $p^d$ elements.

What if $x^{d-1} + x^{d-2} + ... + x + 1$ was reducible? (This can happen, for example, when $p = 2$.) In this case, we must replace the formal symbol $e_2 = \overline{m}^{1/d}$ above by a root, call it $e_2$, of an arbitrary but fixed irreducible polynomial $g(x)$ of degree $d$ having coefficients in $\mathbb{F}_p$. (Such a polynomial exists but there is no simple general formula for writing one down.) Again, we let $F = \{x_1e_1 + ... + x_de_d \mid x_i \in \mathbb{F}_p\}$, where $e_i = e_2^{i-2}$ for $i = 3, ..., d$. This is a field for the same reason as before.

Let $\mathbb{F}_p[x]$ denote the collection of all polynomials having coefficients in $\mathbb{F}_p$. This set is closed under ordinary polynomial addition and multiplication. Given any fixed polynomial $g(x)$ which is not a constant, let $\mathbb{F}_p[x]/g(x)\mathbb{F}_p[x]$ denote the set of all 'cosets' $f(x) + g(x)\mathbb{F}_p[x]$. Define $+, \cdot$ on $\mathbb{F}_p[x]/g(x)\mathbb{F}_p[x]$ by $(f_1(x) + g(x)\mathbb{F}_p[x]) + (f_2(x) + g(x)\mathbb{F}_p[x]) = (f_1(x) + f_2(x)) + g(x)\mathbb{F}_p[x]$ and $(f_1(x) + g(x)\mathbb{F}_p[x]) \cdot (f_2(x) + g(x)\mathbb{F}_p[x]) = (f_1(x) \cdot f_2(x)) + g(x)\mathbb{F}_p[x]$.

**Lemma 12.2.3.** *If $g(x)$ is an irreducible polynomial in $\mathbb{F}_p[x]$ then $\mathbb{F}_p[x]/g(x)\mathbb{F}_p[x]$, which addition and multiplication defined above, is a field.*

See Artin [Ar], chapter 13, for a proof and more details on finite fields.
In fact, the field constructed above is isomorphic (as a field) to $\mathbb{F}_p[x]/g(x)\mathbb{F}_p[x]$.

## 12.3  $PGL(2, \mathbb{F}_5)$ and two faces of the cube

This section is devoted to 'determining' the **two-face group** generated by only
two basic moves, $\langle F, U \rangle$. D. Singmaster [Si] has shown that

$$\langle F, U \rangle \cong S_7 \times PGL_2(\mathbb{F}_5),$$

where $PGL_2(\mathbb{F}_5)$ is a group of order 120 which is defined below. Here $S_7$ arises
from the action of the Rubik's Cube group on the edges and $PGL_2(\mathbb{F}_5)$ arises
from the action on the corners. In this section, we focus on the action on the
corners.

### 12.3.1  Möbius transformations

**Definition 12.3.1.** *The* **projective plane**

$$\mathbb{P}^1(\mathbb{F}_p) = \{\overline{0}, \overline{1}, ..., \overline{p-1}, \infty\}$$

*is defined to be the set of lines through the origin in the Cartesian plane* $\mathbb{F}_p^2$,
*associating each number (including $\infty$) with the slope of the corresponding line.*

If $a, b, c, d \in \mathbb{F}_p$ are given numbers (not all equal to zero) then we define the
**Möbius transformation** $f$ by:

$$f : \mathbb{P}^1(\mathbb{F}_p) \to \mathbb{P}^1(\mathbb{F}_p)$$
$$x \longmapsto \frac{ax+b}{cx+d}.$$

This function is named after August Möbius (1790-1868), best known for his
work in topology, especially for his conception of the Möbius strip, a two dimensional surface with only one side. (Can't resist the riddle: What's one-sided
and swims in the sea? Möbius Dick.)

**Theorem 12.3.1.** $f$ *is a bijection if and only if* $\det \begin{pmatrix} a & b \\ c & d \end{pmatrix} \neq 0.$

Before proving this, we need the following

**Definition 12.3.2.** *Define*

$$GL(2, \mathbb{F}_5) = \{ \begin{pmatrix} a & b \\ c & d \end{pmatrix} \mid a, b, c, d \in \mathbb{F}_5, \ ad - bc \neq 0\}.$$

*This set is a group under ordinary matrix multiplication and, furthermore, acts
on the set* $\mathbb{P}^1(\mathbb{F}_5)$ *by means of Möbius transformations thus defining a function*

$$\begin{pmatrix} a & b \\ c & d \end{pmatrix} : \mathbb{P}^1(\mathbb{F}_5) \to \mathbb{P}^1(\mathbb{F}_5)$$

**Lemma 12.3.1.** *(a) The center of $GL(2, \mathbb{F}_p)$ (i.e., the subgroup of all elements which commute with every element in $GL(2, \mathbb{F}_p)$) is given by*

$$Z(GL(2, \mathbb{F}_p)) = \{ \begin{pmatrix} a & 0 \\ 0 & a \end{pmatrix} \mid a \in \mathbb{F}_p, \ a \neq 0 \}.$$

*(b) This subgroup is normal in $GL(2, \mathbb{F}_p)$.*

*(c) There is an isomorphism*

$$Z(GL(2, \mathbb{F}_p)) \cong \mathbb{F}_p^{\times}.$$

**Proof:** (a) Since

$$\begin{pmatrix} r & 0 \\ 0 & r \end{pmatrix} \begin{pmatrix} a & b \\ c & d \end{pmatrix} = \begin{pmatrix} r & 0 \\ 0 & r \end{pmatrix} \begin{pmatrix} a & b \\ c & d \end{pmatrix}$$

we can conclude that

$$\{ \begin{pmatrix} a & 0 \\ 0 & a \end{pmatrix} \mid a \in \mathbb{F}_p, \ a \neq 0 \} \subset Z(GL(2, \mathbb{F}_p)).$$

To show that

$$Z(GL(2, \mathbb{F}_p)) \subset \{ \begin{pmatrix} a & 0 \\ 0 & a \end{pmatrix} \mid a \in \mathbb{F}_p, \ a \neq 0 \},$$

assume that

$$\begin{pmatrix} r & s \\ u & v \end{pmatrix} \begin{pmatrix} a & b \\ c & d \end{pmatrix} = \begin{pmatrix} r & s \\ u & v \end{pmatrix} \begin{pmatrix} a & b \\ c & d \end{pmatrix}$$

for all $a, b, c, d$. This implies $bu = cs$ for all $b, c$. This is impossible unless $u = s = 0$. This in turn forces $cr = cv$, for all $c$. This implies $r = v$. This proves the desired inclusion.

The proof of parts (b) and (c) are left as an exercise for the reader. $\square$

**Definition 12.3.3.** *The quotient group, denoted*

$$PGL(2, \mathbb{F}_p) = GL(2, \mathbb{F}_p)/Z(GL(2, \mathbb{F}_p)),$$

*is called the **projective linear group**. (This is a group since the center is a normal subgroup by the lemma above.)*

*The quotient group, denoted*

$$PSL(2, \mathbb{F}_p) = SL(2, \mathbb{F}_p)/Z(SL(2, \mathbb{F}_p)),$$

*is called the **projective special linear group**.*

**Lemma 12.3.2.** *This group $PGL(2, \mathbb{F}_p)$ acts on the set $\mathbb{P}^1(\mathbb{F}_p)$ by means of the linear fractional transformations.*

**Remark 12.3.1.** *In fact, the action of $PGL(2, \mathbb{F}_p)$ on the set $\mathbb{P}^1(\mathbb{F}_p)$ is 3-transitive. This not hard to prove but we left it to the interested reader to look it up in [R] (see Theorem 9.48).*

**Proof:** First, we show that the $GL(2, \mathbb{F}_p)$ acts on the set $\mathbb{P}^1(\mathbb{F}_p)$ by means of the linear fractional transformations. In other words, if

$$\phi \begin{pmatrix} a & b \\ c & d \end{pmatrix} (x) = \frac{ax + b}{cx + d}$$

then

(a) $\phi \begin{pmatrix} 1 & 0 \\ 0 & 1 \end{pmatrix} (x) = x$, for all $x$ (i.e., the linear fractional transformation

$\phi \begin{pmatrix} 1 & 0 \\ 0 & 1 \end{pmatrix}$ is the identity map),

(b) $\phi(A) \circ \phi(B) = \phi(AB)$, for all $A, B \in GL(2, \mathbb{F}_p)$.

We leave the verification of (a) to the reader, if he or she wishes, and only check (b). Let $A = \begin{pmatrix} a & b \\ c & d \end{pmatrix}$ and $B = \begin{pmatrix} r & s \\ u & v \end{pmatrix}$. Then

$$AB = \begin{pmatrix} ar + bu & as + bv \\ cr + du & cs + dv \end{pmatrix},$$

so

$$\phi(AB)(x) = \frac{(ar + bu)x + as + bv}{(cr + du)x + cs + dv}.$$

On the other hand, $\phi(A)(\phi(B)(x))$ is equal to

$$\phi(A)(\frac{rx + s}{ux + v}) = \frac{a(\frac{rx+s}{ux+v}) + b}{c(\frac{rx+s}{ux+v}) + d}.$$

Simplifying this, we see that the last two displayed equations are equal. This verifies (b).

Therefore, $GL(2, \mathbb{F}_p)$ acts on the set $\mathbb{P}^1(\mathbb{F}_p)$.

Let $Z = Z(GL(2, \mathbb{F}_p))$. Since $\phi \begin{pmatrix} a & b \\ c & d \end{pmatrix} = \phi \begin{pmatrix} ra & rb \\ rc & rd \end{pmatrix}$, for all non-zero $r$, it follows that we may define an action of $PGL(2, \mathbb{F}_p)$ (still denoted by $\phi$) on the set $\mathbb{P}^1(\mathbb{F}_p)$ by $\phi(A \cdot Z) = \phi(A)$, for all $A \in GL(2, \mathbb{F}_p)$.
$\square$

**Proof:** (of Theorem 12.3.1) Let $\phi$ be as above and let $f$ be as in the statement of the theorem.

($\Leftarrow$): Since $GL(2, \mathbb{F}_p)$ acts on the set $\mathbb{P}^1(\mathbb{F}_p)$, we have $1 = \phi(AA^{-1}) = \phi(A)\phi(A^{-1})$, so $\phi(A)$ is invertible. This implies that $f$ is a bijection.

($\Rightarrow$): We prove the contrapositive. Suppose that $\det \begin{pmatrix} a & b \\ c & d \end{pmatrix} = 0$. By a result in linear algebra (see any text book, for example [JN]), the row vectors

197

of this matrix are linearly dependent. This implies that there is an $r \in \mathbb{F}_p$ such that either $(a, b) = r \cdot (c, d)$ or $(c, d) = r \cdot (a, b)$. In either case, the quotient $f(x) = \frac{ax+b}{cx+d}$ is a constant independent of $x$, so cannot be surjective. This proves that $f$ is not a bijection, which verifies the contrapositive.

$\square$

## 12.3.2    The main isomorphism

Let $G$ denote the Rubik's Cube group. Let $H$ be the subgroup generated by $F$ and $U$:

$$H = \langle F, U \rangle.$$

**Ponderable 12.3.1.** *Show the group $H$ acts on the set of vertices above (via the Rubik's Cube group).*

We describe how to label the six vertices on the 'up' and 'front' faces of the cube,

$$fru, flu, dfl, dfr, bru, blu,$$

with the elements in the projective plane

$$\mathbb{P}^1(\mathbb{F}_5) = \{\overline{0}, \overline{1}, \overline{2}, \overline{3}, \overline{4}, \infty\}$$

in a certain way. More precisely, we will label the six vertices above with elements of $\mathbb{P}^1(\mathbb{F}_5)$ in such a way that (a), (b) of the following theorem hold true.

**Theorem 12.3.2.** *There are $a_0, a_1, b_0, b_1, c_0, c_1, d_0, d_1 \in \mathbb{F}_5$ (given explicitly below) such that*

(a) *the action of $F$ (the usual rotation of the front face) on these vertices is the same as the action of some linear fractional transformation*

$$f_F(x) = \frac{a_0 x + b_0}{c_0 x + d_0}$$

(b) *the action of $U$ (the usual rotation of the up face) on these vertices is the same as the action of some linear fractional transformation*

$$f_U(x) = \frac{a_1 x + b_1}{c_1 x + d_1}$$

In other words, the basic moves $F$, $U$ may be regarded as linear fractions transformations over a finite field!

**Theorem 12.3.3.** $PGL(2, \mathbb{F}_5) = \langle f_F, f_U \rangle.$

**Remark 12.3.2.** $PGL(2, \mathbb{F}_5)$ *is isomorphic to $S_5$ (this is part of Exercise 9.25 in [R]).*

We shall prove these below.

### 12.3.3 The labeling

Label the up, front, and right vertices as below.

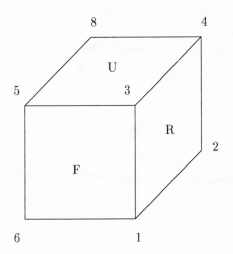

Let

$$f_F(x) = \frac{x-1}{x+1}, \qquad f_U(x) = 3x + 3.$$

The map

$$\phi : F \longmapsto f_F, \quad U \longmapsto F_U,$$

extends to a surjective homomorphism of groups

$$\phi : \langle F, U \rangle \rightarrow \langle f_F, f_U \rangle \subset PGL(2, \mathbb{F}_5).$$

**Ponderable 12.3.2.** *Verify the first theorem above.*

### 12.3.4 Proof of the second theorem

Let

$$\begin{pmatrix} a & b \\ c & d \end{pmatrix}_* \in PGL(2, \mathbb{F}_F)$$

denote the image of

$$\begin{pmatrix} a & b \\ c & d \end{pmatrix} \in GL(2, \mathbb{F}_F)$$

199

under the natural map $GL(2, \mathbb{F}_5) \to PGL(2, \mathbb{F}_5)$, $g \mapsto \mathbb{F}_5^{\times} * g$.
Since

$$f_U^2 = \begin{pmatrix} 0 & -1 \\ 1 & 0 \end{pmatrix}_*$$

we have

$$\begin{pmatrix} 0 & -1 \\ 1 & 0 \end{pmatrix}_* \in \langle f_U, f_F \rangle.$$

Since

$$f_F * f_U^5 = \begin{pmatrix} -1 & 0 \\ -1 & -1 \end{pmatrix}_*$$

it follows that

$$\begin{pmatrix} 1 & 0 \\ 1 & 1 \end{pmatrix}_* \text{ belongs to } \langle f_F, f_U \rangle.$$

Conjugating this matrix by $f_U^2$, we find that

$$\begin{pmatrix} 1 & -1 \\ 0 & 1 \end{pmatrix}_* \text{ belongs to } \langle f_F, f_U \rangle.$$

It is known that $SL(2, \mathbb{F}_5)$ is generated by elementary transvections (see [R]).
Therefore,

$$PSL(2, \mathbb{F}_5) \subset \langle f_F, f_U \rangle \subset PGL(2, \mathbb{F}_5).$$

It is also known (see [R]) that

$$|PSL(2, \mathbb{F}_5)| = 60 \quad \text{and} \quad |PGL(2, \mathbb{F}_5)| = 120.$$

It remains to show that there is an element of $\langle f_F, f_U \rangle$ which does not belong to $PSL(2, \mathbb{F}_5)$. We claim that such an element is $f_U$. Note that $\det(f_U)$ belongs to the set

$$3(\mathbb{F}_5^{\times})^2 = \{3x^2 \mid x \in \mathbb{F}_5^{\times}\}.$$

But an element of $PSL(2, \mathbb{F}_5)$ must have determinant 1. Since $3^{-1} = 2 \pmod 5$ is not a square mod 5, there is no element of $\mathbb{F}_5$ which satisfies $1 = 3x^2$. Thus $f_U$ does not belong to $PSL(2, \mathbb{F}_5)$. $\square$

**Ponderable 12.3.3.** *As an application of theorem 12.3.3, use the example in section 5.7.2 to show that there exists an embedding $D_{12} \hookrightarrow PGL(2, \mathbb{F}_5)$ of the symmetry group of the hexagon into $PGL(2, \mathbb{F}_5)$.*

## 12.4 The cross groups

Define a **cross move** of the cube to be a move of the form $X * Y^{-1}$, where $X, Y \in \{R, L, U, D, F, B\}$. The subgroup of the Rubik's Cube group generated by the cross moves will be called the **cross group**.

The cross moves permute the set $V$ of vertices of the cube and therefore generate a subgroup of $S_V$. This is called the **vertex cross group**. The cross moves permute the set $E$ of edges of the cube and therefore generate a subgroup of $S_E$. This is called the **edge cross group**.

All the enties in the following table are, as far as I am aware, new except for the $M_{12}$ entry.

| Rubik polyhedra | edge cross group | vertex cross group |
|---|---|---|
| tetrahedron | $A_5 \cong PSL_2(\mathbb{F}_5)$ | $C_2 \times C_2$ |
| cube | $A_{12}$ | $PSL_2(\mathbb{F}_7)$ |
| octahedron | $A_{12}$ | $PSL_2(\mathbb{F}_5)$ |
| dodecahedron | $A_{30}$ | $A_{20}$ |
| icosahedron | $A_{30}$ | $A_{12}$ |
| rubicon | $A_{30}$ | $M_{12}$ |

In fact, the subgroup of the dodecahedral edge cross group generated by a subset of the cross moves can yield (smaller but still simple) alternating groups.

**Ponderable 12.4.1.** *(F. Dyson) Work out the analogous cross groups of the 'Rubicized' 4-dimensional regular polyhedra.*

Freeman Dyson (1923-), who has contributed important work to both mathematics and physics, is a Professor at the Institute for Advanced Studies in Princeton, New Jersey.

Regarding Dyson's problem, we conjecture that

- if $E$ denotes the set of edges of a 4-dimensional regular polyhedra then the edge cross group is isomorphic to $A_E$,

- if $V$ denotes the set of vertices of a 4-dimensional regular polyhedra then the vertex cross group is isomorphic to $A_V$,

- if $F$ denotes the set of faces of a 4-dimensional regular polyhedra then the face cross group is isomorphic to $A_F$.

Some cases of this conjecture have been verified using the computer algebra software package GAP [Gap].

Note that these cross groups are, with only one exception so far, all simple.

**Ponderable 12.4.2.** *Are any of the analogous cross groups of the 'Rubicized' 3-dimensional Archimedian polyhedra simple?*

According to some calculations performed using GAP, it appears that the vertex cross group of the regular truncated cube is *not* simple.

## 12.4.1   $PSL(2, \mathbb{F}_7)$ and crossing the cube

Let $C$ denote the vertex cross group of the Rubik's Cube.

**Theorem 12.4.1.** $C \cong PSL_2(\mathbb{F}_7)$.

The first proof of this is by computer!

**Proof:** (number 1) GAP [Gap] gives that $C$ is a simple group of order 168. By the classification of simple groups (or, more simply, Exercise 9.26 in [R]), $C$ must be isomorphic to $PSL_2(\mathbb{F}_7)$. $\square$

The second proof is from [CD].

**Proof:** (number 2) $PSL_2(\mathbb{F}_7)$ can be generated by the three matrices:

$$f_1 = \begin{pmatrix} 0 & -1 \\ 1 & 0 \end{pmatrix}, f_2 = \begin{pmatrix} 2 & 1 \\ 0 & 1 \end{pmatrix}, f_3 = \begin{pmatrix} 2 & 0 \\ 0 & 1 \end{pmatrix}.$$

Label the vertices of the cube as below.

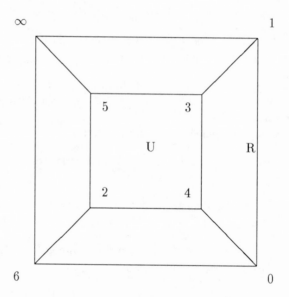

Under this labeling, we can show that

- the image of the move $m_1 = (UD^{-1})^2$ will permute the vertices in this way: $(\infty, 0)(1, 6)(2, 3)(4, 5)$, The same permutation is given by the Möbius transform $(0 \cdot x - 1)/(x + 0)$ acting on $P^1(\mathbb{F}_7)$.

- $m_2 = UR^{-1}$ gives us the permutation: $(0, 1, 3)(2, 5, 4)$, which is given by $(2x + 1)/(0 \cdot x + 1)$ acting of $P^1(\mathbb{F}_7)$.

- $m_3 = BU^{-1}LB^{-1}$ gives us the permutation $(1, 2, 4)(3, 6, 5)$, which is given by $(2x + 0)/(0 \cdot x + 1)$ acting on $P^1(\mathbb{F}_7)$.

You should notice that if the constants in these Möbius transformations $(a, b, c, d)$ are written in matrix form, they correspond to the generators of $PSL_2(\mathbb{F}_7)$. Now we will define a homomorphism $q : C \to PSL_2(\mathbb{F}_7)$, such that $q(m_1) = f_1, q(m_2) = f_2, q(m_3) = f_3$. We want to show that our q is an isomorphism.

To do this we will first show that it is surjective. Let f be a matrix in $PSL_2(\mathbb{F}_7)$, which can be written as a product of generators $f_1, f_2, f_3$ (where $q(m_1) = f_1, q(m_2) = f_2, q(m_3) = f_3$). Now take $f$ as some element of $PSL_2(\mathbb{F}_7)$. $f$ can be broken down as a product of its generators, $f_1, f_2, f_3$, we'll say

$$f = \prod_{k=1}^{n} f_{i_k}^{e_k}.$$

Since we have a homomorphism, we can write it as a product of the images of the generators of $C$. Again we can rewrite it as $f = q(\prod_{k=1}^{n} m_{i_k}^{e_k})$. Therefore $q$ is surjective.

To show that $q$ is one to one we need to know that $PSL_2(\mathbb{F}_7)$ has order 168 [R], and that the order of the cross group is also 168. (This fact was proven by computer.) We will prove by contradiction that $q$ is one to one.

Now we assume that $c_1$ and $c_2$ are elements of $C$, such that $q(c_1) = q(c_2)$, and $c_1$ is not equal to $c_2$. $|PSL_2(\mathbb{F}_7)| = |q(C)|$. We now subtract $c_2$ from $C$, and $|q(C)| = |q(C - c_2)|$ because $q(c_1) = q(c_2)$. Now we can say that $|q(C - c_2)| < or = |C - c_2|$ because we know that q is surjective.

Since we have taken $c_2$ out of $C$, we know $|C - c_2| < |C|$, which by transitivity implies $|PSL_2(\mathbb{F}_7)| < |C|$. This is a contradiction because we know $|PSL_2(\mathbb{F}_7)| = |C|$. Therefore $q$ is injective.

Now that we have shown that $q$ is both surjective and injective, it is bijective and an isomorphism. $\square$

The above proof of the theorem tells us explicitly that there exists a labeling of the vertices $V$ of the cube by the elements of the projective line

$$\mathbb{P}^1(\mathbb{F}) = \{\infty, \overline{0}, \overline{1}, \overline{2}, \overline{3}, \overline{4}, \overline{5}, \overline{6}\},$$

with the property that there is a move $c : V \to V$ in $C$ if and only if there is a Möbius transformation $f : \mathbb{P}^1(\mathbb{F}) \to \mathbb{P}^1(\mathbb{F})$ in $PSL_2(\mathbb{F}_7)$.

Because the group of Möbius transformations in $PSL_2(\mathbb{F}_7)$ acts 2-transitively on the projective line $\mathbb{P}^1(\mathbb{F})$ (see [R], Theorem 9.45), it follows that we have the following

**Corollary 12.4.1.** *C acts 2-transitively on V. In other words, for any ordered pairs* $(v_1, v_2)$, $(v'_1, v'_2)$ *of distinct vertices there is an element* $c \in C$ *sending* $v_i$ *to* $v'_i$, *for* $i = 1, 2$.

## 12.4.2 Klein's 4-group and crossing the Pyraminx

We leave the main result of this section as an exercise, actually more of a project, for the interested reader.

**Ponderable 12.4.3.** *Show that the subgroup of $S_V$ generated by the twist-untwist moves of the Pyraminx is isomorphic to the Klein 4-group $C_2 \times C_2$.*

The determination of the cross group for the Megaminx is due to J. Conway. It will be presented in chapter 14.

# Chapter 13

# Other Rubik-like puzzle groups

'The youthful impulse is rather to prowl about the problem for awhile, looking perhaps not for an open window or a door with a weak lock, for if these were available some earlier malefactor would have discovered them, but for some wall that can be scaled or some unsuspected underground access.'

*Robert Langlands*

We've spent a lot of time thinking about the Rubik's Cube and how to solve it - we scaled that wall using group theory. What about the Megaminx, or other puzzles? This chapter shall say something about the group-theoretical structure of some of the other permutation puzzle groups.

## 13.1  A uniform approach

Here we give a *uniform* discussion of the Pyraminx, the $3 \times 3$ Rubik's Cube, and the Megaminx.

**Notation:** Let

- $G_p$ (resp., $G_R$, $G_m$) denote the permutation puzzle group generated by the basic moves of the Pyraminx (resp., the Rubik's Cube, Megaminx),

- $V_p$ (resp., $V_R$, $V_m$) denote the set of vertex pieces of the Pyraminx (resp., the Rubik's Cube, Megaminx),

- $E_p$ (resp., $E_R$, $E_m$) denote the set of edge pieces of the Pyraminx (resp., the Rubik's Cube, Megaminx),

- $F_p$ (resp., $F_R$, $F_m$) denote the set of facets of the movable pieces of the Pyraminx (resp., the Rubik's Cube, Megaminx).

### 13.1.1 General remarks

Let $G, V, E, F$ (resp.) denote either $G_p, E_p, V_p, F_p$ (resp.), or $G_R, E_R, V_R, F_R$ (resp.), or $G_m, E_m, V_m, F_m$ (resp.).

**Lemma 13.1.1.** *$G$ acts on the set $V$, (resp., $E, F$).*

If $g$ is any move in $G$ then, since $g$ acts on the sets $V$, $E$, and $F$, we may regard $g$

- as an element of the symmetric group $S_V$ of $V$,

- as an element of the symmetric group $S_E$ of $E$, or

- as an element of the symmetric group $S_F$ of $F$.

These groups $S_V$, $S_E$, and $S_F$ are different, so to distinguish these three ways of regarding $g$, let us write

- $g_V$ for the element of $S_V$ corresponding to $g$,

- $g_E$ for the element of $S_E$ corresponding to $g$,

- $g_F$ for the element of $S_F$ corresponding to $g$.

What is the kernel of $f_V$? What is its image? To answer this question (actually, we shall not answer this precise question but one similar to it) we introduce a certain subgroup of the symmetric group.

Recall the alternating group $A_X$ is the subgroup of all even permutations of $X$ (in the sense of Example 5.6.2 above).

### 13.1.2 Parity conditions

Consider the function
$$f_{VE} : G \to S_V \times S_E$$
$$g \longmapsto (g_V, g_E)$$
It is easy to check that this is a homomorphism.

**Theorem 13.1.1.** *The image $f_{VE}(G)$ of $f_{VE}$ is isomorphic to*

$$\begin{cases} A_V \times A_E, & \text{for the Pyraminx, Megaminx} \\ \{(x,y) \in S_V \times S_E \mid x, y \text{ both even or both odd}\}, & \text{for Rubik's Cube.} \end{cases}$$

This is a consequence of a result proven below in the case of the Rubik's Cube. To see what this theorem means, we look at an example.

**Example 13.1.1.** *Let $G = G_R$.*

*Question: Can you find a move of the Rubik's Cube which flips a single edge subcube over, leaving the rest of the puzzle pieces unmoved?*

*If so, then the image of $f_{EV}$ would have to contain an element $(x, y)$ with $x = 1$ (since moving an edge only does not effect the vertices) and where $y$ is a 2-cycle. But $x = 1$ is even and a 2-cycle is odd. This contradicts the theorem, which says that $x, y$ are either both even or both odd. Therefore, the answer is no: a single edge flip is impossible.*

Next, some more **notation**: let

$$K = ker(f_{VE}) \lhd G$$

denote the kernel of the map $f_{VE}$ introduced above. This is a normal subgroup of G.

**Example 13.1.2.** *In the case of the Rubik's Cube, this subgroup $K$ is the set of moves which may reorient (i.e., flip or rotate) a subcube but does not swap it with some other subcube. For example, the move*

$$(R^{-1} * D^2 * R * B^{-1} * U^2 * B)^2,$$

*which twists the ufr corner clockwise and the bld corner counterclockwise, belongs to $K$.*

**Theorem 13.1.2.** *(Gold, Turner [GT]) G is a semi-direct product of $K$ with $f_{VE}(G)$.*

This is a consequence of a result proven above in the case of the Rubik's Cube. In the case of the $3 \times 3$ Rubik's Cube, some more details are given in chapter 11. See also [GT], chapter 2 of [B], or [NST], chapter 19.

# 13.2 On the group structure of the Skewb

**Notation**: We fix an orientation of the cube and label the sides by $R, L, U, D, F, B$ as in the case of the Rubik's Cube. The 120 degree clockwise rotation of a corner is denoted by a 3-letter juxtaposition of the letters abbreviating the 3 faces which the corner meets. (When you twist a corner of the Skewb you must permute three other corners but the opposite side of the Skewb is unaffected.) Such a move will be called a **basic move** - there are 8 of them, though twisting about a corner and twisting about the antipodal opposite corner is basically the same move (up to a rotation of the entire cube.) For example, FRU denotes the 120 degree clockwise rotation of the front-right-up corner, leaving the rest of the cube alone.

Let $C$ denote the set of square center facets and $V$ the set of vertices of the cube.

Let

$$G = \langle FRU, FLU, BRU, BLU, DFR, DFL, BDR, BDL \rangle$$

denote the group of all (legal) Skewb moves. Let $G^*$ denote the group of all legal and 'illegal moves' (where disassembly then reassembly is allowed).

On each square center facet of the Skewb we may choose a vertex with the following property: if we draw an arrow pointing from the chosen vertex of the square to the diametically opposite vertex on the square then the moves of the Skewb permute these arrows amongst themselves, except that some arrows may possibly be reversed. This determines an orientation of each center facet. Call this puzzle the **super Skewb**. For this new puzzle, let

$$G_{super} = \langle FRU, FLU, BRU, BLU, DFR, DFL, BDR, BDL \rangle$$

denote the group of all (legal) super Skewb moves. Let $G^*_{super}$ denote the group of all legal and 'illegal moves' (where disassembly is allowed).

We orient the corners as in the case for the Rubik's Cube. Let $y(g) \in C_3^8 = \{0, 1, 2\}^8$ denote the orientation for the corners.

For the superSkewb, we orient the center facets similarly. Let $z(g) \in C_4^6 = \{0, 1, 2, 3\}^6$ denote the orientation for the centers.

Let $S_C$ denote the symmetric group on the set $C$, $S_V$ the symmetric group on the set $V$.

Claim: There are homomorphisms

$$\rho : G \to S_C, \qquad \sigma : G \to S_V,$$

given, for each move $g \in G$, by

$$\rho(g) = \text{permutation of the center facets associated to } g,$$

and

$$\sigma(g) = \text{permutation of the vertices associated to } g.$$

Let

$$H = C_3^8 \times S_C \times S_V$$

and define $* : H \times H \to H$ by

$$(\vec{y}, r, s) * (\vec{y}', r', s') = (rr', ss', r^{-1}(\vec{y}') + \vec{y}).$$

Let

$$H_{super} = C_3^8 \times S_V \times C_4^6 \times S_C$$

and define $* : H_{super} \times H_{super} \to H_{super}$ by

$$(\vec{y}, r, \vec{z}, s) * (\vec{y}', r', \vec{z}', s') = (r^{-1}(\vec{y}') + \vec{y}, rr', s^{-1}(\vec{z}') + \vec{z}, ss').$$

Observation: There is an embedding of $G$ into $H$ and an embedding of $G_{super}$ into $H_{super}$.

Let $G_C$ be the group that acts only on the center facets of the Skewb, and $G_V$ the group that acts only on the vertices. Now,

$$G = G_C \times G_V.$$

Every generator of G is a 3-cycle on the center facets. This means that r is an element of $A_C$. It is a fact that the elements $(i, j, k)$ of $S_n$ generate $A_n$ (for $i, j, k$ elements of $\{1, 2, ..., n\}$). Therefore, $G_C = A_6$.

The group that acts on the vertices of the Skewb is slightly more complicated. Unlike the Rubik's Cube, there is no condition like conservation of twists which applies to the entire vertex set. Instead, we must split the vertices of the Skewb into two 4-corner orbits. This idea is borrowed from Bandelow's booklet on Mickey's Challenge (a puzzle similar to the Skewb). An orbit is constructed by starting with one corner and including the opposite corner of each face that meets at the first corner. Referring back to our original labeling of the Skewb, the orbits are the odd corners $\{1, 3, 5, 7\}$, and the even corners $\{2, 4, 6, 8\}$. Let the orbit of odd corners be denoted by $V(odd)$, and let $V(even)$ denote the orbit of even corners. We now partition $G_V$ so that

$$G_V = G_{V(odd)} \times G_{V(even)}.$$

We know that each orbit maps to a permutation on 4 vertices and an orientation on 4 vertices. So

$$G_V \text{ is a subgroup of } (C_3^4 \rtimes S_4) \times (C_3^4 \rtimes S_4).$$

Let $h = (s, u(h), t, v(h))$ be an element of $G_V$.

First we will examine the permutations of the vertices of both orbits. Each generator produces a 3-cycle on the vertices, whether they are in the odd or even orbit. Therefore, we can say that the permutations of each orbit generate $A_4$ by the same argument used for the center permutations. Now,

$$G_V \text{ is a subgroup of } (C_3^4 \rtimes A_4) \times (C_3^4 \rtimes A_4).$$

**Claim 1:** There exist $h$, such that $s$ is an element of $A_4$, $u(h) = (0, 0, 0, 0)$, $t$ is an element of $A_4$, and $v(h) = (0, 0, 0, 0)$. We know this is true because there are clean Skewb moves which only permute 3 vertices.

**Claim 2:** Given any permutation $u'$ of $(1, 2, 0, 0)$ by an element of $A_4$ and any permutation $v'$ of $(1, 2, 0, 0)$ by an element of $A_4$, there exist $h$, such that $s = 1$, $u(h) = u'$, $t = 1$, and $v(h) = v'$. This is true because there are clean Skewb moves which only twist vertices.

If we combine the moves of Claims 1 and 2, we should generate all of the possible moves of $G_V$. The condition on each of the vertex 4-tuples will drop them in dimension to elements of $C_3^3$. So we can conclude that $G_V$ is a subgroup of index 9 of

$$(C_3^4 \rtimes A_4) \times (C_3^4 \rtimes A_4).$$

Note: This claim is verified by GAP.

Since $G_C = A_6$, and $G_V = (C_3^3 \rtimes A_4) \times (C_3^3 \rtimes A_4)$, we can conclude that

$$G = A_6 \times (C_3^3 \rtimes A_4) \times (C_3^3 \rtimes A_4).$$

and

$$|G| = (6!/2)(4!/2)^2 3^6 = 37,791,360.$$

In conclusion, it is interesting to note that if we let $G'$ denote the illegal Skewb group - where reassembly is permitted - then

$$G' = S_6 \times S_8 \times C_3^8.$$

and

$$|G|/|G'| = .0001984127...$$

This means that if you could take apart the Skewb and reassemble it however you wanted (leaving the stickers intact however), then only about .02 percent of all possible reassemblies would be solvable. The analogous percentage for the Rubik's Cube is about 8.3 percent. As M. Schönert points out in a post to [CL], this makes the Skewb harder to solve that the Rubik's Cube, in some sense.

### Permutation and Orientation Tables

| Move | Center Permutation | Vertex Permutation |
|------|--------------------|--------------------|
| UFR | (1, 5, 2) | (2, 6, 4) |
| UFL | (1, 4 ,5) | (1, 7, 3) |
| DFR | (1, 2, 6) | (1, 5, 7) |
| DFL | (1, 6, 4) | (4, 6, 8) |
| BRU | (2, 5, 3) | (1, 3, 5) |
| BLU | (3, 5, 4) | (2, 4, 8) |
| DBR | (2, 3, 6) | (2, 8, 6) |
| DBL | (3, 4, 6) | (3, 7, 5) |

| Move | Vertex Orientation |
|------|--------------------|
| UFR | (1 2 0 2 0 2 0 0) |
| UFL | (2 0 2 1 0 0 2 0) |
| DFR | (2 0 0 0 2 1 2 0) |
| DFL | (0 0 0 2 0 2 1 2) |
| BRU | (2 1 2 0 2 0 0 0) |
| BLU | (0 2 1 2 0 0 0 2) |
| DBR | (0 2 0 0 1 2 0 2) |
| DBL | (0 0 2 0 2 0 2 1) |
| UFR*UFL | (0 2 2 2 0 0 2 0) |
| DFR*DFL | (2 0 0 2 0 0 2 2) |

Note: The orientations for the generator moves contain two repeated orbits - permutations of (1 0 0 0) and permutations of (2 2 2 0).

This section was inspired by Martin Schönert's contribution on this topic to [CL].

## 13.3  Mathematical description of the 2 × 2 cube

This section derives the group structure of the 2 × 2 Rubik's Cube.

A position on the 2 × 2 cube is determined by

(a) a permutation of the vertices, and

(b) the orientation of the corner sub-cubes.

An **illegal move** on the 2 × 2 cube is a reassembly of the corners.

Let

$$H = \langle R, L, U, D, F, B, \text{ and all the illegal moves} \rangle.$$

This will be called the **illegal 2 × 2 cube group**. Let $G = \langle R, L, U, D, F, B \rangle$. $G$ is contained in $H$.

Let $C_3^8 = \{0, 1, 2\}^8$ be the group of 8-tuples with coordinate-wise addition mod 3. Let $\vec{v} : H \to C_3^8$ be defined as follows: Assume $h \in H$ sends the $i^{th}$ corner to the $j^{th}$ corner. $v_i(h)$ is the number in $C_3 = \{0, 1, 2\}$ which describes the orientation that the standard reference marking of the $i^{th}$ corner is sent to relative to the standard reference marking of the $j^{th}$ corner. The values of $v$ are tabulated in (11.1.1).

Let $S_V$ be the group of permutation of corner sub-cubes. We may identify $S_V$ with $S_8$ since we have labeled the corners $1, ..., 8$. $H$ is a subset of the Cartesian product $S_V \times C_3^8$.

Let $p(h)$ denote the permutation of the vertices of the cube associated to $h \in H$. We have

$$(v, r) * (v', r') = (v + r(v'), r * r')$$
$$(\vec{v}(g), p(g)) * (\vec{v}(h), p(h)) = (\vec{v}(g) + p(g)\vec{v}(h), p(g) * p(h))$$
$$= (\vec{v}(g * h), p(g * h)).$$

It is not hard to show, based on the results of the previous section, that $H = C_3^8 \rtimes S_8 = \{(v, r) \mid r \in S_V, \ v \in C_3^8\}$. In other words, $H$ is the wreath product of $S_8$ and $C_3$.

**Theorem 13.3.1.** *A two-tuple $(\vec{v}, r) \in C_3^8 \times S_V$ corresponds to a legal position if and only if $v_1 + ... + v_8 \equiv 0 \pmod 3$ (conservation of twists).*

**Proof:** PART 1: In this part, we show that any pair $(v, r)$ as in the theorem (where $v$ satisfies conservation of twists) corresponds to a legal move $g$ in such a way that $r = \rho(g)$ and $v = \vec{v}(g)$.

Case 1: Assume $r = 1$ and $\vec{v}$ is arbitrary. From the solved position, any two corners, corner $i$ and corner $j$ say, can be twisted so that corner $i$ has orientation 1, corner $j$ has orientation 2, and all other corners have orientation 0. Call such a move $e_{i,j}$. Example: $(R^{-1} * D^2 * R * B^{-1} * U^2 * B)^2$ is $e_{2,7}$.

Let $y = a_1 * e_{1,8} + ... + a_7 * e_{7,8}$, where $a_i \in \{0, 1, 2\}$. This is a move of the 2 × 2 Rubik's Cube of the form $(v, 1)$ - in other words, it permutes nothing but may twist some corners. By construction, all moves of this form are legal. For each $a_i * e_{i,8}$ there are three different possible positions (independent of all other $a_j * e_{j,8}$). Since there three choices for each $a_i$, there are a total of $3^7$ distinct moves of the form $y$ as above.

On the other hand, there are exactly $3^7$ possible moves of the form $(v, 1)$ which satisfy the conservation of twists. (proof: If $\vec{v} = (v_1, ..., v_8)$, $v_i \in \{0, 1, 2\}$, and $v_1 + ... + v_8 \equiv 0 \pmod 3$, then there are 3 ways to choose each of $v_1, .., v_7$ but then once these are fixed the consevation of twists condition leaves no choice for $v_8$. This leaves a total of $3^7$ choices.) These $3^7$ possible moves include, of course, the legal moves of the form $y$ above. Thus every move of the form $(1, v)$, with $v$ satisfying conservation of twists, is legal.

Case 2: Assume $\vec{v} = \vec{0}$ and $r$ is arbitrary. Recall $S_8$ is generated by the two-cycles (see chapter 3 above, §§3.3-3.4).

Claim 1: Given any pair of corners, there is a 2-cycle move which swaps them. (Example: $F^{-1} * U * B * U^{-1} * F * U^2 * B^{-1} * U * B * U^2 * B^{-1}$). Once two corners have been swapped, you may correct the orientation of any sub-cube by Case 1. Thus any permutation which preserves orientations is a legal moves.

Case 3: Assume $\vec{v}$ and $r$ are both arbitrary but satisfying conservation of twists. By case 2, we may make a legal move that changes $(\vec{v}, r)$ to $(\vec{v}, 1)$. By case 1, $(\vec{v}, 1)$ is a legal move.

PART 2: In this part, we show that any legal move satisfies conservation of twists.

Assume $(\vec{v}, r) \in C_3^8 \times S_V$ is a legal move.

Define the **length** of a move $g \in G$ to be the smallest number $n$ of generators needed to create the move, written $length(g) = n$.

Induction hypothesis: If a move is length $n$, it satisfies conservation of twists.

step $n = 1$: Every $\vec{v}(x)$ where $x \in \{R, L, U, D, F, B\}$ satisfies the conservation of twists.

step $n > 1$: Assume the induction hypothesis is true for all lengths $\leq n-1$. Let $x$ be length $n$ and write $x = x_1 * x_2$, where $length(x_1) \leq n-1$ and $length(x_2) = 1$. Then $\vec{v}(x) = \vec{v}(x_1) + p(x_1)\vec{v}(x_2)$, by the group operation. Furthermore, $v(x_1)$ satisfies conservation of twists by the induction hypothesis. Since $p(x_1)$ simply permutes the coordinates of $\vec{v}(x_2)$, $\vec{v}(x_2)$ still satisfies the conservation of twists. The sum of moves satisfying conservation of twists still satisfies conservation of twists.

Conclusion: by induction, any move $(v, r) \in C_3^8 \times S_V$ that is a legal move satisfies the conservation of twists. $\square$

## 13.4 On the group structure of the Pyraminx

**Notation**: Let

- $V$ denote the vertices of the tetrahedron (which we identify with the set of corner pieces of the Pyraminx),

- $E$ denote the edges of the tetrahedron (which we identify with the set of edge pieces of the Pyraminx),

- $C$ the set of interior pieces of the tetrahedon (ie, movable pieces of the Pyraminx not in $E$ or $V$),

- $S_V$ the permutation group of $V$,

- $A_V$ the alternating group of $V$,

- $S_E$ the permutation group of $E$,

- $A_E$ the alternating group of $E$.

Assume that the tetrahedron is lying on a flat surface in front of you, with the triangle base pointing away from you. The corners are denoted L (left), R (right), U (up), and B (back).

**Basic Moves**: Opposite each corner or vertex there are three **layers**: the **tip**, the **middle layer**, and the **opposite face**. Let

- $l$ denote the 120 degree clockwise rotation of the tip containing the left corner,

- $L$ denote the 120 degree clockwise rotation of the tip/middle layer containing the left corner,

- $r$ denote the 120 degree clockwise rotation of the tip containing the right corner,

- $R$ denote the 120 degree clockwise rotation of the tip/middle layer containing the right corner,

- $u$ denote the 120 degree clockwise rotation of the tip containing the up corner,

- $U$ denote the 120 degree clockwise rotation of the tip/middle layer containing the up corner,

- $b$ denote the 120 degree clockwise rotation of the tip containing the back corner,

- $B$ denote the 120 degree clockwise rotation of the tip/middle layer containing the back corner.

Let $G = \langle R, L, U, B, r, l, u, b \rangle$ denote the **Pyraminx group**.

Each move $g \in G$ induces a permutation of $E$ denoted $\sigma(g)$. Note that $G$ does not permute the vertices. Furthermore, the tip moves $r, l, u, b$ do not effect the edges.

**Lemma 13.4.1.** $\sigma : G \to S_E$ *is a group homomorphism.*

**Example 13.4.1.** $\rho(L)$ *is a 3-cycle in* $S_V$, $\sigma(L)$ *is a 3-cycle in* $S_E$.

## 13.4.1 Orientations

Assume for the moment that the Pyraminx is fixed in space as above and is in the 'solved' position. For each corner or edge piece, choose once and for all one facet on that piece. There are three possible choices for each corner piece and two for the edges. Mark each of these chosen facets with an imaginary '+', leaving the other facets unmarked. For the rest of this section, we shall make the following choices for the marked facets (with reference to the numbering in §4.4 ):

- marked edge facets: $4, 6, 10, 15, 20, 25$

- marked corner facets: $1, 13, 17, 23$

For each edge piece, assign to a move $g \in G$ either

- a '0' if the '+ facet' for that piece when it was in the solved position is sent to the '+ facet' for that piece when it was in the present position,

- a '1' otherwise,

This yields a 6-tuple of 0's and 1's: $\vec{w}(g) = (w_1, w_2, ..., w_6)$.

**Example 13.4.2.** *We compute the effect of the basic twist moves on the edge orientations:*

| $X$ | $\vec{w}(X)$ |
|---|---|
| $B$ | $(0,0,0,1,0,1)$ |
| $R$ | $(0,0,1,0,1,0)$ |
| $L$ | $(1,0,1,0,0,0)$ |
| $U$ | $(0,1,0,1,0,0)$ |
| $R * U^{-1} * R^{-1} * U$ | $(1,1,0,0,0,0)$ |

For each corner piece, assign to a move $g \in G$ either

- a '0' if the '+ facet' for that piece when it was in the solved position is sent to the '+ facet' for that piece in the present position,

- a '1' if the '+ facet' for that piece when it was in the solved position is sent to the facet which is a 120 degrees rotation about its vertex from the '+ facet' for that piece in the present position,

- a '2' otherwise, thus yielding a 4-tuple of 0's, 1's, and 2's: $\vec{v}(g) = (v_1, v_2, v_3, v_4)$.

**Example 13.4.3.** *We compute the effect of the basic twist moves on the corner orientations:*

| $X$ | $\vec{v}(X)$ |
|---|---|
| $B$ | $(0,0,0,2)$ |
| $R$ | $(0,0,2,0)$ |
| $L$ | $(0,2,0,0)$ |
| $U$ | $(2,0,0,0)$ |

**Proposition 13.4.1.** *If $\vec{w}(g) = (w_1, w_2, ..., w_6)$ corresponds to a move $g \in G$ then*

$$w_1 + w_2 + ... + w_6 \equiv 0 \pmod{2}.$$

**Observation:** There is no corresponding condition for the $v_1$, ..., $v_4$, since corner moves move them around freely.

**Proof:** The proof uses the following lemma, but is otherwise essentially the same as the corresponding fact (Theorem 11.2.1 (c)) which we proved for the Rubik's Cube. The modifications required for the proof are left to the student as an exercise to test their understanding of the argument. □

214

**Lemma 13.4.2.** *For $g, h \in G$, we have*

$$\vec{w}(g * h) = \sigma(g)^{-1}(\vec{w}(h)) + \vec{w}(g).$$

**Proof:** The argument for the proof is essentially the same as the corresponding fact (Lemma 11.1.1) which we proved for the Rubik's Cube. The modifications required for the proof are left to the interested reader. □

**Lemma 13.4.3.** *For $g, h \in G$, we have*

$$\vec{v}(g * h) = \vec{v}(h) + \vec{v}(g).$$

**Ponderable 13.4.1.** *Try to verify this lemma.*

Let $H$ denote the **illegal Pyraminx group** generated by $G$ and the 'illegal edge moves' (that is, one may physically remove the edge pieces and reassemble the Pyraminx. Illegal center or corner moves are not allowed in $H$. Let

$$H^* = \{(s, v, w) \mid r \in S_V, s \in S_E, v \in C_3^4, w \in C_2^6\}$$

and define $* : H^* \times H^* \to H^*$ by

$$(s, v, w) * (s', v', w') = (s * s', v + v', s'(w) + w').$$

**Theorem 13.4.1.** • *$H^*$ is, with this operation, a group.*

- *There is an isomorphism*

$$H \cong H^* \cong C_3^4 \times (C_2^6 \rtimes S_E),$$

*and hence between $H$ and the direct product of the tip moves $C_3^4$ and the wreath product*

$$C_3^4 \times (S_E \text{ wr } C_2).$$

- *The map $G \to H^*$ defined by*

$$g \longmapsto (\sigma(g), v(g), w(g)),$$

*is a homomorphism.*

## 13.4.2 Center pieces

Each corner piece has 3 center pieces neighboring it.
Facts:

- The center pieces associated to a corner can never be moved into a middle layer associated to another corner.

- The center pieces associated to a corner can always be color-aligned with the colors of the corner piece by a corner twist move.

The last fact says, in other words, that the center pieces can always be 'solved' by a corner piece.

### 13.4.3  The group structure

**Theorem 13.4.2.** *G is isomorphic to*

$$\{(s, v, w) \in H^* \mid s \text{ even}, w_1 + w_2 + \ldots + w_6 \equiv 0 (\text{mod } 2)\}.$$

**Proof:** The idea to prove this is to show that

- $A_E = \langle \sigma(R), \sigma(L), \sigma(U), \sigma(B) \rangle$,

- the map

$$\begin{aligned} G &\to A_E, \\ g &\longmapsto \sigma(g), \end{aligned}$$

  is surjective, and the map

- $G \to \{(w_1, \ldots, w_6) \in C_2^6 \mid w_1 + \ldots + w_6 \equiv 0(\text{mod } 2)\}$, $g \longmapsto \vec{w}(g)$, is surjective (as a map of sets).

Here's the proof of the first point: We can label (as in §5.10 ) the edges $1, 2, \ldots, 6$ so that the edges on the front face are $1, 2, 3$, resp., where the $fl$ edge is 1, the $fr$ edge is 2, and the $fd$ edge is 3 (here $f$, $r$, $l$, and $d$ denote the front face, right face, left face, and down face, resp.). The move $[R, U^{-1}] = R * U^{-1} * R^{-1} * U$ is the counterclockwise cyclic permutation $(1, 3, 2)$ of the edges on the $f$ face. (This move does not affect any corners but does flip some edges, a fact which we may ignore for now since we are only concerned with the permutations now.) In particular, $(1, 3, 2)$ may be written as a product of the generators in $\{\sigma(R), \sigma(L), \sigma(U), \sigma(B)\}$. Now pick any $i \in \{4, 5, 6\}$ and let $s \in G$ denote a move which sends edge $i$ to edge 2 and does not move edge 1 or 3. The move $s * [R, U^{-1}] * s^{-1}$ is the 3-cycle $(1, 3, i)$. It does not affect any corners or other edges. By Lemma 9.4.4, these permutations generate $A_6 \cong A_E$.

The second point follows immediately from the first point proven above.

Here's the proof of the last point: The move $g = [R, U^{-1}]$ has the following effect on the orientation: $w(g) = (1, 1, 0, 0, 0, 0)$. The group

$$\{(w_1, \ldots, w_6) \in C_2^6 \mid w_1 + \ldots + w_6 \equiv 0(\text{mod } 2)\} \cong C_2^5$$

is a vector space over $\mathbb{F}_2$. The 5 vectors listed in the table for the values for $w$ are all independent. It follows from this and the group law for $G$ proves that the map $g \longmapsto \vec{w}(g)$ is surjective.

The theorem 13.4.2 above is thus proven. $\square$

## 13.5  The homotopy group of the Square 1 puzzle

Here we study the group theoretic properties of the collection $G$ of all 'words' in the basic moves of the Square 1 puzzle which preserve the cube shape. This collection $G$ forms a group which, motivated by Wilson [W], we call the **homotopy group of the Square 1 puzzle**. The list of shapes which the Square 1 puzzle can make is given in [Sn2]. It is not hard to see that the homotopy group of any one of these other shapes is conjugate to $G$.

A bi-product of the proof is an collection of moves which can be used to solve the Square 1 puzzle, once it is put in the cube shape.

## 13.5.1 The main result

Let $S_n$ denote the symmetric group of degree $n$, i.e., the group of permutations of $\{1, 2, ..., n\}$. Let $sgn : S_n \to \{\pm 1\}$ denote the homomorphism which assigns to each permutation its sign, as in Definition 3.1.1 (the sign of a cyclic permutation of length $r$ is $(-1)^{r+1}$, for example).

We shall see that the size of the homotopy group of the Square 1 is about .8 billion.

**Theorem 13.5.1.** *$G$ is isomorphic to the kernel of index 2 in $S_8 \times S_8$ of the homomorphism $f : S_8 \times S_8 \to \{\pm 1\}$ defined by $f(g_1, g_2) = sgn(g_1)sgn(g_2)$. Consequently, $|G| = 2^{13}3^4 5^2 7^2 = 812851200$.*

As a corollary of the proof of this theorem, given below, we shall see that any even permutation of the corners is possible and any even permutation of the wedges is possible.

Let $H$ denote the **illegal Square 1 group** generated by all legal moves preserving the cube shape and all illegal moves (i.e., disassembly and reassembly is allowed) preserving the cube shape. It is clear that

$$H \cong S_8 \times S_8.$$

## 13.5.2 Some notation

We shall assume that the puzzle is in the solved position with the 'Square 1' side in front, right-side up. Let

- $u$ denote rotation of the up face by $30^o$ clockwise,

- $d$ denote rotation of the up face by $30^o$ clockwise,

- $R$ denote rotation of the cube by $180^o$ though one of the skew-diagonal cuts (in a given position, at most one such move is possible, so this is unambiguous).

Like the 15 Puzzle, and unlike the Rubik's Cube, not any sequence of $u, d$, and $R$'s is possible.
Let

$$T(x, y) = u * R * x * y * R * u^{-1}, B(x, y) = d^{-1} * R * x * y * R * d,$$

where $x, y$ are moves of the Square 1 puzzle.

back

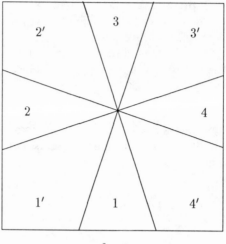

front

The up facets of the 'Square 1' puzzle.

front

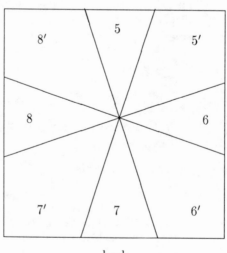

back

The down facets of the 'Square 1' puzzle.

In the notation of these diagrams, we have

$$uRu^{-1}d^{-1}Rd = (2,8)(4,6)$$
$$T(u^3,1) = (1',6',7',4')(1,6,7,4)$$
$$T(1,d^3) = (2',3',8',5')(2,3,8,5)$$
$$B(u^3,1) = (1,2,7,8)(1',6',7',4')$$
$$B(1,d^3) = (3,4,5,6)(2',3',8',5').$$

### 13.5.3 Two subgroups

Let

$$G_u = \langle T(u^3,1), T(1,d^3) \rangle$$

and

$$G_d = \langle B(u^3,1), B(1,d^3) \rangle$$

**Lemma 13.5.1.** $G_u$ and $G_d$ are each isomorphic to $C_4 \times C_4$.

**Proof:** We have $T(u^3,1)T(1,d^3) = T(1,d^3)T(u^3,1)$. Moreover, $T(u^3,1)$ and $T(1,d^3)$ are each of order 4. Since

$$C_4 \times C_4 = \langle a,b \mid a^4 = 1, b^4 = 1, ab = ba \rangle,$$

the lemma follows. $\square$

The homotopy group of the Square 1 puzzle is defined to be

$$G = \langle d^3, u^3, B(u^3,1), B(1,d^3), T(u^3,1), T(1,d^3) \rangle$$

We shall use the following labelings to describe the moves of the Square 1 puzzle

### 13.5.4 Proof of the theorem

We shall prove the theorem in the following steps:

- Show that the wedge 3-cycle $(1,2,3)$ and the corner 3-cycle $(1',2',3')$ each belong to $G$.

- Show that any wedge 3-cycle $(1,2,i)$ and each corner 3-cycle $(1',2',i')$ belong to $G$.

- Show that there is a injective homomorphism $\phi : G \to S_8 \times S_8$ where the image $\phi(G)$ contains $A_8 \times A_8$.

- Conclude that $G \cong S_8 \times S_8/\{\pm 1\}$.

Step 1: First, we claim that $(1,2,3)$ belongs to $G$. In fact, the 3-cycle $(1,2,3)$ is obtained from the move $M_1$ defined to be:

$$(B(u^3,1)*d^3)*((B(u^3,1)*d^{-3})*(B(u^{-3},1)*T(1,d^{-3})*d^6)))^4*(B(u^3,1)*d^3)^{-1}.$$

219

(Incidently, this 80 move long manuever may be verified using GAP [Gap]. See also [Sn2].)

Next, we claim that $(1', 2', 3')$ belongs to $G$. In fact,

$$M_2 = Ru^3 Rd^{-3} Ru^3 (Ru^{-3})^2 d^3 Ru^{-3}$$

is the product of 2-cycles $(2', 3')(3, 4)$. (This move was found in [Sn2].) Therefore, $u^3 M_2 u^{-3}$ is the product of 2-cycles $(1', 2')(2, 3)$. The product of these is $(1', 2', 3')(2, 3, 4)$. Since $(2, 3, 4)$ is obtained from $u^{-3} M_1 u^3$, we see that $(1', 2', 3')$ is in $G$. (This may also be verified using GAP.)

Step 2: Let $g$ be any move in $G$ which sends wedges 3 to wedge $i$, resp., and does not move wedges $1, 2$ (it may permute other wedges and corners). Then $(1, 2, i) = g * (1, 2, 3) * g^{-1}$. Thus $(1, 2, i) \in G$.

The proof that each $(1', 2', i') \in G$ is similar.

Step 3: It is clear from our definition that there is an injection $G \to S_8 \times S_8$ as sets. The verification that this is a homomorphism is straightforward.

Step 4: The group $A_8$ is generated by the 3-cycles $(1, 2, i)$ (see Lemma 9.4.4). Since these all belong to $G$, all even wedge permutations are possible. Similarly, all even corner permutations are possible. Thus $A_8 \times A_8 \subset G$.

Let $p_1 : S_8 \times S_8 \to S_8$ denote the projection onto the first factor. Let $p_2$ denote the projection onto the second factor. For each generator $g \in \{d^3, u^3, B(u^3, 1), B(1, d^3), T(u^3, 1), T(1, d^3)\}$ of $G$ we have $sgn(p_1(g)) = sgn(p_2(g))$. Thus the image $\phi(G)$ is strictly contained in $S_8 \times S_8$. In fact, this shows that $\phi(G)$ is contained in the kernel $ker(f)$ of the homomorphism $f : S_8 \times S_8 \to \{\pm 1\}$ defined in the statement of the theorem. Since

$$A_8 \times A_8 \subset G \subset ker(f),$$

$[ker(f) : A_8 \times A_8] = 2$, and $T(u^3, 1) \notin A_8 \times A_8$, the theorem follows. $\square$

## 13.6  Hockeypuck groups

Recall the basic moves of the Hockeypuck puzzle: For the $i^{th}$ radial segment, let $f_i$ be the transformation which flips over that half of the circle which contains the $i^{th}$ pie piece. Let $r$ denote the rotation of this circle by $30^\circ$ radians counterclockwise, so $r = (1, 2, ..., 12)$.

### 13.6.1  A more general puzzle

Consider a more general puzzle, made as follows.

Slice a disk with n diametrical cuts into 2n congruent 'pie pieces'. Color the side of the circle facing up red and the other side blue. You may make the following moves with this puzzle:

- you may flip over any half of the disk bounded by a cut, leaving the other half unmoved (of course this will change the colors of those pie slices which were flipped);

- you may rotate the disk by any multiple of $\pi/n$.

Label the pie pieces $1, 2, ..., 2n$ counterclockwise (as viewed from the front) in some fixed way and label the $2n$ radial segments extending away from the center with the same number as the region counterclockwise from it. (Thus for each pie piece the front (red) side and back (blue) side share the same label.)

For the $i^{th}$ radial segment, let $f_i$ be the transformation which flips over that half of the circle which contains the $i^{th}$ pie piece. Let $r$ denote the rotation of this circle by $\pi/n$ radians counterclockwise, so $r = (1, 2, ..., 2n)$. Let

$$F_0 = \langle f_1, ..., f_{2n} \rangle$$

and let

$$F_1 = \langle F_0, r \rangle.$$

We shall determine $F_1$.

**Lemma 13.6.1.** $F_1 = S_{2n}$.

**Proof:** Clearly, $F_1 \subset S_{2n}$. We have

$$(1, 2n) = r^{-1}(f_1 ... f_n)^2, \qquad r^j(1, 2n)r^{-j} = (2n - j, 2n + 1 - j),$$

for $j = 1, 2, ..., 2n - 1$. Since $S_{2n}$ is generated by the 2-cycles, this implies $S_{2n} \subset F_1$. $\square$

## 13.7 The Masterball group

Recall the notation from §4.3. The Masterball group,

$$G = \langle f_1, f_2, f_3, f_4, f_5, f_6, f_7, f_8, r_1, r_2, r_3, r_4 \rangle,$$

is a subset of $S_{32}$. We can provide a simple proof to show that $G$ is a subset of $S_{32}$, by showing the following result.

**Theorem 13.7.1.** $G = S_{16} \times S_{16}$.

**Proof:** Since the Masterball is made up of 32 facets, the permutation group $G$, on these facets, must be a subset of $S_{32}$. If we agree that we can take the ball apart and put the pieces back in any position, we have generated $S_{32}$, saying that any swap is legal. However, as stated before equator facets can only be swapped with equator facets and polar facets can only be swapped with polar facets. Since there are sixteen polar facets, the group of polar swaps, $G_p$ is a subset of $S_{16}$. Similarly, since there are 16 equatorial facets, the group of equatorial swaps, $G_e$ is a subset of $S_{16}$. To prove that $G_p$ and $G_e$ do each generate $S_{16}$, we need only to recall Steinhaus' Theorem 3.4.1.

If then we can provide a two cycle which swaps any two polar facets and a two cycle which swaps any two equatorial facets (neglecting set up moves), we have shown that $G_p$ and Ge, both products of two cycles, have generated $S_{16}$. The following polar swap can be represented by the two cycle, $(18, 48)$:

$$r_4 r_1 f_1 r_1 f_1 r_4 f_1 r_1^{-1} f_1 r_4^{-2} f_1 r_1 f_1 r_4 f_1 r_1^{-1} f_1 r_4^4 f_1 r_1 f_1 r_4 \times$$
$$\times f_1 r_1^{-1} f_1 r_4^{-2} f_1 r_1 f_1 r_4^2 f_1 r_1^{-1} f_1 r_4^2 f_1 r_1 f_1 r_4^{-1} f_1 r_1^{-1} f_1 r_4^{-1} f_1 \times$$
$$\times r_1 f_1 r_4 f_1 r_1^{-1} f_1 r_1 r_4^{-2} f_1 r_1^{-1} f_1 r_4^{-1} f_1 r_1 f_1 r_1^{-1} r_4 f_1 r_1 f_1 r_1^{-1} \times$$
$$\times f_1 r_1 f_1 r_4^2 f_1 r_1^{-1} f_1 r_1 f_1 r_1^{-1} f_1 r_1 r_4^{-1} f_1 r_1^{-1} f_1 r_4 f_1 r_1 f_1 r_1^{-2} r_4^{-2}.$$

The following equator swap can be represented by the two cycle, $(28, 38)$:

$$r_3 r_2 f_1 r_2 f_1 r_3 f_1 r_2^{-1} f_1 r_3^{-2} f_1 r_2 f_1 r_3 f_1 r_2^{-1} f_1 r_3^4 f_1 r_2 f_1 r_3 f_1 r_2^{-1} \times$$
$$\times f_1 r_3^{-2} f_1 r_2 f_1 r_3^2 f_1 r_2^{-1} f_1 r_3^2 f_1 r_2 f_1 r_3^{-1} f_1 r_2^{-1} f_1 r_3^{-1} f_1 r_2 f_1 r_3 \times$$
$$\times f_1 r_2^{-1} f_1 r_2 r_3^{-2} f_1 r_2^{-1} f_1 r_3^{-1} f_1 r_2 f_1 r_2^{-1} r_3 f_1 r_2 f_1 r_2^{-1} f_1 r_2 f_1 r_3^2 \times$$
$$\times f_1 r_2^{-1} f_1 r_2 f_1 r_2^{-1} f_1 r_2 r_3^{-1} f_1 r_2^{-1} f_1 r_3 f_1 r_2 f_1 r_2^{-2} r_3^{-2}.$$

Since any polar facet can be placed in the 18 and 48 positions, and any equator facet can be placed in the 28 and 38 positions (with the aid of set up moves) we have shown that $G_p$ and $G_e$ can be written as products of two-cycles and therefore generate $S_{16}$. Then it follows, since $G = G_e \times G_p$, then $G = S_{16} \times S_{16}$ and is therefore a subgroup of $S_{32}$. $\square$

# Chapter 14

# Crossing the Rubicon

'Mathematical structures are among the most beautiful discoveries by the human mind. The best of these discoveries have tremendous metaphorical and explanatory power.'
*Douglas Hofstadter,* **Metamathematical themas, 1985**

The title of this chapter is, however, 'borrowed' from a similarly worded title of an article by D. Hofstadter [H].

This chapter shall be a little more advanced than some of the others. The reader will be assumed to be familiar with some topics covered in a course in linear algebra and elementary number theory or coding theory. We shall also assume some results from Rotman [R], though for the understanding of the material in this chapter the reader may simply take them on faith.

Let $g_1, ..., g_{12}$ denote the basic moves of the Rubik isocahedron. A surprising result of Conway states that the group generated by $g_i * g_j^{-1}$ is the simple 'sporadic' group $M_{12}$. (This is stated more precisely below.) We shall describe, in this chapter, what $M_{12}$ is and some of its remarkable properties. They form a basis for my opinion, which I hope you will agree with, that $M_{12}$ is one of the most interesting objects in mathematics.

Let $p$ be a prime unless otherwise stated and let $q$ be a power of $p$. **We shall assume that $p > 3$.**

## 14.1   Doing the Mongean shuffle

Consider a deck of 12 cards labeled 0, 1, ..., 11. Let $r, s$ be the permutations

$$r(t) = 11 - t, \qquad s(t) = min(2t, 23 - 2t).$$

The permutation $r$ that reverses the cards around and the permutation $s$ is called the 'Mongean shuffle', named after Gaspard Monge (1746-1818), a French mathematician known mostly for his work in geometry. Monge also worked on the calculus of variations, partial differential equations, and combinatorics. To perform the reverse shuffle, simply take a stack of cards (face down, say) in your left hand and put them in your right hand one-at-a-time (face down). To perform the Mongean shuffle, take the same stack of cards and, one-at-a-time, put them alternately into one of two piles: the first card face up into the first pile, the second card face down into the second pile, the third card face up into the first pile, the fourth card face down into the second pile, and so on until the pile is exhausted. Now pick up the first pile of face up cards, flip the entire pile over so that they are all face down and put it on top of the second pile.

| cards | reverse shuffle | Mongean shuffle |
|-------|-----------------|-----------------|
| 0     | 11              | 0               |
| 1     | 10              | 2               |
| 2     | 9               | 4               |
| 3     | 8               | 6               |
| 4     | 7               | 8               |
| 5     | 6               | 10              |
| 6     | 5               | 11              |
| 7     | 4               | 9               |
| 8     | 3               | 7               |
| 9     | 2               | 5               |
| 10    | 1               | 3               |
| 11    | 0               | 1               |

**Definition 14.1.1.** *The* **Mathieu group** $M_{12}$ *is defined to be the permutation group* $M_{12} = <r, s> < S_{12}$.

Emile Mathieu (1835-1890) actually discovered five 'sporadic' simple groups, of which $M_{12}$ is only one, in his PhD thesis. (Most simple groups belong to an infinite family of simple groups - 'sporadic' simply means that it is not from one of these infinite families.) Mathieu's main work was in mathematical physics and special (hypergeometric) functions.

## 14.2  Background on $PSL_2$

We need a few basic facts about the projective special linear group of degree 2. We have already discussed the related group $GL(2, \mathbb{F}_p)$ in the previous chapter, so we refer to there for more details.

Möbius transformations, introduced in §12.3.1 are bijections from the projective line to itself, so we may interpret each Möbius transformation as an element of $S_X$ , where $X = \mathbb{P}^1(\mathbb{F}_q)$ (and therefore also of $S_n$, where $n = |\mathbb{P}^1(\mathbb{F}_q)| = q+1$).

**Example 14.2.1.** *Let* $p = 11$ *and let* $f(x) = -1/x$. *Then*

| $x$    | $\infty$ | $0$      | $1$  | $2$ | $3$ | $4$ | $5$ | $6$ | $7$ | $8$ | $9$ | $10$ |
|--------|----------|----------|------|-----|-----|-----|-----|-----|-----|-----|-----|------|
| $f(x)$ | $0$      | $\infty$ | $10$ | $5$ | $7$ | $8$ | $2$ | $9$ | $3$ | $4$ | $6$ | $1$  |

224

*Therefore, as a permutation,* $f = (\infty, 0)(1, 10)(2, 5)(3, 7)(4, 8)(6, 9).$

The following facts are known about the projective special linear group.

**Theorem 14.2.1.** *If $q > 3$ then $PSL_2(\mathbb{F}_q)$ is a simple group. Moreover, for all prime powers $q$,*
$$|PSL_2(\mathbb{F}_q)| = (q^2 - 1)q/gcd(2, q - 1).$$

(Recall a simple group was defined in Definition 9.5.1.) This theorem is over 100 years old. It is proven, for example, in [R].

**Theorem 14.2.2.** *Choose a $k \in \mathbb{F}_q$ such that $< k >= \mathbb{F}_q^\times$. Let*

$$f_1(x) = x + 1, \quad f_2(x) = k \cdot x, \quad f_3(x) = -1/x.$$

*Then $PSL_2(\mathbb{F}_q)$ is generated by $f_1, f_2$, and $f_3$. In particular, the action of $PSL_2(\mathbb{F}_q)$ on the projective line $X = \mathbb{P}^1(\mathbb{F}_q)$ yields an injective homomorphism $PSL_2(\mathbb{F}_q) \to S_X$.*

Basically, this is proven in [R] as well.

## 14.3   Galois' last dream

Supposedly, the night before he died in a duel, Galois wrote a letter to a friend stating the following remarkable theorem:

**Theorem 14.3.1.** *(Galois) Assume $p > 11$. Then $PSL_2(\mathbb{F}_p)$ has no embedding into a symmetric group $S_n$ with $n \leq p$.*

The following isomorphisms (for $q \leq 11$) are known:

$$PSL_2(\mathbb{F}_q) \cong \begin{cases} A_4, & q = 3, \\ A_5, & q = 5, \\ A_6, & q = 9. \end{cases}$$

If $p = 7$ or $p = 11$ then explicit embeddings of $PSL_2(\mathbb{F}_p)$ into $A_8$ ($p = 7$), $A_{12}$ ($p = 11$) are known (see [CS], ch 10, or [K] for an excellent discussion of this).

## 14.4   The $M_{12}$ generation

One of the most amazing aspects about $M_{12}$ is its close relationship with other 'interesting' groups.

**Definition 14.4.1.** *Define the permutation $f_4$ of the set $\mathbb{P}^1(\mathbb{F}_p) = \{\infty, 0, 1, ..., p-1\}$, for $3 \leq p \leq 11$, as follows*

$$f_4 = \begin{cases} 1, & p = 3, \\ (1, 2)(3, 4), & p = 5, \\ (1, 2)(3, 6), & p = 7, \\ (2, 10)(3, 4)(5, 9)(6, 7), & p = 11. \end{cases}$$

We have run across the group $S_6$ before, when studying the symmetries of the icosahedron. We have also seen that $S_6$ is rather an interesting group because it is the only non-abelian symmetric group $S_n$ which has an outer automorphism. One rather interesting connection between $M_{12}$ and $S_6$ is given by the following

**Theorem 14.4.1.** *(a) If $p = 5$ then $S_6 =< f_1, f_2, f_3, f_4 >$.*
*(b) If $p = 11$ then $M_{12} =< f_1, f_2, f_3, f_4 >$.*

We shall see another interpretation of $M_{12}$ below using coding theory!

**Definition 14.4.2.** *Let*

$$\delta(x) = \begin{cases} x^3/9, & x \in (\mathbb{F}_{23})^2 - 0, \\ 9x^3, & x \in \mathbb{P}^1(\mathbb{F}_{23}) - (\mathbb{F}_{23})^2. \end{cases}$$

*This is an element of $S_X$, where $X = \mathbb{P}^1(\mathbb{F}_{23})$. Define the* **Mathieu group** $M_{24}$ *by*

$$< f, \delta \mid f \in PSL_2(\mathbb{F}_{23}) > .$$

*This is a permutation group in $S_X$, where $X = \mathbb{P}^1(\mathbb{F}_{23})$.*

By the way,
(a) $(\mathbb{F}_{23})^2 = \{0, 1, 2, 3, 4, 6, 8, 9, 12, 13, 16, 18\}$,
(b) $M_{12} = 8 \cdot 9 \cdot 10 \cdot 11 \cdot 12 = 95040$,
(c) $|M_{24}| = 244823040$.

# 14.5 Coding the Golay way

Codes are used in everyday life, from ISBN numbers on books to barcodes on food products to music CDs to satellite transmissions. There are many types of codes, some more efficient than others, some with better error correcting ability than others, some more practical than others, and so on. We shall concern ourselves only with aspects which are related to (in one way or another) permutation puzzles.

Let $q = p^k$ be a prime power.

**Definition 14.5.1.** *A* **q-ary code** *is a subset $C$ of a finite dimensional vector space $V$ over the finite field $\mathbb{F}_q$. A* **code word** *is an element of $C$. The number of coordinates (i.e., the dimension of $V$) is called the* **length** *of the code word. If $q = 2$ then the code is called* **binary** *(instead of 2-ary) and if $q = 3$ then the code is called* **ternary** *(instead of 3-ary).*

A code $C$ which is a vector space over $\mathbb{F}_q$ is called a **linear code**.

Two linear codes $C, C' \subset \mathbb{F}_q^n$ are **equivalent** if there is a bijection $F; C \to C'$ which is obtained by composing a permutation of the positions of the code words $c = (c_1, c_2, ..., c_n)$ (via an element $p$ of $S_n$ sending $c$ to $(c_{p(1)}, c_{p(2)}, ..., c_{p(n)})$) with a map of the form $c = (c_1, c_2, ..., c_n) \longmapsto c' = (a_1 c_1, a_2 c_2, ..., a_n c_n)$, for some $a_1, ..., a_n \in \mathbb{F}_q^\times$.

Imagine we have a long message to send: 'Hi Mom! I just learned how to solve the Rubik's Cube ...' First, chop this message into blocks of a specific size, say each block contains $k$ consecutive letters. (This depends on the code you want to use.) Next, each letter, space and punctuation mark can be translated into a number. (It doesn't matter how.) There are only finitely many letters and punctuation marks, so we can choose $q$ larger than every number arising this way. Now associate to each number arising this way a corresponding element of $\mathbb{F}_q$, so that different numbers are associated to different elements. In this way, our message can be regarded as a sequence of elements of $\mathbb{F}_q^k$. For this reason, the vector space $\mathbb{F}_q^k$ over $\mathbb{F}_q$ is often called the **information space**.

Suppose that we use a linear code to transmit our message. A linear code of length $n$ may be regarded as the image of an injective linear transformation $G : \mathbb{F}_q^k \to \mathbb{F}_q^n$. Give $\mathbb{F}_q^k$ and $\mathbb{F}_q^n$ their standard bases. The encoding map $G$, with respect to these bases, is called the **generating matrix** of $C$. The rows of $G$ form a basis for the vector space $C$ over $\mathbb{F}_q$.

One other interesting thing about linear codes is that there is a surjective linear map $H : \mathbb{F}_q^n \to \mathbb{F}_q^{n-k}$ with the property that $Hv = \vec{0}$ if and only if $v \in C$. In other words, $H$ checks whether or not your received vector $v$ contains any errors (i.e., if $v$ is actually a code word). The map $H$ with respect to the standard basis is called the **parity check matrix**.

**Ponderable 14.5.1.** *Show that for all $w \in \mathbb{F}_q^k$, $HGw = \vec{0}$.*

**Example 14.5.1.** $V = \mathbb{F}_q^n = \mathbb{F}_q \times ... \times \mathbb{F}_q$ *(n times) is a code.*

If $v = (v_1, v_2, ..., v_n)$, $w = (w_1, w_2, ..., w_n)$ are vectors in $V$ then we define

$$d(v, w) = |\{i \mid 1 \leq i \leq n, \ v_i \neq w_i\}|$$

to be the **Hamming distance** between $v$ and $w$.

**Definition 14.5.2.** *We say that $C$ is $e$-**error correcting** if the following property always holds: given any $w \in F^n$ whose Hamming distance from some $c \in C$ is $\leq e$ then any $c' \in C$, $c \neq c'$, must satisfy $d(w, c') > e$ (in other words, a code word at most distance $e$ from $w$ is unique if $w$ has at most $e$ errors).*

**Definition 14.5.3.** *Let $Mon_n(\mathbb{F}_q)$ denote the group of all $n \times n$ matrices which have exactly one non-zero entry from $\mathbb{F}_q$ per row and per column. An element of $Mon_n(\mathbb{F}_q)$ is called a **monomial matrix**.*

**Ponderable 14.5.2.** *(a) Show $Mon_n(\mathbb{F}_q)$ is a group.*
*(b) Show $|Mon_n(\mathbb{F}_q)| = (q-1)^n \cdot n!$.*

**Definition 14.5.4.** *The set of all $A \in Mon_n(\mathbb{F}_q)$ such that $A * C = C$ (i.e., are the same code) is called the **automorphism group** of $C$, denoted $Aut(C)$.*

We shall see some examples below.

**Definition 14.5.5.** *If $w$ is a code word in $\mathbb{F}_q^n$ then the number of non-zero coordinates of $w$ is called the* **weight** *of $w$, denoted $wt(w)$.*

*If $C$ is a linear code then the smallest non-zero weight of any code word is called the* **minimum distance** *of $C$.*

*A* **cyclic code** *is a code which has the property that whenever $(c_0, c_1, ..., c_{n-1})$ is a code word then so is $(c_{n-1}, c_0, ..., c_{n-2})$. If $c = (c_0, c_1, ..., c_{n-1})$ is a non-zero code word in a cyclic code $C$ with the property that the degree of*

$$g_c(x) = c_0 + c_1 x + ... + c_{n-1} x^{n-1}$$

*is minimum (amongst all such polynomials, as $c$ ranges over all the non-zero code words in $C$) then we call $g_c$ a* **generator polynomial for $C$** .

**Example 14.5.2.** *The code with generator matrix*

$$\begin{pmatrix} 1 & 0 & 0 & 0 & 0 & 1 & 1 \\ 0 & 1 & 0 & 0 & 1 & 0 & 1 \\ 0 & 0 & 1 & 0 & 1 & 1 & 0 \\ 0 & 0 & 0 & 1 & 1 & 1 & 1 \end{pmatrix}$$

*is a binary code of length 7 and minimum distance 3. It is equivalent to a cyclic code having generator polynomial*

$$g(x) = 1 + x + x^3.$$

*(For the proof of this statement, see for example chapter 12 in Hill [Hi].) This code is called a 'Hamming code' and has many interesting properties which, to describe, would take us too far afield. The interested reader is refered to [CS], ch. 3, and [Hi], chapter 8.)*

**Ponderable 14.5.3.** *Check that the code in Example 14.5.2 is 1-error correcting.*

**Definition 14.5.6.** *Let $n$ be a positive integer relatively prime to $q$ and let alpha be a primitive n-th root of unity. Each generator polynomial $g$ of a cyclic code $C$ of length $n$ has a factorization of the form*

$$g(x) = (x - \alpha^{k_1})...(x - \alpha^{k_r}),$$

*where $\{k_1, ..., k_r\} \subset \{0, ..., n-1\}$. The numbers $\alpha^{k_i}$, $1 \le i \le r$, are called the* **zeros** *of the code $C$. They do not depend on the choice of $g$.*

**Definition 14.5.7.** *Let $p$ and $n$ be distinct primes and assume that $p$ is a square mod $n$. The* **quadratic residue code** *of length $n$ over $\mathbb{F}_p$ is the cyclic code whose generator polynomial has zeros*

$$\{\alpha^k \mid k \text{ is a square mod } n\}.$$

*The* **binary Golay code** *$GC_{23}$ is the quadratic residue code of length 23 over $\mathbb{F}_2$. The* **binary Golay code** *$GC_{24}$ is the code of length 24 over $\mathbb{F}_2$ obtained by appending onto $GC_{23}$ a zero-sum check digit.*

228

*The* **ternary Golay code** $GC_{11}$ *is the quadratic residue code of length 11 over* $\mathbb{F}_3$. *The* **ternary Golay code** $GC_{12}$ *is the code of length 12 over* $\mathbb{F}_3$ *obtained by appending onto* $GC_{11}$ *a zero-sum check digit.*

A generator matrix of $GC_{11}$ is

$$
\begin{pmatrix}
1 & 0 & 0 & 0 & 0 & 0 & 2 & 0 & 1 & 2 & 1 \\
0 & 1 & 0 & 0 & 0 & 0 & 1 & 2 & 2 & 2 & 1 \\
0 & 0 & 1 & 0 & 0 & 0 & 1 & 1 & 1 & 0 & 1 \\
0 & 0 & 0 & 1 & 0 & 0 & 1 & 1 & 0 & 2 & 2 \\
0 & 0 & 0 & 0 & 1 & 0 & 2 & 1 & 2 & 2 & 0 \\
0 & 0 & 0 & 0 & 0 & 1 & 0 & 2 & 1 & 2 & 2
\end{pmatrix}
$$

and a generator matrix of $GC_{23}$ is

$$
\begin{pmatrix}
1 & 0 & 0 & 0 & 0 & 0 & 0 & 0 & 0 & 0 & 0 & 0 & 1 & 0 & 1 & 0 & 1 & 1 & 1 & 0 & 0 & 0 & 1 \\
0 & 1 & 0 & 0 & 0 & 0 & 0 & 0 & 0 & 0 & 0 & 0 & 1 & 1 & 1 & 1 & 1 & 0 & 0 & 1 & 0 & 0 & 1 \\
0 & 0 & 1 & 0 & 0 & 0 & 0 & 0 & 0 & 0 & 0 & 0 & 1 & 1 & 0 & 1 & 0 & 0 & 1 & 0 & 1 & 0 & 1 \\
0 & 0 & 0 & 1 & 0 & 0 & 0 & 0 & 0 & 0 & 0 & 0 & 1 & 1 & 0 & 0 & 0 & 1 & 1 & 1 & 0 & 1 & 1 \\
0 & 0 & 0 & 0 & 1 & 0 & 0 & 0 & 0 & 0 & 0 & 0 & 1 & 1 & 0 & 0 & 1 & 1 & 0 & 1 & 1 & 0 & 0 \\
0 & 0 & 0 & 0 & 0 & 1 & 0 & 0 & 0 & 0 & 0 & 0 & 0 & 1 & 1 & 0 & 0 & 1 & 1 & 0 & 1 & 1 & 0 \\
0 & 0 & 0 & 0 & 0 & 0 & 1 & 0 & 0 & 0 & 0 & 0 & 0 & 0 & 1 & 1 & 0 & 0 & 1 & 1 & 0 & 1 & 1 \\
0 & 0 & 0 & 0 & 0 & 0 & 0 & 1 & 0 & 0 & 0 & 0 & 1 & 0 & 1 & 1 & 0 & 1 & 1 & 1 & 1 & 0 & 0 \\
0 & 0 & 0 & 0 & 0 & 0 & 0 & 0 & 1 & 0 & 0 & 0 & 1 & 0 & 1 & 1 & 0 & 1 & 1 & 1 & 1 & 0 & 0 \\
0 & 0 & 0 & 0 & 0 & 0 & 0 & 0 & 0 & 1 & 0 & 0 & 0 & 0 & 1 & 0 & 1 & 1 & 0 & 1 & 1 & 1 & 1 \\
0 & 0 & 0 & 0 & 0 & 0 & 0 & 0 & 0 & 0 & 1 & 0 & 1 & 0 & 1 & 0 & 1 & 1 & 1 & 0 & 0 & 1 & 0 \\
0 & 0 & 0 & 0 & 0 & 0 & 0 & 0 & 0 & 0 & 0 & 1 & 0 & 1 & 0 & 1 & 1 & 1 & 0 & 0 & 0 & 1 & 1
\end{pmatrix}
$$

(These were obtained using MAGMA [Magma].)

According to MacWilliams and Sloane [MS], the Golay codes are among the the most important of all error-correcting codes. They are named after Marcel Golay (1902-1989), a physicist born in Switzerland, educated in Zürich and then at the University of Chicago. Right after receiving his diploma, he worked for the U. S. Army Signal Corps Labs for almost 25 years. It was during this time that he discovered, in 1949, the Golay codes and the non-binary generalization of the Hamming codes. (Golay knew only of the $(7, 4)$- binary Hamming code, discovered the binary Hamming codes independently, and then generalized them to the non-binary case.)

The following result illustrates how the Matheiu groups arise in coding theory.

**Theorem 14.5.1.** *(a) There is a normal subgroup $N$ of $Aut(GC_{12})$ of order 2 such that $Aut(GC_{12})/N$ is isomorphic to $M_{12}$.*
*(b) $Aut(GC_{24}) = M_{24}$.*

Since the Mathieu groups are so large, this theorem above indicates that the Golay codes $GC_{12}$ and $GC_{24}$ have a lot of symmetry.

It is a basic rule of thumb in mathematics that whenever you find something displaying a lot of symmetry then it will quite often have other interesting properties. With this philosophy spurring us on, let us turn to some of the other properties of these codes.

**Lemma 14.5.1.** *Any two code words in $GC_{24}$ differ by 8 bits. The code $GC_{24}$ detects 4 errors (per 24 bits) and corrects 3 errors.*

When you compare that with the correcting ability of bar-codes or ISBN codes (which have a check-digit), $GC_{24}$ is much better.

**Lemma 14.5.2.** *If $w$ is a code word in $GC_{24}$ then $wt(w)$ is either $0, 8, 12, 16, 24$.*

**Definition 14.5.8.** *The code words of weight 12 in $GC_{24}$ are called* **dodecads**.

We may identify a code word $w = (c_0, c_1, ..., c_{23})$ with the set of indices i of the non-zero coordinates $c_i \neq 0$.

**Theorem 14.5.2.** *$M_{12}$ is the stabilizer in $M_{24}$ of a dodecad, regarded as a set of indices.*

## 14.6 $M_{12}$ is crossing the Rubicon

The result of this section was mentioned briefly in §12.3 above.

Let $f_1, f_2, ..., f_{12}$ denote the basic moves ($2\pi/5$ degree turns of a pentagon about a vertex) of the Rubik isocahedron, regarded as elements of $S_V$, where $V$ denotes the set of 12 vertices of the Rubik isocahedron ('Rubicon').

The following remarkable result is due to John Conway [CS]. John Conway (1937-) is a very interesting character in the evolution of group theory. In the late 1960's, at a time when many experts thought they had classified all finite simple groups, Conway found another one, of order 8,315,553,613,086,720,000. This group he found had the nice property that two of its subgroups were also undiscovered finite simple groups!

**Theorem 14.6.1.** $M_{12} = \langle x * y^{-1} \mid x, y \in \{f_1, ..., f_{12}\} \rangle$.

In other words, the Mathieu group $M_{12}$ (which, recall, is a finite simple group discovered by Mathieu) is generated by the twist-untwist moves of the Rubik isocahedron, or Rubicon. If we call a 'twist-untwist' move of the form $x * y^{-1}$ (with $x, y$ as in the theorem above) a **cross** move then, with apologies to Caeser, the theorem above says that $M_{12}$ is generated by the crosses of the Rubicon.

In fact, $C$ acts 5-transitively on the set of vertices of the Rubicon (this is implicit in [CS]).

## 14.7 An aside: A pair of cute facts

It's hard to resist stating some more interesting facts about the Mathieu groups.

## 14.7.1 Hadamard matrices

Let $A = (a_{ij})_{1 \leq i,j \leq n}$ denote a real $n \times n$ matrix. The following question seems quite natural in a course in advanced vector calculus or real analysis:

**Question**: What is the maximum value of $|\det(A)|$, where the entries of $A$ range over all real numbers $|a_{ij}| \leq 1$?

From vector calculus we know that the absolute value of the determinant of a real square matrix equals the volume of the parallelpiped spanned by the row (or column) vectors of the matrix. The volume of a parallelpiped with sides of a *fixed* length depends on the angles the row vectors make with each other. This volume is maximized when the row vectors are mutually orthogonal, i.e., when the parallelpiped is a cube in $\mathbb{R}^n$. Suppose now that the row vectors of $A$ are all orthogonal. The row vectors of $A$, $|a_{ij}| \leq 1$, are longest when each $a_{ij} = \pm 1$, which implies that the length of each row vector is $\sqrt{n}$. Suppose, in addition, that the row vectors of $A$ are all of length $\sqrt{n}$. Such a matrix is called a **Hadamard matrix of order** $n$. Then $|\det(A)| = (\sqrt{n})^n = n^{n/2}$, since the cube has $n$ sides of length $\sqrt{n}$. Now, if $A$ is any matrix as in the above question then we must have $|\det(A)| \leq n^{n/2}$. This inequality is called **Hadamard's inequality**.

Jacques Hadamard (1865-1963) was a prolific mathematician who worked in many areas but he is most famous for giving the first proof of the prime number theorem (in 1896). (The prime number theorem, 'known' to Gauss though not proven, roughly states that the number of primes less than $N$ is about $N/\log(N)$, as $N$ grows to infinity.)

What might be surprising at first sight is that there does not always exist a Hadamard matrix - for some $n$'s they exist and for other $n$'s they don't. For example, there is a $2 \times 2$ Hadamard matrix but not a $3 \times 3$ one. What is perhaps even more suprising is that, in spite of the fact that the above question (which is unsolved for arbitrary $n$) arose from an analytic perspective, Hadamard matrices are related more to coding theory, number theory, and combinatorics [vLW]!

**Example 14.7.1.** *Let*

$$A := \begin{pmatrix}
1 & 1 & 1 & 1 & 1 & 1 & 1 & 1 & 1 & 1 & 1 & 1 \\
-1 & 1 & 1 & 1 & 1 & 1 & 1 & -1 & -1 & -1 & 1 & -1 \\
-1 & -1 & 1 & 1 & -1 & 1 & 1 & 1 & -1 & -1 & -1 & 1 \\
-1 & 1 & -1 & 1 & 1 & -1 & 1 & 1 & 1 & -1 & -1 & -1 \\
-1 & -1 & 1 & -1 & 1 & 1 & -1 & 1 & 1 & 1 & -1 & -1 \\
-1 & -1 & -1 & 1 & -1 & 1 & 1 & -1 & 1 & 1 & 1 & -1 \\
-1 & -1 & -1 & -1 & 1 & -1 & 1 & 1 & -1 & 1 & 1 & 1 \\
-1 & 1 & -1 & -1 & -1 & 1 & -1 & 1 & 1 & -1 & 1 & 1 \\
-1 & 1 & 1 & -1 & -1 & -1 & 1 & -1 & 1 & 1 & -1 & 1 \\
-1 & 1 & 1 & 1 & -1 & -1 & -1 & 1 & -1 & 1 & 1 & -1 \\
-1 & -1 & 1 & 1 & 1 & -1 & -1 & -1 & 1 & -1 & 1 & 1 \\
-1 & 1 & -1 & 1 & 1 & 1 & -1 & -1 & -1 & 1 & -1 & 1
\end{pmatrix}$$

*This is a Hadamard matrix of order* 12.

**Ponderable 14.7.1.** *Show that*

231

(a) if you swap two rows or columns of a Hadamard matrix, you will get another Hadamard matrix,

(b) if you multiply any row or column of a Hadamard matrix by $-1$, you will get another Hadamard matrix,

(c) if you multiply any Hadamard matrix on the left by a signed permutation matrix (that is, a matrix with exactly one $\pm 1$ per row and column) then you will get another Hadamard matrix,

(d) if you multiply any Hadamard matrix on the left by a signed permutation matrix (that is, a matrix with exactly one $\pm 1$ per row and column) then you will get another Hadamard matrix.

**Ponderable 14.7.2.** Let $A, B$ be two Hadamard matrices of order $n$. Call $A$, $B$ **left equivalent** if there is an $n \times n$ signed permutation matrix $P$ such that $A = PB$. Show that this defines an equivalence relation on the set of all Hadamard matrices of order $n$.

**Ponderable 14.7.3.** Let $A$ be a Hadamard matrix of order $n$. Let $Aut(A)$ denote the set of all $n \times n$ signed permutation matrices $Q$ such that $A$ is left equivalent to $AQ$. Show that $Aut(A)$, called the **automorphism group of** $A$, is a group under matrix multiplication.

The following result is yet another indication of the unique role of these Mathieu groups in mathematics:

**Theorem 14.7.1.** (Assmus-Mattson [AM]) Let $A$ be the Hadamard matrix of order $12$ in the above example. Then $Aut(A) \cong M_{12}$.

## 14.7.2  5-transitivity

The following result exemplifies once more the unique role of these Mathieu groups in group theory:

**Theorem 14.7.2.** If $G$ is a subgroup of $S_X$ for some finite set $X$ and if $G$ acts 5-transitively on $X$ then exactly one of the following must be true:

(a) $G \cong S_n$, for some $n > 4$,

(b) $G \cong A_m$, for some $m > 6$,

(c) $G \cong M_{12}$,

(d) $G \cong M_{24}$.

Furthermore, each of the groups in (a)-(d) acts 5-transitively on some finite set.

For a proof of this, see [CS] and [R], ch 9.

# Chapter 15

# Some solution strategies

'I had a feeling once about Mathematics - that I saw it all. Depth beyond depth was revealed to me - the Byss and Abyss. I saw - as one might see the transit of Venus or even the Lord Mayor's Show - a quantity passing through infinity and changing its sign from plus to minus. I saw exactly why it happened and why the tergiversation was inevitable but it was after dinner and I let it go.'

*Sir Winston Churchill*

It is a person far better than I who can look at the Rubik's Cube and solve it behind their back (there are such people, I believe). They remind me of Churchill's amusing quote - they can 'see it all'. Speaking for myself, I need strategies, and the simpler the better. This chapter includes some strategies for solving the Rubik's Cube, the Masterball, the Skewb, the Pyraminx, and a few others. This is good hands-on practice: in the words of Aristotle, 'What we have to learn to do, we learn by doing.' In addition, we discuss some of the mathematical ideas behind the algorithms used to solve the Rubik's Cube - my favorite method involves a little bit of group theory (see §15.2 below).

For unexplained notation used in some of the sections below, see chapter 4. A good online source of solutions is Jaap Scherphuis' web site [Sch2].

## 15.1   A strategy for solving the Rubik's Cube

I recommend that the person who has never 'played' with the Rubik's Cube before and wants to follow the strategy indicated below first learn how to 'play a move' without looking at the cube. Why? Think of the cube as a musical instrument like a guitar. To play music on the guitar, you should know how to look at the song sheet and play without having to constantly keep looking at your fingers. The cube is similar: in the beginning, you want to be able to make a move such as $F^2 * L^2 * U^2 * (F^2 * L^2)^3 * U^2 * L^2 * F^2$ without having to constantly keep looking at your fingers. This will help make you faster and cut down on frustrating mistakes.

One way to solve it is the **corner-edge method**. The basic idea is as follows.

- Ignoring orientations, i.e., twists and flips, first solve the corner subcubes. This means, make sure that the corners will, up to possible twisting, match up the center subcubes.

- Ignoring orientations, i.e., twists and flips, second solve the edge subcubes. This means, make sure that the edges will, up to possible flipping, match up the center subcubes.

- Fix the corner orientations, i.e., twists of the corner subcubes. Now all the corners are completely solved.

- Fix the edge orientations, i.e., flips of the edge subcubes. Now the cube is solved.

For completeness, we mention another method. The **layer method** solution strategy is composed of 3 stages.

- Solve the top face and top edges.
- Solve the middle edges (and bottom edges as best as possible).
- Solve the bottom corners (and bottom edges if necessary).

There is nothing wrong with this method but we shall not discuss this method further. (The reader who wishes to become a fast Rubik's Cube solver should learn both methods.)

## 15.1.1  Strategy for solving the cube

Remember, the basic moves are denoted $R, L, U, D, F, B$. Let $M_R$ denote clockwise (with respect to **R**ight side) quarter turn of the **M**iddle slice parallel to the right side.

To 'solve' the corners (ignoring orientations) we first describe a simple method, which I call 'fishing', for achieving this. (Mathematically, this amounts to performing some carefully choosen commutators.) Hold the cube in front of you. Suppose you want the $lfd$ corner to be in the $urf$ position without messing up too much of the rest of the cube. First perform $R^{-1}$ (this is 'baiting the hook'), then $D$ ('setting the hook'), $R$ ('reeling in the hook'), and finally $D^{-1}$. This last move can be omitted since after $R^{-1}DR$ you have moved $lfd$ to $urf$.

By fishing you can either (a) solve all the corners (if you are lucky), (b) solve all except two which need to be swapped, or (c) all but 3 which need to be cyclically permuted. To finish, we need a corner swap move in case (b) and a corner 3-cycle in case (c). In case (b), the move

$$U * F * [R, U]^3 * F^{-1} =$$
$$U * F * R * U * R^{-1} * U^{-1} * R * U * R^{-1} * U^{-1} * R * U * R^{-1} * U^{-1} * F^{-1}$$

swaps the corners $(ubr, ufl)$ and permutes the edges $(uf, ul, ub, ur)$, leaving the rest of the cube alone. (If the corners you need to 'fix' are not in the $ubr, ufl$ positions then you first need the following trick. Rotate the cube so that one of them is in the $ubr$ position say then move ('bait') the other into the $ufl$ position, make the move above, then unmove ('cut bait').) In case (c), the move

$$[R * D * R^{-1}, U] = R * D * R^{-1} * U * R * D^{-1} * R^{-1} * U^{-1}$$

is a corner 3-cycle $(brd, urb, ulb)$, leaving the rest of the cube alone. (Again, if the corners you need to 'fix' are not in the $brd, urb, ulb$ positions, you will need the 'bait/cut bait' trick.)

Once the corners have been solved, to 'solve' the edges (ignoring orientations), one simply has to use the edge 3-cycle

$$M_R^2 * U^{-1} * M_R^{-1} * U^2 * M_R * U^{-1} * M_R^2$$

repeatedly. This move is a 3-cycle on $(uf, ul, ur)$ and leaves the rest of the cube alone. (This is not quick and it is not at all obvious that only this move (along with the 'bait/cut bait' trick) works, but it can be proven using that fact that 3-cycles generate the alternating group.)

To twist the corners correctly, you need a corner twisting move. The move

$$(R^{-1} * D^2 * R * B^{-1} * U^2 * B)^2$$

twists the $ufr$ clockwise and the $bld$ corner counterclockwise, leaving the rest of the cube alone. This move (along with the 'bait/cut bait' trick) will solve all the corners.

Finally, to flip the edges correctly, use

$$(M_R * U)^3 * U * (M_R^{-1} * U)^3 * U$$

which flips the top edges $uf, ub$ (leaving the rest of the cube alone), and the move

$$(M_R * U)^4$$

which flips $ub, ul$ and flips $df, db$ (leaving the rest of the cube alone).

Now practice!

To shorten the notation and emphasize the group-theoretical symmetry of some of the moves given below, let $x^y = y^{-1} * x * y$ denote conjugation and $[x, y] = x * y * x^{-1} * y^{-1}$ denote the commutator, for $x, y$ group elements.

A table of 'clean' edge and corner moves (i.e., moves which perform as advertised and don't mess up the rest of your cube!):

| | |
|---|---|
| $M_R^2 * U^{-1} * M_R^{-1} * U^2 * M_R * U^{-1} * M_R^2$ | edge 3-cycle (uf,ul,ur) |
| $(M_R * U)^3 * U * (M_R^{-1} * U)^3 * U$ | flips the top edges uf, ub |
| $(R^2 * U^2)^3$ | permutes (uf,ub)(fr,br) |
| $(M_R * U)^4$ | flips ub,ul and flips df,db |
| $(R^{-1} * D^2 * R * B^{-1} * U^2 * B)^2$ | $ufr+, bld++$ |
| $[R, U]^3 = (R * U * R^{-1} * U^{-1})^3$ | permutes $(ufr, dfr)(ubr, ubl)$ |
| $F^2 * L^2 * U^2 * (F^2 * L^2)^3 * U^2 * L^2 * F^2$ | permutes $(uf, ub)(ur, ul)$ |
| $(D^2 * R^2 * D^2 * (F^2 * R^2)^2 * U)^2$ | permutes $(ufl, ubr)(dfr, dbl)$ |
| $(M_R^2 * U * M_R^2 * U^2)^2$ | permutes $(ufl, ubr)(ufr, ubl)$ |
| $R * D * R^{-1} * U * R * D^{-1} * R^{-1} * U^{-1}$ | corner 3-cycle $(brd, urb, ulb)$ |

These moves were compiled with help from the books [Si], [B], and [Sn1].

### 15.1.2  Catalog of $3 \times 3$ Rubik's 'supercube' moves

The supercube is the Rubik's Cube with each center facet marked with a short line through it and an adjoining edge (i.e., the center facets are oriented).

\* $(M_R^2 * U^{-1} * M_R^{-1} * U^2 * M_R * U^{-1} * M_R^2)^2$ is the top edge 3-cycle $(uf, ur, ul)$,

\* $(R^{-1} * D^2 * R * B^{-1} * U^2 * B)^2$ twists the $ufr$ corner clockwise and the $bld$ corner counterclockwise (and does not twist any centers).

\* $M_R^{-1} * M_D^{-1} * M_R * U^{-1} * M_R^{-1} * M_D * M_R * U$ is the center twist $u+, r-$ (for these last three moves, see [Si], [Sn1])

## 15.2  The subgroup method

One approach to solve the Rubik's Cube using a computer has been to construct a certain sequence of subgroups

$$G_n = \{1\} \subset G_{n-1} \subset \ ... \ \subset G_1 \subset G_0 = G,$$

where $G = \langle R, L, F, B, U, D \rangle$ is the Rubik's Cube group, which allows the following strategy to be implemented:

- represent a given position of the Rubik's Cube by an element $g_0 \in G$,

- determine a complete set of coset representatives of $G_{k+1}/G_k$:

$$G_{k+1}/G_k = \cup_{i=1}^{r_k} g_{k+1,i} G_{k+1}, \qquad \text{some } r_k > 1, \forall 0 \le k < n$$

(note $m_{n-1} = 1, g_{n,1} = 1$),

- (step 1) if $g_0 \in g_{1,i}G_1$ (where $i \in \{1, ..., n_1\}$) then let $g_1 = g_{1,i}$ and $g_1' = g_1^{-1}g_0$ (note $g_1' \in G_1$),

- (inductive step) if $g_k' \in G_k$ has been defined and if $g_k' \in g_{k+1,j}G_k$ (where $j \in \{1, ..., n_1\}$) then let $g_{k+1} = g_{k+1,j}$ and $g_{k+1}' = g_{k+1}^{-1}g_k'$ (note $g_{k+1}' \in G_{k+1}$),

- putting all these together, we obtain $1 = g_n^{-1} g_{n-1}^{-1} g_{n-2}^{-1} \cdots g_1^{-1} g_0$, so

$$g_0 = g_1 g_2 \cdots g_{n-1} g_n.$$

The hope is to be able to choose the sequence of subgroups $G_i$ in such a way that the coset representatives are short, relatively simple moves on the Rubik's Cube so that the 'solution' $g_0 = g_1 g_2 \cdots g_{n-1} g_n$ is not too long.

### 15.2.1 Example: the corner-edge method

We now present an example - a fairly unsophisticated one but you will get the idea.

Let $G_1$ denote the subgroup which does not permute any corners, let $G_2$ denote the subgroup which does not permute any corners or edges, let $G_3$ denote the subgroup which does not permute any corners or edges and does not reorient any corners, and let $G_4 = \{1\}$:

$$G_4 = \{1\} \subset G_3 \subset G_2 \subset G_1 \subset G_0 = G.$$

This choice of subgroups crudely models the 'corner-edge method' due to Singmaster [Si].

The idea is simple.

1. Represent a given position of the Rubik's Cube by an element $g_0 \in G$.

2. Let $g_1$ denote the move which moves all the corners into the correct positions (i.e., permutes them into the solved position and possibly twists them), so $g_1^{-1} g_0 \in G_1$. Let $g_1' = g_1^{-1} g_0$.

3. Let $g_2$ denote the move which moves all the edges into the correct positions (i.e., permutes them into the solved position and possibly reorients corners and edges) and leaves all other pieces unpermuted, so $g_2^{-1} g_1' \in G_2$. Let $g_2' = g_2^{-1} g_1'$.

4. Let $g_3$ denote the move which 'solves' all the corners (i.e., twists them all into the correct orientation and may flip some edges) but does not permute any pieces, so $g_3^{-1} g_2' \in G_3$. Let $g_3' = g_3^{-1} g_2'$.

5. Let $g_4$ denote the move which 'solves' all the edges (i.e., flips them all into the correct orientation) and leaves all other facets alone.

6. The 'solution' is $g_0 = g_1 g_2 g_3 g_4$.

### 15.2.2 Example: Thistlethwaite's method

Morwen Thistlethwaite (a mathematician formerly a colleague of David Singmaster, now at the Univ. of Tennessee) developed one of the best subgroup methods for solving the cube [FS]. He takes

$$G_1 = \langle R, L, F, B, U^2, D^2 \rangle, \quad G_2 = \langle R, L, F^2, B^2, U^2, D^2 \rangle,$$
$$G_3 = \langle R^2, L^2, F^2, B^2, U^2, D^2 \rangle, \quad G_4 = \{1\}.$$

$G_2$ is isomorphic to the 'Rubik's $3 \times 3 \times 2$-domino' group. Its order is $(8!)^2 \cdot 12$, according to [FS], §7.6. $G_3$ is the 'squares' group. Its order is $2^{13} \cdot 3^4$, according to [FS], §7.6.

Thistlethwaite has shown (using a computer to help with some of the work) the following facts.

- There is a complete set of coset representatives $\{g_{1,i} \mid 1 \leq i \leq n_1\}$ of $G/G_1$ such that each $g_{1,i}$ is at most 7 moves long (and $n_1 = 2048$). This set of moves flips edges only.

- There is a complete set of coset representatives $\{g_{2,i} \mid 1 \leq i \leq n_2\}$ of $G_1/G_2$ such that each $g_{2,i}$ is at most 13 moves long (and $n_2 = 1082565$). This set of moves twists corners only.

- There is a complete set of coset representatives $\{g_{3,i} \mid 1 \leq i \leq n_3\}$ of $G_2/G_3$ such that each $g_{3,i}$ is at most 15 moves long (and $n_3 = 29400$). This set of moves puts all the edge subcubes and corner subcubes in the correct position.

- there is a complete set of coset representatives $\{g_{4,i} \mid 1 \leq i \leq n_4\}$ of $G_3/G_4$ such that each $g_{4,i}$ is at most 17 moves long (and $n_4 = 663552$).

Therefore, the Rubik's Cube can be solved in at most $7 + 13 + 15 + 17 = 52$ moves.

More recent improvements on this method have gotten this number down to 45 in the 'face-turn metric' (see [Lo] for details and recent updates).

## 15.3 The hockeypuck puzzle

The Hockeypuck is in the shape of a circular disk sliced into 12 equal pieces, as one might slice up a pie. These 6 slices give 12 radial segments, each emanating from the center of the Hockeypuck. Label these 12 segments $1, 2, ..., 12$.

Basic moves: For the $i^{th}$ radial segment, let $f_i$ be the transformation which flips over that half of the Hockeypuck which contains the $i^{th}$ pie piece. Let $r$ denote the rotation of this circle by $30^o$ radians counterclockwise, so $r = (1, 2, ..., 12)$.

Solution strategy:

Step 1 : Solve half of the puzzle by 'fishing', using moves of the form $f_n r^{-1} f_n r$, where $n$ is choosen appropriately.

Step 2 : Use moves of the form

$$(f_1 * f_2 * f_3 * f_4 * f_5 * f_6)^2$$

to swap adjacent puzzle pieces (in this case, this move swaps 1 with 12 and flips over every piece).

# 15.4    Rainbow Masterball

The solution strategy

Step 1 : The idea is to first get all the middle bands aligned first, so you get the Masterball corresponding to a matrix of the form

$$
\begin{array}{cccccccc}
* & * & * & * & * & * & * & * \\
1 & 2 & 3 & 4 & 5 & 6 & 7 & 8 \\
1 & 2 & 3 & 4 & 5 & 6 & 7 & 8 \\
* & * & * & * & * & * & * & *
\end{array}
$$

Here, $*$ denotes any color. We have labeled the colors on the Masterball as $1, 2, ..., 8$ in order of occurrence.

We describe a method, which I call 'fishing', for achieving this. (Mathematically, this amounts to performing some carefully choosen commutators.) Without too much trouble you can always assume that we have one column aligned. You may need to flip or rotate the ball a little bit to do this. Call this aligned column 'column 1' and call the color in column 1, 'color 1'. We want to get the middle two entries in column 2 aligned. Call the color in the (2,3)-entry 'color 2'.

We want to get color 2 in the (2,2)-entry. The remaining large color 2 tile is what we will 'fish' for. Hold the ball in front of you in such a way that column 2 is slightly to the left of center and column 3 is slightly to the right of center. There are 4 facets in the right upper middle band, 4 facets in the left upper middle band, 4 facets in the right lower middle band and 4 facets in the left lower middle band. A flip about the center on the right half (i.e., perform $f_2$) exchanges these. We may assume that color 2 is on one of the four facets in the right lower middle band. (If it isn't you need to apply $f_2$ first). Now perform $r_2^{-1} * f_2^{-1} * r_2 * f_2$: first perform $r_2^{-1}$ (this is 'baiting the hook'), then $f_2^{-1}$ ('putting the hook in the water'), then $r_2$ ('setting the hook'), and finally $f_2$ ('reeling in the hook'). You may or may not have color 2 in the (2,2) place like you want but the color 1 stripe is intact. If necessary, try again. After at most 4 tries you'll be successful.

Step 2 : Repeat this 'fishing' strategy to get color 2 in the (1,2) position (using $r_1^{-1} * f_2^{-1} * r_1 * f_2$ in place of $r_2^{-1} * f_2^{-1} * r_2 * f_2$). Now, by turning the ball over if necessary, repeat this idea to get color 2 in the (4,2) position. Now you have two 'aligned' stripes on your ball - color 1 in column 1 and color 2 in column 2. We say, in this case, that columns 1 and 2 have been 'solved'.

Step 3 : Repeat this for columns 3 and 4.

Step 4 : Use the moves in the catalog below to finish the puzzle. (I believe the only moves needed are the 'equator2swap36' and the 'polar2swap36' below, along with suitable cleverly choosen 'set-up moves'.)

## 15.4.1 A catalog of Masterball moves

**Column moves**

We number the columns as 1,...,8. We will use a signed cycle notation to denote an action of a move on the columns of the Masterball.

**Example 15.4.1.** *A move which switches the 1st and 3rd column but flips both of them over will be denoted by* $(1,3)_-$.

*A move which sends the 4th column to the 6th column, the 6th column to the 5th column, and switches the 2nd and 3rd column but flips both of them over will be denoted by* $(2\ 3)_-(6\ 5\ 4)$.

| move | cycle |
|:---:|:---:|
| $f_1$ | $(1,4)_-(2,3)_-$ |
| $f_2$ | $(2,5)_-(3,4)_-$ |
| $f_3$ | $(3,6)_-(4,5)_-$ |
| $f_4$ | $(4,7)_-(5,6)_-$ |
| $f_5$ | $(5,8)_-(6,7)_-$ |
| $f_6$ | $(1,6)_-(7,8)_-$ |
| $f_7$ | $(2,7)_-(1,8)_-$ |
| $f_8$ | $(3,8)_-(1,2)_-$ |
| $f_1 * f_2 * f_1$ | $(1,2)_-(3,5)$ |
| $f_1 * f_2 * f_1 * f_2$ | $(5,4,3,2,1)_-$ |
| $f_1 * f_3 * f_1$ | $(1,5)(2,6)$ |
| $f_2 * f_3 * f_2$ | $(2,3)_-(6,5,4)$ |
| $f_1 * f_4 * f_1$ | $(1,7)(5,6)$ |
| $f_1 * f_5 * f_1$ | $(5,8)_-(6,7)_-$ |
| $f_1 * f_8 * f_1$ | $(2,8)(3,4)_-$ |
| $f_8 * f_1 * f_8$ | $(1,8)_-(4,3,2)$ |
| $f_2 * f_1 * f_2$ | $(1,3)(4,5)_-$ |
| $f_3 * f_1 * f_3$ | $(1,3)(4,8)_-$ |
| $f_8 * f_1 * f_2$ | $(1,4)_-(2,3,8,5)_-$ |

Finally, $(f_1 * f_2 * f_3 * f_4)^2 * r_1 * r_2 * r_3 * r_4$ swaps the 7,8 columns and leaves all the others fixed but flipped over.

**Some products of 2-cycles on the facets**

The polar2swap and equator2swap were obtained by trying variations of some moves of Andrew Southern's on a MAPLE implementation of the Masterball [J] (see also [JS]).

We number the facets in the $i^{th}$ column, north-to-south, as $i1, i2, i3, i4$ (where $i = 1, 2, ..., 8$).

| move | cycle |
|---|---|
| $x = r_1 * f_4 * r_1^{-1} * r_4 * f_4 * r_4^{-1}$ | $(41, 84)(44, 81)$ |
| $x * r_1^4 * x * r_4^4$ | $(41, 81)(44, 84)$ |
| $f_1 * r_1 * f_4 * r_1^{-1} * r_4 * f_4 * r_4^{-1} * f_1$ | $(14, 84)(11, 81)$ |
| polar2swap36 | $(11, 14)(31, 61)$ |
| polar2swap18 | $(61, 64)(11, 81)$ |
| equator2swap36 | $(12, 13)(32, 62)$ |
| equator2swap18 | $(62, 63)(12, 82)$ |

where polar2swap36 is:

$$f_1 * r_3^{-1} * r_4^{-1} * f_1 * f_2 * r_1 * r_4^{-1} * f_2 * r_4^4 * f_2 * r_1^{-1} * r_4 * f_2 * r_4^4 * f_1 * r_3 * r_4 * f_1$$

(moreover, if you replace $r_3$ by $r_2$ both times in this move you get the same effect); polar2swap18 is:

$$f_1 * r_3^{-1} * r_4^{-1} * f_3 * f_4 * r_1 * r_4^{-1} * f_4 * r_4^4 * f_4 * r_1^{-1} * r_4 * f_4 * r_4^4 * f_3 * r_3 * r_4 * f_1;$$

equator2swap36 is:

$$f_1 * r_4^{-1} * r_3^{-1} * f_1 * f_2 * r_2 * r_3^{-1} * f_2 * r_3^4 * f_2 * r_2^{-1} * r_3 * f_2 * r_3^4 * f_1 * r_4 * r_3 * f_1;$$

and equator2swap18 is:

$$f_1 * r_4^{-1} * r_3^{-1} * f_3 * f_4 * r_2 * r_3^{-1} * f_4 * r_3^4 * f_4 * r_2^{-1} * r_3 * f_4 * r_3^4 * f_3 * r_4 * r_3 * f_1.$$

For further details on the rainbow puzzle, see [JS], [J].

# 15.5 The Skewb

## 15.5.1 Strategy

The goal here is to collect enough moves to support the following solution strategy: fix the centers and solve the corners using 'clean corner moves' (i.e., moves which do not effect the centers).

The basic moves are twists by 120 degrees clockwise about each of the six corners $FRU, FLU, BRU, BLU, BDR, BDL, DFR, DFL$.

## 15.5.2 A catalog of Skewb moves

The descriptions below were discovered with the help of a simulation of the Skewb written for MAPLE [Jwww].

1. $FRU * BLU * FRU^{-1}$ is order 3.

2. $[FRU * FLU]^3$ twists 6 corners clockwise by 120 degrees. The 2 corners not twisted are those opposite the $FRU, FLU$ corners: the $BDR, BDL$ corners. The centers are all fixed. $(FRU * FLU)^3 = (FLU * FRU)^3$ rotates all the corners except for the bd corners. It does not permute any facets.

3. The move, $[FRU*BLU]^5$, fixes all the centers and the 2 'opposite' corners: $DFL, BDR$. It twists the 3 corners $FLU$, $BLU$, and $FRU$. On the remaining 3 corners, it acts as the permutation $(DFR, BRU, BDL)$.

4. $FRU = BDL$ (actually, they are only equal up to a rotation of the entire cube). In general, a corner move is equal to the opposite corner move up to a rotation of the entire cube.

5. $(FRU * FLU)^3 * (BDL * BDR)^6$ rotates all the corners except for the $bu$ and the $df$ corners. The $uf$ corners are rotated clockwise and the $bd$ corners counterclockwise. It does not permute any facets. $(FRU*FLU)^6 * (BDL * BDR)^3$ is the same move, but rotates in the opposite direction.

6. $(FRU * FLU)^6 * (BDL * BDR)^6$ rotates the corners as follows:

   - the $uf$ corners counterclockwise,
   - the $db$ corners counterclockwise,
   - the $df$ corners clockwise,
   - the $ub$ corners clockwise.

   It does not permute any facets.

7. Let $bottomspin = (FRU*FLU)^6*(BDL*BDR)^3*(DFR*DFL)^3$. This move rotates the 4 bottom corners (the $df$ corners clockwise and the $db$ counterclockwise). It does not permute any facets.

8. $(BRU * FLU)^9$ is a 5 cycle on the center facets $(F, R, B, U, L)$. It fixes the bottom and does not affect any corners. $(BLU * FRU)^9$ is a 5 cycle on the center facets. It fixes the bottom and does not affect any corners.

9. $(BLU * FRU)^9 * (BRU * FLU)^9$ is a product of 2 transpositions on the center facets, swapping front/back and up/right. It fixes the bottom and does not affect any corners.

10. Let $U$ denote the clockwise (with respect to the up face) rotation of entire cube by 90 degrees. Then $bottomspin * U * bottomspin$ rotates but does not swap 2 corners (the DFR and BDL) and does not affect any other corners or faces.

## 15.6 The Pyraminx

Assume that the tetrahedron is lying on a flat surface in front of you, with the triangle base pointing away from you. The corners are denoted $L$ (left), $R$ (right), $U$ (up), and $B$ (back).

Basic Moves: Let

- L denote the 120 degree clockwise rotation of the 2-level subtetrahedron containing the left corner,

- R denote the 120 degree clockwise rotation of the 2-level subtetrahedron containing the right corner,

- U denote the 120 degree clockwise rotation of the 2-level subtetrahedron containing the up corner,

- B denote the 120 degree clockwise rotation of the 2-level subtetrahedron containing the back corner.

First, get the 'center' facets solved, then twist the corner tips to solve them and the center facets. Finally, to solve the edge facets, use the following moves (given in [EK]):

- $[R, U^{-1}]$ is a 3-cycle of edge pieces on the $URL$ face,

- $[R, U^{-1}] * [R^{-1}, L]$ is a flip of two edges ($UR$ edge and $UL$ edge) on the $URL$ face.

## 15.7 The Megaminx

The strategy here is the same as for the $3 \times 3$ Rubik's Cube:

- place the corners correctly first (ignoring correct corner orientation),

- place the edges correctly first (ignoring correct edge orientation),

- twist the corners if necesary,

- flip the edges if necessary.

Moves useful for carrying out these steps are included in the following catalog.

### 15.7.1 Catalog of moves

First, some notation. We label the faces $f_1, f_2, ..., f_6$ on top and label the bottom faces $f_7, f_8, ..., f_{12}$ as in chapter 4. The same notation is used to indicate the move of the Megaminx given by rotating that face of the Megaminx by 72 degrees clockwise.

- $f_1^{-1} * f_2^{-1} * f_1 * f_2 * f_1^{-1} * f_2^{-1} * f_1 * f_2 * f_1^{-1} * f_2^{-1} * f_1 * f_2 = [f_1, f_2]^3$ - swaps the $f_1.f_3$ and the $f_2.f_6$ corners: $(f_1.f_2.f_3, f_1.f_3.f_4)(f_1.f_2.f_6, f_2.f_6.f_7)$

- $m = f_3 * f_6^{-1} * f_4 * f_2^{-1} * f_5 * f_3^{-1} * f_6 * f_4^{-1} * f_2 * f_5^{-1}; f_6 * f_1 * m^6 * f_1^{-1} * f_6^{-1}$ - swaps 2 pairs of corners on the $f_1$ face: $(f_1.f_2.f_6, f_1.f_3.f_4)(f_1.f_2.f_3, f_1.f_5.f_6)$

- $f_1 * f_6 * f_1^{-1} * f_2 * f_1 * f_6^{-1} * f_1^{-1} * f_2^{-1}$ 3 cycle on corners and 3-cycle on edges $(f_1.f_2.f_6, f_11.f_7.f_6, f_2.f_6.f_7)(f_1.f_2, f_6.f_7, f_2.f_6)$

- $M2 = f_6 * f_2 * f_1 * f_2^{-1} * f_1^{-1} * f_6^{-1} * f_3^{-1} * f_1^{-1} * f_2^{-1} * f_1 * f_2 * f_3$ (Mark Longridge) - edge 3-cycle $(f_1.f_2, f_2.f_3, f_2.f_6)$

- $(f_6^{-1} * f_2^{-1} * f_3^{-1} * f_6 * f_2 * f_3)^6$ - triple corner twister, ccw twists of $f_1 \cdot f_2 \cdot f_3, f_1 \cdot f_2 \cdot f_6, f_1 \cdot f_5 \cdot f_6$

- $M3 = f_3^{-2} * f_6^2 * f_2 * f_1^{-1} * f_6 * f_1 * f_3^2 * f_6^2 * f_1 * f_6^{-2} * f_3^{-2} * f_1^{-1} * f_6^{-1} * f_1 * f_2^{-1} * f_6^{-2} * f_3^2 * f_1^{-1}$ (Mark Longridge) - edge 2-flip of $f_1 \cdot f_2, f_1 \cdot f_6$

- $M3a = f_6^{-1} * f_2^{-1} * f_1 * f_3^{-1} * f_1^{-1} * f_3 * f_2 * f_6 * f_3 * f_2 * f_1^{-1} * f_6 * f_1 * f_6^{-1} * f_2^{-1} * f_3^{-1}$ (Mark Longridge) - edge 2-flip of $f_1 \cdot f_2, f_1 \cdot f_3$

For further details, see the internet sites [J] or [Lo].

## 15.8 Solution of Lights Out

Solution ($5 \times 5$ version only):

1. For each light on row 5, press the button above it on row 4.

2. Repeat step a for rows 2-4, so that now you only have lights on row 1.

3. The only 'row 1 only' states which are solvable are the following (0 for off and 1 for on): $s_{15}, s_{134}, s_{235}, s_{1245}$,

$$
s_{24} = \begin{pmatrix} 0 & 1 & 0 & 1 & 0 \\ 0 & 0 & 0 & 0 & 0 \\ 0 & 0 & 0 & 0 & 0 \\ 0 & 0 & 0 & 0 & 0 \\ 0 & 0 & 0 & 0 & 0 \end{pmatrix}, \quad
s_{123} = \begin{pmatrix} 1 & 1 & 1 & 0 & 0 \\ 0 & 0 & 0 & 0 & 0 \\ 0 & 0 & 0 & 0 & 0 \\ 0 & 0 & 0 & 0 & 0 \\ 0 & 0 & 0 & 0 & 0 \end{pmatrix},
$$

and

$$
s_{345} = \begin{pmatrix} 0 & 0 & 1 & 1 & 1 \\ 0 & 0 & 0 & 0 & 0 \\ 0 & 0 & 0 & 0 & 0 \\ 0 & 0 & 0 & 0 & 0 \\ 0 & 0 & 0 & 0 & 0 \end{pmatrix}.
$$

The solutions of the last states are

$$
x_{24} = \begin{pmatrix} 0 & 0 & 1 & 1 & 1 \\ 0 & 0 & 0 & 0 & 0 \\ 0 & 0 & 1 & 1 & 1 \\ 0 & 1 & 0 & 1 & 0 \\ 1 & 1 & 1 & 0 & 0 \end{pmatrix}, \quad
x_{123} = \begin{pmatrix} 1 & 0 & 1 & 1 & 1 \\ 0 & 1 & 1 & 1 & 0 \\ 0 & 0 & 0 & 1 & 0 \\ 0 & 1 & 0 & 0 & 1 \\ 1 & 1 & 1 & 0 & 1 \end{pmatrix},
$$

and

$$
x_{345} = \begin{pmatrix} 0 & 0 & 1 & 1 & 0 \\ 0 & 1 & 1 & 1 & 0 \\ 1 & 0 & 0 & 1 & 1 \\ 1 & 0 & 0 & 1 & 0 \\ 0 & 1 & 1 & 0 & 0 \end{pmatrix},
$$

respectively. The others, we leave to the reader.

For another solution strategy, see [Sch1].

# 15.9 Solution of Deluxe Lights Out

Solution (6 × 6 version only):

1. For each light on row 6, press the button above it on row 5.

2. Repeat step a for rows 2-5, so that now you only have lights on row 1.

3. Use the following table to solve the $2^6 - 1 = 63$ possible 'row 1' possibilities (only 12 are included for reasons of space).

row 1: [1 0 0 0 0 0]

$$\begin{bmatrix} 1 & 0 & 1 & 0 & 1 & 1 \\ 0 & 0 & 1 & 0 & 0 & 0 \\ 1 & 1 & 0 & 1 & 1 & 1 \\ 0 & 0 & 1 & 0 & 1 & 0 \\ 1 & 0 & 1 & 1 & 0 & 0 \\ 1 & 0 & 1 & 0 & 0 & 0 \end{bmatrix}$$

row 1: [0 1 0 0 0 0]

$$\begin{bmatrix} 0 & 0 & 0 & 0 & 1 & 1 \\ 0 & 1 & 0 & 1 & 0 & 0 \\ 1 & 1 & 0 & 1 & 0 & 1 \\ 0 & 1 & 0 & 0 & 0 & 1 \\ 0 & 0 & 1 & 1 & 1 & 0 \\ 0 & 0 & 0 & 1 & 0 & 0 \end{bmatrix}$$

row 1: [0 0 1 0 0 0]

$$\begin{bmatrix} 1 & 0 & 1 & 1 & 0 & 0 \\ 1 & 0 & 1 & 0 & 1 & 0 \\ 0 & 0 & 0 & 1 & 1 & 1 \\ 1 & 0 & 0 & 0 & 0 & 0 \\ 1 & 1 & 0 & 1 & 1 & 1 \\ 1 & 0 & 0 & 0 & 1 & 0 \end{bmatrix}$$

row 1: [0 0 0 1 0 0]

$$\begin{bmatrix} 0 & 0 & 1 & 1 & 0 & 1 \\ 0 & 1 & 0 & 1 & 0 & 1 \\ 1 & 1 & 1 & 0 & 0 & 0 \\ 0 & 0 & 0 & 0 & 0 & 1 \\ 1 & 1 & 1 & 0 & 1 & 1 \\ 0 & 1 & 0 & 0 & 0 & 1 \end{bmatrix}$$

row 1: [0 0 0 0 1 0]

$$\begin{bmatrix} 1 & 1 & 0 & 0 & 0 & 0 \\ 0 & 0 & 1 & 0 & 1 & 0 \\ 1 & 0 & 1 & 0 & 1 & 1 \\ 1 & 0 & 0 & 0 & 1 & 0 \\ 0 & 1 & 1 & 1 & 0 & 0 \\ 0 & 0 & 1 & 0 & 0 & 0 \end{bmatrix}$$

row 1: [0 0 0 0 0 1]

$$\begin{bmatrix} 1 & 1 & 0 & 1 & 0 & 1 \\ 0 & 0 & 0 & 1 & 0 & 0 \\ 1 & 1 & 1 & 0 & 1 & 1 \\ 0 & 1 & 0 & 1 & 0 & 0 \\ 0 & 0 & 1 & 1 & 0 & 1 \\ 0 & 0 & 0 & 1 & 0 & 1 \end{bmatrix}$$

row 1: [1 1 0 0 0 0]

$$\begin{bmatrix} 1 & 0 & 1 & 0 & 0 & 0 \\ 0 & 1 & 1 & 1 & 0 & 0 \\ 0 & 0 & 0 & 0 & 1 & 0 \\ 0 & 1 & 1 & 0 & 1 & 1 \\ 1 & 0 & 0 & 0 & 1 & 0 \\ 1 & 0 & 1 & 1 & 0 & 0 \end{bmatrix}$$

row 1: [1 0 1 0 0 0]

$$\begin{bmatrix} 0 & 0 & 0 & 1 & 1 & 1 \\ 1 & 0 & 0 & 0 & 1 & 0 \\ 1 & 1 & 0 & 0 & 0 & 0 \\ 1 & 0 & 1 & 0 & 1 & 0 \\ 0 & 1 & 1 & 0 & 1 & 1 \\ 0 & 0 & 1 & 0 & 1 & 0 \end{bmatrix}$$

row 1: [1 0 0 1 0 0]

$$\begin{bmatrix} 1 & 0 & 0 & 1 & 1 & 0 \\ 0 & 1 & 1 & 1 & 0 & 1 \\ 0 & 0 & 1 & 1 & 1 & 1 \\ 0 & 0 & 1 & 0 & 1 & 1 \\ 0 & 1 & 0 & 1 & 1 & 1 \\ 1 & 1 & 1 & 0 & 0 & 1 \end{bmatrix}$$

row 1: [1 0 0 0 1 0]

$$\begin{bmatrix} 0 & 1 & 1 & 0 & 1 & 1 \\ 0 & 0 & 0 & 0 & 1 & 0 \\ 0 & 1 & 1 & 1 & 0 & 0 \\ 1 & 0 & 1 & 0 & 0 & 0 \\ 1 & 1 & 0 & 0 & 0 & 0 \\ 1 & 0 & 0 & 0 & 0 & 0 \end{bmatrix}$$

row 1: [1 0 0 0 0 1]

$$\begin{bmatrix} 0 & 1 & 1 & 1 & 1 & 0 \\ 0 & 0 & 1 & 1 & 0 & 0 \\ 0 & 0 & 1 & 1 & 0 & 0 \\ 0 & 1 & 1 & 1 & 1 & 0 \\ 1 & 0 & 0 & 0 & 0 & 1 \\ 1 & 0 & 1 & 1 & 0 & 1 \end{bmatrix}$$

row 1: [0 1 1 0 0 0]

$$\begin{bmatrix} 1 & 0 & 1 & 1 & 1 & 1 \\ 1 & 1 & 1 & 1 & 1 & 0 \\ 1 & 1 & 0 & 0 & 1 & 0 \\ 1 & 1 & 0 & 0 & 0 & 1 \\ 1 & 1 & 1 & 0 & 0 & 1 \\ 1 & 0 & 0 & 1 & 1 & 0 \end{bmatrix}$$

245

# 15.10 Solutions to the Orbix

We shall only solve the Orbix (in the easiest mode of play) in the case of 1 or 2 lights. The puzzle can usually be reduced to this after a few minutes by simply trial and error.

| State | Solution |
|---|---|
| [1, 0, 0, 0, 0, 0, 0, 0, 0, 0, 0, 0, 0] | [0, 1, 1, 1, 1, 1, 1, 0, 0, 0, 0, 0, 0] |
| [0, 1, 0, 0, 0, 0, 0, 0, 0, 0, 0, 0, 0] | [1, 0, 1, 0, 0, 1, 1, 0, 0, 0, 1, 0] |
| [0, 0, 1, 0, 0, 0, 0, 0, 0, 0, 0, 0, 0] | [1, 1, 0, 1, 0, 1, 1, 1, 0, 0, 0, 0] |
| [0, 0, 0, 1, 0, 0, 0, 0, 0, 0, 0, 0, 0] | [1, 0, 1, 0, 1, 0, 0, 1, 1, 0, 0, 0] |
| [0, 0, 0, 0, 1, 0, 0, 0, 0, 0, 0, 0, 0] | [1, 0, 0, 1, 0, 1, 0, 0, 1, 1, 0, 0] |
| [0, 0, 0, 0, 0, 1, 0, 0, 0, 0, 0, 0, 0] | [1, 1, 0, 0, 1, 0, 0, 0, 0, 1, 1, 0] |
| [0, 0, 0, 0, 0, 0, 1, 0, 0, 0, 0, 0, 0] | [0, 1, 1, 0, 0, 0, 0, 1, 0, 0, 1, 1] |
| [0, 0, 0, 0, 0, 0, 0, 1, 0, 0, 0, 0, 0] | [0, 0, 1, 1, 0, 0, 1, 0, 1, 0, 0, 1] |
| [0, 0, 0, 0, 0, 0, 0, 0, 1, 0, 0, 0, 0] | [0, 0, 0, 1, 1, 0, 0, 1, 0, 1, 0, 1] |
| [0, 0, 0, 0, 0, 0, 0, 0, 0, 1, 0, 0, 0] | [0, 0, 0, 0, 1, 1, 0, 0, 1, 0, 1, 1] |
| [0, 0, 0, 0, 0, 0, 0, 0, 0, 0, 1, 0, 0] | [0, 1, 0, 0, 0, 1, 1, 0, 0, 0, 1, 0] |
| [0, 0, 0, 0, 0, 0, 0, 0, 0, 0, 0, 1, 0] | [0, 0, 0, 0, 0, 0, 1, 1, 1, 1, 1, 0] |
| [0, 0, 0, 0, 0, 0, 0, 0, 0, 0, 0, 0, 1] | [0, 0, 0, 0, 0, 0, 1, 1, 1, 1, 1, 0] |
| [1, 1, 0, 0, 0, 0, 0, 0, 0, 0, 0, 0, 0] | [1, 1, 0, 1, 1, 0, 1, 0, 0, 0, 0, 0] |
| [1, 0, 1, 0, 0, 0, 0, 0, 0, 0, 0, 0, 0] | [1, 0, 1, 0, 1, 1, 1, 1, 0, 0, 0, 0] |
| [1, 0, 0, 1, 0, 0, 0, 0, 0, 0, 0, 0, 0] | [1, 1, 0, 1, 0, 1, 0, 1, 1, 0, 0, 0] |
| [1, 0, 0, 0, 1, 0, 0, 0, 0, 0, 0, 0, 0] | [1, 1, 0, 1, 0, 1, 0, 1, 1, 1, 0, 0] |
| [1, 0, 0, 0, 0, 1, 0, 0, 0, 0, 0, 0, 0] | [1, 1, 1, 0, 1, 0, 0, 0, 1, 1, 0, 0] |
| [1, 0, 0, 0, 0, 0, 1, 0, 0, 0, 0, 0, 0] | [1, 0, 1, 1, 0, 1, 0, 0, 0, 1, 1, 0] |
| [1, 0, 0, 0, 0, 0, 0, 1, 0, 0, 0, 0, 0] | [0, 0, 0, 1, 1, 1, 0, 1, 0, 0, 1, 1] |
| [1, 0, 0, 0, 0, 0, 0, 0, 1, 0, 0, 0, 0] | [0, 1, 0, 0, 1, 1, 1, 0, 1, 0, 0, 1] |
| [1, 0, 0, 0, 0, 0, 0, 0, 0, 1, 0, 0, 0] | [0, 1, 1, 0, 0, 1, 0, 1, 0, 1, 0, 1] |
| [1, 0, 0, 0, 0, 0, 0, 0, 0, 0, 1, 0, 0] | [0, 1, 1, 1, 1, 0, 0, 0, 1, 0, 1, 1] |
| [1, 0, 0, 0, 0, 0, 0, 0, 0, 0, 0, 1, 0] | [0, 0, 1, 1, 1, 0, 1, 0, 0, 1, 0, 1] |
| [1, 0, 0, 0, 0, 0, 0, 0, 0, 0, 0, 0, 1] | [0, 1, 1, 1, 1, 1, 1, 1, 1, 1, 1, 0] |
| [0, 1, 1, 0, 0, 0, 0, 0, 0, 0, 0, 0, 0] | [0, 1, 1, 1, 1, 0, 1, 0, 1, 0, 0, 1, 0] |
| [0, 1, 0, 1, 0, 0, 0, 0, 0, 0, 0, 0, 0] | [0, 0, 0, 0, 0, 1, 1, 1, 1, 1, 1, 0] |
| [0, 1, 0, 0, 1, 0, 0, 0, 0, 0, 0, 0, 0] | [0, 0, 1, 1, 1, 0, 0, 1, 0, 1, 1, 1, 0] |
| [0, 1, 0, 0, 0, 1, 0, 0, 0, 0, 0, 0, 0] | [0, 1, 1, 0, 1, 1, 1, 0, 0, 1, 0, 0] |
| [0, 1, 0, 0, 0, 0, 1, 0, 0, 0, 0, 0, 0] | [1, 1, 0, 0, 0, 1, 1, 1, 0, 0, 0, 1] |
| [0, 1, 0, 0, 0, 0, 0, 1, 0, 0, 0, 0, 0] | [1, 0, 0, 1, 0, 1, 0, 0, 1, 0, 1, 1] |
| [0, 1, 0, 0, 0, 0, 0, 0, 1, 0, 0, 0, 0] | [1, 0, 1, 1, 1, 1, 1, 1, 0, 1, 1, 1] |
| [0, 1, 0, 0, 0, 0, 0, 0, 0, 1, 0, 0, 0] | [1, 0, 1, 0, 1, 0, 1, 0, 1, 0, 0, 1] |
| [0, 1, 0, 0, 0, 0, 0, 0, 0, 0, 1, 0, 0] | [1, 1, 1, 0, 0, 0, 0, 0, 0, 1, 1, 1] |
| [0, 1, 0, 0, 0, 0, 0, 0, 0, 0, 0, 0, 1] | [1, 0, 1, 0, 0, 0, 1, 0, 1, 1, 1, 0, 0] |
| [0, 0, 1, 1, 0, 0, 0, 0, 0, 0, 0, 0, 0] | [0, 1, 1, 1, 1, 1, 0, 1, 0, 1, 0, 0, 0] |
| [0, 0, 1, 0, 1, 0, 0, 0, 0, 0, 0, 0, 0] | [0, 1, 0, 0, 0, 1, 1, 1, 1, 1, 1, 0, 0] |
| [0, 0, 1, 0, 0, 0, 1, 0, 0, 0, 0, 0, 0] | [1, 0, 1, 1, 0, 0, 1, 0, 0, 0, 1, 1] |
| [0, 0, 1, 0, 0, 0, 0, 1, 0, 0, 0, 0, 0] | [1, 1, 1, 0, 0, 0, 0, 1, 1, 0, 0, 1] |
| [0, 0, 1, 0, 0, 0, 0, 0, 1, 0, 0, 0, 0] | [1, 1, 0, 0, 0, 1, 0, 1, 0, 0, 1, 1] |
| [0, 0, 1, 0, 0, 0, 0, 0, 0, 1, 0, 0, 0] | [1, 1, 0, 1, 1, 1, 1, 1, 1, 1, 0, 1, 1] |
| [0, 0, 1, 0, 0, 0, 0, 0, 0, 0, 1, 0, 0] | [1, 0, 0, 1, 0, 1, 0, 1, 0, 1, 0, 1] |
| [0, 0, 1, 0, 0, 0, 0, 0, 0, 0, 0, 0, 1] | [1, 1, 0, 1, 0, 0, 0, 0, 0, 1, 1, 1, 0] |
| [0, 0, 0, 1, 1, 0, 0, 0, 0, 0, 0, 0, 0] | [0, 0, 1, 1, 1, 1, 1, 0, 1, 0, 1, 0, 0] |
| [0, 0, 0, 1, 0, 1, 0, 0, 0, 0, 0, 0, 0] | [0, 1, 1, 0, 0, 0, 1, 1, 1, 1, 1, 0] |
| [0, 0, 0, 1, 0, 0, 1, 0, 0, 0, 0, 0, 0] | [1, 1, 0, 0, 1, 0, 0, 0, 1, 0, 1, 1] |
| [0, 0, 0, 1, 0, 0, 0, 1, 0, 0, 0, 0, 0] | [1, 0, 0, 1, 1, 0, 1, 1, 0, 0, 0, 1] |
| [0, 0, 0, 1, 0, 0, 0, 0, 1, 0, 0, 0, 0] | [1, 0, 1, 1, 0, 0, 0, 0, 1, 1, 0, 1] |
| [0, 0, 0, 1, 0, 0, 0, 0, 0, 1, 0, 0, 0] | [1, 0, 1, 0, 0, 1, 0, 0, 1, 0, 0, 1, 1] |
| [0, 0, 0, 1, 0, 0, 0, 0, 0, 0, 1, 0, 0] | [1, 1, 1, 0, 1, 1, 1, 1, 1, 1, 0, 1, 1] |
| [0, 0, 0, 1, 0, 0, 0, 0, 0, 0, 0, 1, 0] | [1, 0, 0, 1, 0, 1, 0, 1, 0, 0, 1, 1, 0] |
| [0, 0, 0, 1, 0, 0, 0, 0, 0, 0, 0, 0, 1] | [1, 1, 0, 1, 0, 0, 0, 0, 0, 1, 1, 1, 0] |
| [0, 0, 0, 0, 1, 1, 0, 0, 0, 0, 0, 0, 0] | [0, 1, 0, 1, 1, 1, 1, 0, 0, 1, 0, 1, 0] |
| [0, 0, 0, 0, 1, 0, 1, 0, 0, 0, 0, 0, 0] | [1, 1, 1, 1, 1, 0, 1, 0, 1, 1, 1, 1, 1] |
| [0, 0, 0, 0, 1, 0, 0, 1, 0, 0, 0, 0, 0] | [1, 0, 1, 0, 1, 0, 0, 1, 1, 0, 0, 1, 0] |
| [0, 0, 0, 0, 1, 0, 0, 0, 1, 0, 0, 0, 0] | [1, 0, 0, 0, 1, 1, 0, 1, 1, 0, 0, 0, 1] |
| [0, 0, 0, 0, 1, 0, 0, 0, 0, 1, 0, 0, 0] | [1, 0, 0, 1, 1, 0, 0, 0, 0, 1, 1, 1] |
| [0, 0, 0, 0, 1, 0, 0, 0, 0, 0, 1, 0, 0] | [1, 1, 0, 1, 0, 0, 1, 0, 1, 0, 0, 1] |
| [0, 0, 0, 0, 1, 0, 0, 0, 0, 0, 0, 1, 0] | [1, 0, 0, 1, 1, 0, 1, 1, 1, 1, 0, 0, 1] |
| [0, 0, 0, 0, 1, 0, 0, 0, 0, 0, 0, 0, 1] | [1, 1, 1, 1, 1, 1, 0, 1, 0, 1, 1, 1, 1] |
| [0, 0, 0, 0, 0, 1, 1, 0, 0, 0, 0, 0, 0] | [1, 1, 0, 1, 0, 0, 0, 1, 0, 0, 0, 1] |
| [0, 0, 0, 0, 0, 1, 0, 1, 0, 0, 0, 0, 0] | [1, 1, 1, 1, 1, 0, 1, 0, 1, 1, 1, 1, 1] |
| [0, 0, 0, 0, 0, 1, 0, 0, 1, 0, 0, 0, 0] | [1, 1, 0, 1, 0, 0, 0, 0, 1, 0, 0, 1, 1] |
| [0, 0, 0, 0, 0, 1, 0, 0, 0, 1, 0, 0, 0] | [1, 0, 0, 0, 0, 1, 1, 0, 1, 1, 1, 0, 0] |
| [0, 0, 0, 0, 0, 1, 0, 0, 0, 0, 1, 0, 0] | [1, 0, 0, 0, 1, 1, 0, 1, 1, 1, 0, 0, 1] |
| [0, 0, 0, 0, 0, 1, 0, 0, 0, 0, 0, 1, 0] | [1, 0, 0, 1, 1, 0, 0, 0, 0, 1, 1, 1] |
| [0, 0, 0, 0, 0, 1, 0, 0, 0, 0, 0, 0, 1] | [1, 1, 0, 1, 0, 0, 1, 0, 1, 1, 0, 0, 1] |
| [0, 0, 0, 0, 0, 0, 1, 1, 0, 0, 0, 0, 0] | [1, 0, 0, 1, 0, 1, 1, 1, 1, 0, 0, 0, 1] |
| [0, 0, 0, 0, 0, 0, 1, 0, 1, 0, 0, 0, 0] | [1, 1, 1, 1, 1, 1, 0, 1, 0, 1, 1, 1, 1] |
| [0, 0, 0, 0, 0, 0, 1, 0, 0, 1, 0, 0, 0] | [1, 1, 0, 1, 0, 0, 0, 0, 1, 0, 0, 1, 1] |
| [0, 0, 0, 0, 0, 0, 1, 0, 0, 0, 0, 0, 0] | [0, 1, 0, 1, 0, 0, 1, 1, 1, 0, 1, 0] |
| [0, 0, 0, 0, 0, 0, 1, 0, 1, 0, 0, 0, 0] | [0, 1, 1, 1, 1, 0, 0, 0, 0, 1, 1, 0] |

246

| State | Solution |
|---|---|
| [0, 0, 0, 0, 0, 0, 1, 0, 0, 1, 0, 0] | [0, 1, 1, 0, 1, 1, 0, 1, 1, 0, 0, 0] |
| [0, 0, 0, 0, 0, 0, 1, 0, 0, 0, 1, 0] | [0, 0, 1, 0, 0, 1, 1, 1, 1, 0, 1, 1, 0] |
| [0, 0, 0, 0, 0, 0, 1, 0, 0, 0, 0, 1] | [0, 1, 1, 0, 0, 0, 1, 0, 1, 1, 0, 1] |
| [0, 0, 0, 0, 0, 0, 0, 1, 1, 0, 0, 0] | [0, 0, 1, 0, 1, 0, 1, 1, 1, 1, 0, 0] |
| [0, 0, 0, 0, 0, 0, 0, 1, 0, 1, 0, 0] | [0, 0, 1, 1, 1, 1, 1, 1, 0, 0, 0, 1, 0] |
| [0, 0, 0, 0, 0, 0, 0, 1, 0, 0, 1, 0] | [0, 1, 1, 1, 0, 1, 0, 0, 1, 1, 0, 0] |
| [0, 0, 0, 0, 0, 0, 0, 1, 0, 0, 0, 1] | [0, 0, 1, 1, 0, 0, 0, 1, 0, 1, 1, 1] |
| [0, 0, 0, 0, 0, 0, 0, 0, 1, 1, 0, 0] | [0, 0, 0, 1, 0, 1, 0, 1, 1, 1, 1, 0] |
| [0, 0, 0, 0, 0, 0, 0, 0, 1, 0, 1, 0] | [0, 1, 0, 1, 1, 1, 1, 1, 1, 0, 0, 0] |
| [0, 0, 0, 0, 0, 0, 0, 0, 1, 0, 0, 1] | [0, 0, 0, 1, 1, 0, 1, 0, 1, 0, 1, 1] |
| [0, 0, 0, 0, 0, 0, 0, 0, 0, 1, 1, 0] | [0, 1, 0, 0, 1, 0, 1, 0, 1, 1, 1, 0] |
| [0, 0, 0, 0, 0, 0, 0, 0, 0, 1, 0, 1] | [0, 0, 0, 0, 1, 1, 1, 1, 0, 1, 0, 1] |
| [0, 0, 0, 0, 0, 0, 0, 0, 0, 0, 1, 1] | [0, 1, 0, 0, 0, 1, 0, 1, 1, 0, 1, 1] |

See also Jaap's Orbix page [Sch1] or Leonard Campbell's Orbix hint page [Cam].

# 15.11  Solutions to the Keychain Lights Out

For brevity, we shall only solve the Keychain Lights Out in the case of states of 1 light. The puzzle can often be reduced to this after a few minutes by simply trial and error.

| State | Solution |
|---|---|
| $\begin{pmatrix} 1 & 0 & 0 & 0 \\ 0 & 0 & 0 & 0 \\ 0 & 0 & 0 & 0 \\ 0 & 0 & 0 & 0 \end{pmatrix}$ | none (other states with no solutions will be omitted) |
| $\begin{pmatrix} 0 & 1 & 0 & 0 \\ 0 & 0 & 0 & 0 \\ 0 & 0 & 0 & 0 \\ 0 & 0 & 0 & 0 \end{pmatrix}$ | $\begin{pmatrix} 1 & 1 & 1 & 0 \\ 0 & 1 & 0 & 0 \\ 0 & 0 & 0 & 0 \\ 0 & 1 & 0 & 0 \end{pmatrix}$ |
| $\begin{pmatrix} 0 & 0 & 0 & 1 \\ 0 & 0 & 0 & 0 \\ 0 & 0 & 0 & 0 \\ 0 & 0 & 0 & 0 \end{pmatrix}$ | $\begin{pmatrix} 1 & 0 & 1 & 1 \\ 0 & 0 & 0 & 1 \\ 0 & 0 & 0 & 0 \\ 0 & 0 & 0 & 1 \end{pmatrix}$ |
| $\begin{pmatrix} 0 & 0 & 0 & 0 \\ 0 & 1 & 0 & 0 \\ 0 & 0 & 0 & 0 \\ 0 & 0 & 0 & 0 \end{pmatrix}$ | $\begin{pmatrix} 0 & 1 & 0 & 0 \\ 1 & 1 & 1 & 0 \\ 0 & 1 & 0 & 0 \\ 0 & 0 & 0 & 0 \end{pmatrix}$ |
| $\begin{pmatrix} 0 & 0 & 0 & 0 \\ 0 & 0 & 0 & 1 \\ 0 & 0 & 0 & 0 \\ 0 & 0 & 0 & 0 \end{pmatrix}$ | $\begin{pmatrix} 0 & 0 & 0 & 1 \\ 1 & 0 & 1 & 1 \\ 0 & 0 & 0 & 1 \\ 0 & 0 & 0 & 0 \end{pmatrix}$ |

| State | | | | Solution | | | |
|---|---|---|---|---|---|---|---|
| 0 | 0 | 0 | 0 | 0 | 0 | 0 | 0 |
| 0 | 0 | 0 | 0 | 1 | 0 | 0 | 0 |
| 1 | 0 | 0 | 0 | 1 | 1 | 0 | 1 |
| 0 | 0 | 0 | 0 | 1 | 0 | 0 | 0 |
| 0 | 0 | 0 | 0 | 0 | 1 | 0 | 0 |
| 0 | 0 | 0 | 0 | 0 | 0 | 1 | 0 |
| 0 | 0 | 1 | 0 | 0 | 0 | 1 | 1 |
| 0 | 0 | 0 | 0 | 1 | 1 | 0 | 0 |
| 0 | 0 | 0 | 0 | 0 | 0 | 0 | 0 |
| 0 | 0 | 0 | 0 | 0 | 0 | 0 | 1 |
| 0 | 0 | 0 | 1 | 1 | 0 | 1 | 1 |
| 0 | 0 | 0 | 0 | 0 | 0 | 0 | 1 |
| 0 | 0 | 0 | 0 | 1 | 0 | 0 | 0 |
| 0 | 0 | 0 | 0 | 0 | 0 | 0 | 0 |
| 0 | 0 | 0 | 0 | 1 | 0 | 0 | 0 |
| 1 | 0 | 0 | 0 | 1 | 1 | 0 | 1 |
| 0 | 0 | 0 | 0 | 0 | 1 | 1 | 0 |
| 0 | 0 | 0 | 0 | 0 | 0 | 0 | 0 |
| 0 | 0 | 0 | 0 | 0 | 1 | 1 | 0 |
| 0 | 0 | 1 | 0 | 1 | 0 | 0 | 1 |
| 0 | 0 | 0 | 0 | 0 | 1 | 0 | 1 |
| 0 | 0 | 0 | 0 | 0 | 0 | 0 | 0 |
| 0 | 0 | 0 | 0 | 0 | 1 | 0 | 1 |
| 0 | 0 | 0 | 1 | 0 | 1 | 0 | 1 |

# Chapter 16

# Coda: Questions and other directions

'[Lefschetz and Einstein] had a running debate for many years. Lefschetz insisted that there was difficult mathematics. Einstein said that there was no difficult mathematics, only stupid mathematicians. I think that the history of mathematics is on the side of Einstein.'
*Richard Bellman*, **Eye of the hurricane**, *1984*

In the spirit of Galois' quote in chapter 7, this book will conclude with some of the (many) things which I don't know related to the mathematics of the Rubik's Cube-like puzzles. Some of these have already appeared previously in the book.

## 16.1 Coda

Let $G$ be the group of all legal moves of a permutation puzzle such as the Rubik's Cube, Skewb, Pyraminx, Megaminx, or Masterball.

1. (God's algorithm, Singmaster) Find the diameter of the Cayley graph of $G$.

   Is it 26 (in the "quarter turn metric", i.e., relative to the usual generators $R, L, U, D, F, B$)? Is Mike Reid's move "superflip composed with the four-spot" the longest move of the Rubik's Cube?

2. (Schwenk) Is the Cayley graph of $G$ a Hamiltonian graph?

3. With the exception of the vertex cross group of the Pyraminx, are all the cross groups simple?

4. (Dyson) What are the cross groups of the 4-dimensional regular polyhedra?

5. Are all the cross groups of the 4-dimensional regular polyhedra alternating groups?

6. Let $H$ be a group in the table in §9.4. If $H$ is a subgroup of $G$, find explicit puzzle moves which satisfy the relations given for $H$ in the table.

7. (Singmaster) Find the number of elements of each order in $G$.

8. Compute the Poincare polynomial of $G$ with respect to the generators $\{R, R^{-1}, L, L^{-1}, U, U^{-1}, D, D^{-1}, F, F^{-1}, B, B^{-1}\}$.

9. Compute the generating polynomial of $G$.

10. Find the "simplest" presentation of $G$.

11. If $G$ is the Rubik's Cube group generated by $\{U, D, R, L, F, B\}$, what is the average length of a word in $G$?

There are other group-theoretical questions related to puzzles and games which we have not touched on. The reader interested in pursuing these notions further is referred (for example) to **Winning Ways** [BCG], Conway and Sloane's book [CS], their paper on lexicographic codes [CS2], their paper on Gray codes [CSW], M. Gardner [Gar2], Luers' thesis [Lu], Stiller's thesis [Sti], Elkies' paper on the combinatorical game theory aspects of chess [El], and Eriksson's paper [E].

# Bibliography

[AF]  M. Anderson and T. Feil, "Turning Lights Out with linear algebra," Math. Magazine vol 71(1998)300-303.

[Ar]  M. Artin, **Algebra**, Prentice-Hall, 1991

[AM]  E. Assmus, Jr. and H. Mattson, "On the automorphism groups of Paley-Hadamard matrices", in **Combinatorial mathematics and its applications**, ed. R. Bose, T. Dowling, Univ of North Carolina Press, Chapel Hill, 1969

[Ba]  J. Baez, "Some thoughts on the number 6", internet newsgroup sci.math article, posted May 22, 1992,

   http://math.ucr.edu/home/baez/README.html

[B]  C. Bandelow, **Inside Rubik's cube and beyond**, Birkhäuser, Boston, 1980

[BH]  R. Banerji and D. Hecker, "The slice group in Rubik's cube", Math. Mag. 58(1985)211-218

[BCG]  Berlekamp, J. Conway, R. Guy, **Winning ways, II**, Academic Press, 1982

[BFR]  E. Bonsdorff, K. Fabel, O. Riihimaa, **Schach und zahl**, Walter Rau Verlag, Düsseldorf, 1966

[BLS]  A. Björner, L. Lovasz, and P. Shor, "Chip-firing games on graphs," European J. of Combin. 12(1991)283-291

[Bu]  G. Butler, **Fundamental algorithms for permutation groups**, Springer-Verlag, Lecture Notes in Computer Science, 559, 1991

[Ca]  R. Calinger, **Classics of mathematics**, Prentice Hall, NJ, 1982

[Cam]  L. Campbell, "Orbix hints page,"
   http://web.usna.navy.mil/~wdj/orbix_hints2.htm

[CD] M. Conrady and M. Dunivan, "The Cross Group of the Rubik's Cube", SM485C project, April,1997

[CS] J. Conway and N. Sloane, **Sphere packings, lattices, and groups**, Springer-Verlag, 1993

[CS2] ——, "Lexicographic codes: error-correcting codes from game theory", IEEE Transactions on Information Theory, $\underline{32}$(1986)337-348

[CSW] —— and A. Wilks, "Gray codes and reflection groups", Graphs and combinatorics $\underline{5}$(1989)315-325

[CFS] G. Cooperman, L. Finkelstein and N. Sarawagi, "Applications of Cayley graphs", in **Applied algebra ...**, Springer-Verlag, Lecture Notes in Computer Science, $\underline{508}$, 1990

[Cox] H. S. M. Coxeter, **Regular polytopes**, Dover, 1973

[C] J. Crossley, et al, **What is mathematical logic?**, Dover, 1972

[CL] ftp archives of the "cube-lovers" list at ftp://ftp.ai.mit.edu/pub/cube-lovers/

[CG] S. Curran and J. Gallian, "Hamiltonian cycles and paths in Cayley graphs and diagraphs - survey", Discrete Math. $\underline{156}$(1996)1-18

[DM] J. Davies and A. O. Morris, "The schur multiplier of the generalized symmetric group", J. London Math. Soc. $\underline{8}$(1974)615-620

[El] N. Elkies, "On numbers and endgames: Combinatorial game theory in chess endgames," pages 135-150 in **Games of no chance**, 151–192, Math. Sci. Res. Inst. Publ., 29, Cambridge Univ. Press, Cambridge, 1996.)

[E] K. Eriksson, "The numbers game and Coxeter groups," Discrete Math. $\underline{139}$(1995)155-166

[EK] J. Ewing and C. Kosniowski, **Puzzle it out, cubes, groups, and puzzles**, Cambridge Univ Press, 1982

[FS] A. Frey and D. Singmaster, **Handbook of cubik math**, Enslow Pub., 1982

[G] A. Gaglione, **An introduction to group theory**, NRL, 1992
http:\\web.usna.navy.mil\~wdj\tonybook\index.html

[Gap] Martin Schönert et al, **GAP manual**, Lehrstuhl D für Mathematik, RWTH Aachen
http:\\www.ccs.neu.edu\mirrors\GAP\

[Gar1] M. Gardner, "Combinatorial card problems" in **Time travel and other mathematical bewilderments**, W. H. Freeman, New York, 1988

[Gar2] ——, "The binary Gray code", in **Knotted donuts and other mathematical entertainments**, F. H. Freeman and Co., NY, 1986

[GJ] M. Garey and D. Johnson, **Computers and intractibility**, W. H. Freeman, New York, 1979

[GT] K. Gold, E. Turner, "Rubik's group", Amer. Math. Monthly, <u>92</u>(1985)617-629

[GKT] J. Goldwasser and W. Klostermeyer and G. Trapp, "Characterizing switch-setting problems", Congressus Numerantium, vol. 126, 1997, pp. 99-111 (preprint available at http://www.csee.wvu.edu/~wfk/fib.html )

[GKTZ] —– and —– and —- and C. Zhang, "Setting switches in a grid, " 1995 preprint available at http://www.csee.wvu.edu/~wfk/fib.html

[Gr] R. Grimaldi, **Discrete and combinatorial mathematics**, $4^{th}$ ed., Addison-Wesley-Longman, 1999

[Hi] R. Hill, **A first course in coding theory**, Oxford Univ. Press, 1986

[H] D. Hofstater, **Metamagical themas**, Basics Books, 1985 (Mostly a collection of Scientific American columns he wrote; the articles referred to here were also published in Scientific American, March 1981, July 1982)

[Hum] J. Humphreys, **Reflection groups and coxeter groups**, Cambridge Univ Press, 1990

[I] J. Isbell, "The Gordon game of a finite group", Amer. Math. Monthly <u>99</u>(1992)567-569

[J] D. Joyner, "Rainbow Masterball page", internet www page

http://web.usna.navy.mil/~wdj/mball/rainbow.html

[Jwww] ——, "Permutation puzzle page", internet www page

http://web.usna.navy.mil/~wdj/rubik.html

[JN] —– and G. Nakos, **Linear algebra and applications**, Brook-Cole, 1998.

[JKT] ——, R. Kreminski, J. Turisco, **Applied Abstract Algebra**, lecture notes, to be published, www page

http://web.usna.navy.mil/~wdj/book/index.html

[JS] —— and A. Southern, "The Masterball puzzle", preprint (at [J])

[K]    B. Kostant, "The graph of the truncated icosahedron and the last letter of Galois", Notices of the A.M.S. 42(1995)959-968

[Lo]   M. Longridge, "God's algorithm calculations for Rubik's Cube, Rubik's subgroups, and related puzzles", internet www page

`http://cubeman.org/fullcube.txt`

(see also `http://cubeman.org/` and click on "Cube Notes" for more information)

[Lu]   A. Luers, "The group structure of the pyraminx and the dodecahedral faces of $M_{12}$", USNA Honors thesis, 1997 (Advisor W. D. Joyner)

`http://web.usna.navy.mil/~wdj/m_12.htm`

[MT]   The MacTutor History of Mathematics archive (maintained by John J O'Connor and Edmund F Robertson at the School of Mathematical and Computational Sciences, University of St Andrews, Scotland),

`http://www-groups.dcs.st-and.ac.uk/~history/index.html`

[MS]   F. MacWilliams and N. Sloane, **The theory of error-correcting codes**, North-Holland, 1977

[Magma] W. Bosma, J. Cannon, C. Playoust, "The MAGMA algebra system, I: The user language," J. Symb. Comp., 24(1997)235-265.
(See also the MAGMA homepage at
`http://www.maths.usyd.edu.au:8000/u/magma/` or the MAGMA html documentation.)

[MKS]  W. Magnus, A. Karrus and D. Solitar, **Combinatorial group theory**, 2nd ed, Dover, 1976

[Maple] Commercial   computer   algebra   software,   available   from
`http://www.maplesoft.com`

[NST]  P. Neumann, G. Stoy and E. Thompson, **Groups and geometry**, Oxford Univ. Press, 1994

[Os]   P. Osterlund, the `abstab` package for GAP,

`http://www.math.umn.edu/~osterlu/`

and

`http://www-gap.dcs.st-and.ac.uk/~gap/Info/deposit.html`

[P]    D. Pelletier, "Merlin's magic square," Amer. Math. Monthly 94(1987)143-150.

[Pe]  M. Petković, **Mathematics and chess**, Dover, 1997

[Ro]  T. Rothman, "Genius and biographers: the fictionalization of Évariste Galois," Amer. Math. Monthly 89 (1982)84–106

[R]  J. J. Rotman, **An introduction to the theory of groups**, 4th ed, Springer-Verlag, Grad Texts in Math 148, 1995

[Ru]  E. Rubik, et al, **Rubik's cubic compendium**, Oxford Univ Press, 1987

[Rus]  D. Rusin's web pages on Lights Out,
    `http://www.math.niu.edu/~rusin/papers/uses-math/games/other/lights`

[Sa]  Dorothy Sayer, **The nine tailors**, Harcourt Brace and Co., 1962

[Sch1]  Jaap Scherphuis' Lights Out puzzle page, on the WWW at the URL:
    `http://www.org2.com/jaap/puzzles/lights.htm`

[Sch2]  Jaap Scherphuis' puzzle page, on the WWW at the URL:
    `http://www.org2.com/jaap/puzzles/`

[Se]  J.-P. Serre, **Linear representations of finite groups**, Springer-Verlag, 1977

[Ser]  ——, **Trees**, Springer-Verlag, 1980

[Si]  D. Singmaster, **Notes on Rubik's magic cube**, Enslow, 1981

[Sn1]  R. Snyder, **Get cubed**, booklet, 1990

[Sn2]  ——, **Turn to square 1**, booklet, 1993

[Sti]  Lewis Benjamin Stiller, **Exploiting symmetries on parallel architecture**, PhD Dissertation, CS Dept, John Hopkins University, 1995 (see also "Multilinear algebra and chess endgames", in **Games of no chance (Berkeley, CA, 1994)**, 151–192, Math. Sci. Res. Inst. Publ., 29, Cambridge Univ. Press, Cambridge, 1996)

[Sto]  D. Stock, "Merlin's magic square revisited," Amer. Math. Monthly 96(1989)608-610

[St]  R. Stoll, **Set theory and logic**, Dover, 1963

[Ta]  J. Tawney, "Turning the Lights Out in three dimensions," Rose-Hulman Institute of Technology Undergraduate Mathematics Journal, Volume 1, 2000

[TW]  A. D. Thomas and G. V. Wood, **Group tables**, Shiva Publishing Ltd, Kent, UK, 1980

[Th]  J. G. Thompson, "Rational functions associated to presentations of finite groups," J. of Algebra 71(1981)481-489

[vLW] J. van Lint and R. M. Wilson, **A course in combinatorics**, Cambridge Univ. Press, 1992

[We] H. Weyl, **Symmetry**, Princeton Univ Press, 1952

[Wh] White, Arthur, "Fabian Stedman: The First Group Theorist?", American Mathematical Monthly, Nov. 1996, pp771-778

[W] R. M. Wilson, "Graph puzzles, homotopy, and the alternating group", J. of Combin. Theory, 16 (1974)86-96

# Index